Localization and Its Discontents

Localization and Its Discontents

A Genealogy of Psychoanalysis and the Neuro Disciplines

KATJA GUENTHER

THE UNIVERSITY OF CHICAGO PRESS   CHICAGO AND LONDON

Katja Guenther is assistant professor of the history of science at Princeton University. She lives in Princeton, New Jersey.

The University of Chicago Press, Chicago 60637
The University of Chicago Press, Ltd., London

Printed in the United States of America

24 23 22 21 20 19 18 17 16 15 1 2 3 4 5

ISBN-13: 978-0-226-28820-8(cloth)
ISBN-13: 978-0-226-28834-5(e-book)
DOI: 10.7208/chicago/9780226288345.001.0001

This book was supported by funds made available by the Cultural Foundations of Social Integration Center of Excellence at the University of Konstanz, established within the framework of the German Federal and State Initiative for Excellence.

Library of Congress Cataloging-in-Publication Data
Guenther, Katja, author.
Localization and its discontents : a genealogy of psychoanalysis and the neuro disciplines / Katja Guenther.
pages ; cm
Includes bibliographical references and index.
ISBN 978-0-226-28820-8 (cloth : alk. paper) — ISBN 978-0-226-28834-5 (e-book)
1. Psychiatry—History. 2. Neurology—History. 3. Neurosciences—History.
4. Psychoanalysis—History. I. Title.
QP360.G84 2015
612.8—dc23

2015011295

⊗ This paper meets the requirements of ANSI/NISO Z39.48-1992 (Permanence of Paper).

Contents

Introduction

In early 2013, the European Commission announced the winners of its prestigious Future and Emerging Technologies competition. Of the two projects selected as European "flagships," it was the Human Brain Project that gained most attention. It is designed to run for at least ten years and will receive funding of more than 1 billion euros. Coordinated by the neuroscientist Henry Markram, director of the Brain Mind Institute and the Center for Neuroscience and Technology at the Ecole Polytechnique at Lausanne, the project aims to simulate a human brain in a supercomputer, pushing forward developments in neuroscience, medicine, and computing. Markram is building upon his previous research combining Information and Computing Technology and neurobiology to integrate the knowledge and skills of scientists from different countries and different fields, thus creating a "CERN for the brain."

Just three weeks later, in his State of the Union address, President Barack Obama announced his own billion-dollar investment. The BRAIN project (Brain Research through Advancing Innovative Neurotechnologies) hopes to harness new mapping techniques to examine the interaction of individual neurons. While the Europeans are tying their project to the recent developments in particle physics (not coincidentally, another Swiss success story), the Americans are working on creating a "second Human Genome Project." The comparison speaks not only to the hope for economic return—Obama emphasized in his announcement that every dollar invested in the human genome project returned $140 to the economy—but also to its focus on individual elements, in this case the activity of neurons. The project seeks to combine imaging technologies with single-cell recordings, mapping the activity of every one of the approximately 100 billion neurons in the human brain. For 2014 it was supported with around $100 million, with a similar annual budget projected over its expected lifetime of ten years.

These two announcements have confirmed neuroscience as "big science."[1] The success of neuroscience is a product of new technologies that can examine, track, and model with ever greater precision the workings of the brain. But more broadly, we can attribute its success to the ways in which those technologies have helped confirm a materialist, more specifically a somaticist, viewpoint. Neuroimaging technologies such as fMRI (functional magnetic resonance imaging) and PET (positron emission tomography) couple state-of-the-art detection techniques with powerful computational systems to locate and analyze subtle physiological changes in the brain. When correlated with mental processes, they allow the production of colorful brain scans that strongly suggest the cerebral locations of thought, empathy, emotion, even religiosity. Such images take pride of place in numerous magazines and newspapers and constitute the public face of neuroscience today.

Not surprisingly, these claims have met with criticism, in particular the assumption that the location of physiological changes in the brain can be identified with the site of clearly defined mental functions. The neuroscientist William Uttal argued in his 2001 book *The New Phrenology: The Limits of Localizing Cognitive Processes in the Brain* that this assumption is a holdover from the phrenology of the eighteenth and nineteenth centuries.[2] Phrenologists like Franz Joseph Gall correlated the measurement of the skull with certain psychological functions "seated" in particular areas of the brain, a practice that now appears to us, in Uttal's words, as "nonsense."[3] Insofar as modern neuroscience seems to be doing the same thing, it confronts the same criticism.[4]

While Uttal may be right about these localizing tendencies, especially as they have been nurtured by the successes of neuroimaging, it is important to recognize that today we are witnessing a shift in the neuroscientific landscape. In fact, if we look closely at the two projects I mentioned at the outset, which are representative of broader trends in the field, we can see that the attempt to localize higher functions has taken a backseat to a different endeavor. Both the American project and the European project aim to decode the brain's "connectional architecture," where the main lines on the maps do not divide areas cleanly from each other but rather form a network crossing boundaries and spanning the brain. The key word of this venture is not "center" but "circuit." In the words of *Nature* correspondent Alison Abbott, the scientists want to work out how the "billions of neurons . . .

in the human brain organize themselves into working neural circuits that allow us to fall in love, go to war, solve mathematical theorems or write poetry." What is more, they aim to understand how brain circuitry shifts as synapses, the points of connections between nerve cells, constantly change their form.[5] Scientists are beginning to generate such a "network map," a "comprehensive structural description of the network of elements and connections forming the human brain."[6] The goal is not to create an "exact replica of the connectional anatomy down to the finest ramifications of neurites and individual synaptic boutons." Rather, "like any good road map," the "connectome" of the human brain should afford a "multiscale description of the topological and spatial layout of connectional anatomy."[7] This shift in the goals of contemporary neuroscience has led its practitioners to declare that we have entered a new "connectomic era."[8]

While for most neuroscientists the connective and localizing traditions coexist, that coexistence is an uneasy one. As the German neurobiologist Gerhard Roth has stated, no self-respecting neurobiologist today would seriously claim that a higher function, such as "syntactic-grammatical language," was "'located' in Broca's area." Rather, "these brain centers have these qualities, amongst other things, because they are connected to certain other brain centers as well as with certain sensory inputs and motor outputs."[9] In the connectomic era, function is considered to be the product of highly complex nervous circuits both within and (crucially) crossing anatomical boundaries. When scientists trace the passage of a nervous signal across a brain circuit, it is very difficult to privilege any one part of the path and identify it as the "cerebral location" of a particular function. For this reason the simple identity between location and function, which neuroimaging promoted and Uttal criticized, is challenged by the new paradigm. Insofar as we can identify in it both localizationist and connectomic approaches, modern neuroscience remains caught between contradictory principles.

Localization and Its Discontents

This is not the first time that these principles have coexisted uneasily in the mind and brain sciences. In the 1860s and 1870s, neuropsychiatry, as practiced by figures such as Theodor Meynert in Vienna and Carl Wernicke in Breslau, combined a concern to locate mental faculties in particular centers of the brain with an essentially connectivist model of brain function. In this book, I examine this earlier moment and think through the consequences of the shifting balance between the two principles at the practical, theoretical, and institutional levels.[10]

Like modern neuroscience, nineteenth-century localization participated in a general enthusiasm for somatic explanations. Developments such as the introduction of anesthesia and antiseptic techniques into surgery, and the increasing acceptance of the germ theory of disease, spurred a general optimism about the new "scientific medicine."[11] In Vienna, where much of my story takes place, this drive to a somatic medicine was led by the pathological anatomist Carl von Rokitansky. Rokitansky, after the revolutions of 1848, turned his institute into the center, the hub, of the Vienna General Hospital, and his methods gained traction in a variety of fields.

In psychiatry, the terms of the new paradigm were set with Wilhelm Griesinger's midcentury prescription: "Mental disease is brain disease." Meynert and Wernicke were two of the most prominent figures to heed Griesinger's call. They hoped to translate psychiatric conditions into the language of brain pathology and to identify the cerebral locations of psychological functions.[12] The neuropsychiatric project was able to draw on the support of a number of parallel developments at the time, not least the collaboration between Berlin electrotherapist Eduard Hitzig and anatomist Gustav Fritsch. In their experiments, Fritsch and Hitzig had stimulated parts of a dog's cortex electrically, eliciting the movement of select body parts, which allowed them to map the motor areas of the animal's brain.[13]

But though localization was dominant, during this period it was closely bound up with a connective principle: the reflex. A reflex is made up of the nervous connections between sense organs and muscles, which pass through the central nervous system (the spinal cord or the brain). Over the course of the nineteenth century, this sensory-motor system became the dominant principle for explaining nervous function. It also furnished a clinical practice. The reflex allowed practitioners to engage with the patient's nervous system, and consequently the "reflex test" became the central element of the neurological examination.[14]

Reflex physiology proved useful for neuropsychiatrists like Meynert and Wernicke because it provided an authoritative somatic framework for explaining nervous function and thus underpinned the localization project.[15] Moreover, by furnishing a model for the nervous connections between periphery and center, the reflex helped explain why certain parts of the brain might control certain functions, initiate certain movements, receive sensory experience, or govern particular tasks like speech. In this understanding, the motor center for hand movement, for instance, was connected to the hand by the efferent arc of a reflex. The prominence of the reflex tradition led to one of the great innovations in localization during this period: the division of higher functions into sensory and motor components, each to be located in a different part of the brain.

But it is important to recognize that the connective principle not only supported but also undermined localization, much as in our connectomic era today. From the latter decades of the nineteenth century, as a result of shifting political and social considerations—Bismarck's social insurance, new patient concerns in fin-de-siècle Vienna, changing patient populations during the Great War—the examination and treatment of patients became increasingly central to neuropsychiatrists and neurologists. In this context, physicians like Wernicke prioritized the practical aspects of the reflex. One of the great advantages of the reflex as applied to higher functions was that it provided the tools to structure the clinical psychiatric examination. And the neat demarcations between functions suggested by the localization paradigm showed themselves to be increasingly inadequate for elucidating the unpredictable and confusing symptoms of psychiatric patients. For while reflex physiology as a *theory* supported the division of functions into sensory and motor elements, within reflex *practice* the sensory and motor parts could not be disconnected so easily. According to the localization model, a reflex arc passed through at least two centers: a sensory center at the end of the afferent (inward-conducting) arc, and a motor center at the end of the efferent (outward-conducting) reflex arc. But in the reflex test, a sensory stimulus was always correlated with a motor response, such that it was difficult to see either as independent of the other or to identify neatly compartmentalized sensory and motor centers. That is, the reflex as a *clinical* principle foregrounded its systemic, connective aspects, how different nervous elements functioned together, and this posed a considerable challenge to the atomizing tendencies of the localization project.

A Genealogy of Psychoanalysis and the Neuro Disciplines

There are, of course, marked differences between neuropsychiatry and early-twenty-first-century neuroscience. But the comparison is instructive because it allows us to track how the criticism of localization through a connective principle has played out before, and thus provides valuable means for thinking through our present situation. For what is surprising about the earlier moment is that the prioritization of connectivity, far from heralding a new neuropsychiatric era, introduced considerable disciplinary instability into it. Most importantly, it provided the resources for the construction of two new and diverse fields: psychoanalysis and neurology.

That the histories of neurology and neuropsychiatry cannot be separated, at least in the German context, is simultaneously uncontroversial and understudied.[16] Otfrid Foerster, one of the most prominent figures in German neurology in the interwar period, worked with Wernicke between 1900 and 1904, first as his *Volontärarzt* (unsalaried physician) and then as assistant at the psychiatric clinic and the neurological polyclinic. Foerster also undertook research in Wernicke's neuropathological laboratory, where he was in charge of producing the third volume of a photographic atlas of the brain, a costly and ambitious project funded by the prestigious Königliche Akademie der Wissenschaften.

Scholars are also aware that psychoanalysis can trace roots back to neuropsychiatry.[17] Sigmund Freud joined Meynert's laboratory in 1883, at a time when he was also working as Meynert's *Sekundararzt* at the Vienna General Hospital. Even after Freud had moved on with his clinical rotations, he continued to conduct research in the lab for an additional two years. During this period he wrote a number of the articles submitted for his *Habilitation* in neuropathology.[18]

But despite the intensity of Freud's engagement with the neuropsychiatric tradition, the overwhelming tendency in the scholarly literature has been to consider this early work as part of his "pre-analytic" phase, with little importance for his later studies.[19]

In contrast, I argue that Freud's early participation in the debates over pathological anatomy and reflexes in the 1870s and beyond played a signal role in the emergence of psychoanalysis. By embedding his work in the story of mainstream somatic psychiatry, my book brings Freud into the history of medicine more broadly. The stories of psychoanalysis and pathological anatomy cannot be told independently of each other. From such a perspective, though neurology and psychoanalysis are often seen as opposed in their medical and scientific orientations, their common neuropsychiatric heritage enables me to construct a family tree on which Freud and Foerster appear as close intellectual relatives.[20]

Not only can we trace Freud and Foerster's work back to the same institutional and pedagogical context; in addition they engaged with that context in similar ways. Freud no less than Foerster drew broadly on models of neuronal connections, and they both developed their practices through a criticism of the *Zentrenlehre* (one of the dominant forms of the localization tradition). More specifically, I argue that psychoanalysis and neurology emerged out of Freud and Foerster's break with the localization project as developed in neuropsychiatry, a break enabled by their deployment of connective principles.[21]

My concentration on the shared attributes of psychoanalysis and neurology means that my story cannot be taken as a full account of their emergence, nor indeed of Freud's and Foerster's careers. To provide a common frame for thinking about such diverse disciplines, it privileges particular aspects of their theory and practice, focusing on some strands rather than others. For this reason this is not a history of specialization, though it hopes to contribute to that literature.[22] The processes of specialization are complex, and numerous factors, functioning differently in different national contexts, play roles.[23] The complexity of the task has led most histories of specialization in the mind and brain sciences to concentrate on one discipline or specialty, taking as their object of research a particular institution, an area of study, or a more or less coherent theoretical form and practice.[24]

This study, however, is closer to those that provide dual histories of the brain and mind sciences by choosing a guiding thread—a practice, a function, a disease, or a concept that can be identified in a wide variety of disciplines—to construct their analyses. These histories have been able to think through the relationship between the brain and mind sciences by examining how different medical specialties and academic disciplines responded to and treated these shared elements. Just as their guiding threads are not constrained by disciplinary categories, these histories have been able to escape the limits of a soma-psyche distinction. For instance, Alison Winter and Kurt Danziger have written histories within the mind and brain, and more broadly the human, sciences by analyzing how different disciplines (psychology, neurosurgery, criminology, philosophy) discuss and study memory.[25] And Roger Smith, in his book *Inhibition*, explores how a shared language of "inhibition" cut across the European sciences of the mind and the brain, especially neurophysiology and psychology, from the second third of the nineteenth century.[26]

Such an approach is valuable because it allows me to bring into conversation the sciences of mind and brain, and thus two of the fields that inform modern thinking about subjectivity most authoritatively. In this way, this book contributes to the growing field of research on the history of the self.[27] It offers a way of moving beyond set oppositions between the somatic and the psychological, the cerebral and the psychoanalytic, subjects. Theories of and practices on the nervous system were not necessarily tied to a reductive somaticism, but rather provided resources for widely diverging understandings of subjectivity: self-transparent and self-opaque, learning and fixed, unified and divided, bodily and mental, solipsistic and social.

Localization and Its Discontents, by taking the changing and contradictory relationship between the connective and localizing traditions as its guiding thread, provides a new perspective on the developing medicine of the mind and the brain. Looking at the history of neuropsychiatry and its long shadows—crossing over the Atlantic, reaching across European borders, and especially across disciplinary divisions—we can shed light on the history of psychoanalysis and neurosurgery-neurology, which strongly resists the attempt to present them within the reified categories of "mind" and "brain," the two cultures, the sciences versus the humanities. As the brain today gains in prestige as an object of scientific investigation and is invested with ever more extravagant hopes, such a history becomes particularly relevant. For in turning to the past we become aware of the contingencies, unexpected connections, and ironic narratives that will surely also mark our future.

Chapter Outline

The chapters in this book form a Y-structure. The base of the Y is made up of two chapters discussing the work of neuropsychiatrists Theodor Meynert and Carl Wernicke, which, I argue, provided resources common to both psychoanalysis and neurology. Beginning with chapter 3, I tell these symmetrical histories, which form the two arms of the Y. Chapters 3 and 5 examine the development of psychoanalysis, first Freud's break with the neuropsychiatrists and then the transformation of his ideas in the work of Paul Schilder, an Austrian émigré to the United States. Chapters 4 and 6 treat the development of neurology-neurosurgery, looking first at Otfrid Foerster and then at Wilder Penfield, a Canadian neurosurgeon. The book thus moves through a number of national and institutional contexts: from the morgue of Vienna medicine to the lecture hall in Breslau;

from the couch in Freud's Viennese middle-class practice to Foerster's exercise hall; and from there across the Atlantic to Schilder's clinic-based psychoanalysis and Penfield's operating room.

Though each chapter takes as its primary focus the work of a single practitioner, the book is not simply a set of case studies. As an account of the emergence of neurology-neurosurgery and psychoanalysis as independent medical fields, its examples are constitutive of broader trends. While Meynert and Wernicke were only two of a larger group of neuropsychiatrists in the latter decades of the nineteenth century, they were arguably the most prominent and influential. Freud's career path and intellectual development are not simply exemplary of psychoanalysis but were the single most important factor in its creation. Similarly, Foerster's path from neuropsychiatry to neurology is not merely a useful illustration of the changing face of neurological work in the period; as head of the prestigious Gesellschaft deutscher Nervenärzte from 1924 to 1932 and of the Rockefeller-funded Breslau Neurological Institute, he played a determinative role in the emergence of neurology as distinct from psychiatry. Penfield was one of the best-known neurosurgeons of the twentieth century and a figure of enormous influence in North American medicine. Even Schilder, who had perhaps the least institutional power of all my figures, is useful for considering the broader movement of psychoanalysis, because he became the epicenter of a debate that shook the New York Psychoanalytic Society. In each case, the way in which the doctor rearticulated the relationship between localization and connectivity was central, I argue, for the ongoing development of his field.

At the beginning of my story, and for the neuropsychiatrists, a connective principle (the reflex) was deployed in order to legitimate the localization project, and the mutually reinforcing relationship between the two took center stage. In chapter 1, I show how Meynert's appeal to a version of reflex physiology allowed him to participate in the new enthusiasm for pathological anatomy in Vienna medicine, which had risen to prominence in the political and cultural conditions created by the failure of the 1848 revolutions. To be able to apply the reflex (which previously had been restricted to spinal action) to the localization project (which took as its object the brain), Meynert revised the reflex model to make it appropriate for describing higher functions. Drawing on the language of association psychology, Meynert argued that between the afferent sensory arc and the efferent motor arc, there existed an "association system" that explained the complexity and "plasticity" of brain function. A modified reflex physiology thus emerged as a necessary resource for the localization project. At the same time, the appeal to associationism helped Meynert address a problem resulting from the organization of the Vienna General Hospital. Because pathological anatomy was centralized there, Meynert had to rely on his clinical colleagues for the description of symptoms that he would correlate with anatomical findings. Meynert's recasting of the *physiological* reflex paradigm through an appeal to association *psychology* helped facilitate the communication between somatically inclined pathological anatomists and psychologically oriented psychiatrists.

In chapter 2 I show how the reshaping of the medical landscape after Bismarck's insurance reforms in 1883 opened up space for a challenge to the localization project. Though Carl Wernicke made his name as a neuropsychiatrist, he became increasingly skeptical of the claim that discrete brain centers were responsible for discrete functions. I show this by examining Wernicke's clinical work in Breslau. He originally constructed his clinical demonstrations (*Krankenvorstellungen*) following the principles of neuropsychiatric localization, but the variety and complexity of the patient material led him to sideline that project. Drawing instead on Duchenne de Boulogne's "anatomy of the living," Wernicke began to present disease as the complex interaction of numerous nervous elements, and he reshifted the goals of his patient examination from the identification of a single lesion to the close description and interpretation of the patient's pathology. Wernicke's reflex practice showed itself to be resistant to the localization project for which it had been first deployed.

These tensions were only fully worked out by other figures, leading to two new conceptualizations of the normal and the pathological. In chapters 3 and 4 I show how Sigmund Freud and Otfrid Foerster came to refigure the relationship between reflex and localization, broadly speaking, by emphasizing the systemic aspects of the first over the atomistic aspects of the second. Such a shifting of the relationship, I argue, allowed them to recast the Meynert-Wernicke understanding of nervous action and helped them form the two new disciplines of psychoanalysis and neurology.

Chapter 3 shows how Freud, working in the recesses of his apartment in Berggasse 19 in Vienna, was able to develop a new paradigm of disease and treatment that broke with central tenets of the localization tradition. Freud radicalized the associative elements of Meynert's connectivism in order to challenge the localizationist paradigm for which previously it was a support. First in *On Aphasia*, Freud used elements of Meynert's own system to challenge the then dominant theory of localization. Then in later works, he challenged the lesion model upon which localization theory had been based. In doing so, Freud was able to reevaluate the etiology of mental disturbance, moving from an emphasis on physical to psychological trauma, and he recast the reflex exam as a form of "talk therapy." Freud's mature psychoanalytic practice, I argue, can then be seen as the ultimate rejection of the lesion and pathological anatomical model: because it dispensed with the "cathartic method" and focused on working through resistances, it was no longer structured by the identification and confrontation of an underlying "trauma."

In chapter 4, we see how Foerster made a similar move, again drawing on the systemic aspects of the reflex in order to challenge the assumptions of the localization tradition. Working in the context of the neurological and neurosurgical departments of the Breslau state hospital, Foerster reconceptualized nervous disease by an appeal to a *Reflexgemeinschaft* (reflex community), which allowed him to sever the one-to-one connection between lesion and function that had been assumed in Meynert's pathological anatomy: understanding it within the reflex system, Foerster showed how sensory damage could have motor consequences. That realization had important therapeutic consequences. Confronted with a flood of wounded soldiers after the Great War, Foerster developed a spinal-cord operation in which he would cut the reflex arc to isolate the pathological effects of a lesion. Later on, Foerster extended his operation to the brain in the treatment of epilepsy. Complementary to his surgical interventions, the "plastic" elements of Meynert's reflex system were redeployed in order to draw on the organism's own resources for change, in what Foerster called *Übungstherapie* (lit., exercise therapy). Because the nervous system could be shaped by experience, it was possible to establish a new "organization" of nerves, a process in which certain members of the "reflex community" were recruited for new tasks.

Chapters 5 and 6 foreground one further transformation. Chapter 5 looks into what became of Freudian psychoanalysis when the connective principles upon which it was based were discarded, by following the career of the psychoanalyst Paul Schilder. Schilder, who emigrated from Austria to the United States in 1928, styled himself as a psychoanalyst, but I argue that he remained far from the Freudian orthodoxy. While Wernicke and the early Freud emphasized the functioning of the whole reflex arc, in particular the set of associations connecting sensory to motor arcs, Schilder in his neurological tests treated the two separately. An important effect of this modification of reflex testing is that Schilder had to rely on the patient's report for an account of her sensory experience. The reliance on this report in his clinical practice, I argue, encouraged Schilder's embrace of a self-transparent subject in his psychoanalytic theory, as seen in his rejection of the Freudian unconscious. Further, the debates over Schilder's ideas in the United States, resulting ultimately in his expulsion from the New York Psychoanalytic Society, shed light on deep conflicts within American psychoanalysis.

Chapter 6 charts a similar transformation of Foerster's practice. Wilder Penfield came to study with Foerster in Breslau in 1928 and brought Foerster's epilepsy operation back to North America. There it formed the basis of Penfield's clinical work. In contrast to Foerster, Penfield retasked his surgical technique in order to map the brain. I argue that the reemergence of the localization project, after a thirty-year hiatus, can be explained by Penfield's de-composition of the reflex. Like Schilder, Penfield tested both sides of the reflex arc separately, studying in turn sensory and motor responses. As such, Penfield could sideline the systemic aspects of the reflex that had structured earlier investigations and made localization so unconvincing. And like Schilder, Penfield was thus encouraged to posit a self-transparent patient who could provide insight into sensory states. As I show, the self-transparency of Penfield's introspective patient increasingly became the focus of his research, as in the 1950s he concentrated his efforts on tracking down an ever-elusive "mind."

In the epilogue I return to the theme with which I began this introduction and suggest ways in which this history sheds new light on recent trends in neuroscience—not to offer a full history of the present, but to exploit suggestive structural parallels between the history told in these chapters and neuroscientific developments of the past decade. I argue that a historically grounded understanding of the articulation of the relationship between localization and a principle of "connectivity" helps reframe anxieties about the place of neuroscience in contemporary academic culture, in particular in the humanities.

CHAPTER ONE

In the Morgue

Theodor Meynert, Pathological Anatomy, and the Social Structure of Dissection

In 1913 the German psychiatrist and philosopher Karl Jaspers identified three "prejudices" in psychopathology that "weigh[ed] . . . with paralyzing effect" on the field: the somatic, the philosophical, and the "absolutizing." Of the three prejudices, the first was particularly vicious, producing what had been called "brain mythologies."[1] In making his criticism, Jaspers had a particular form of the somatic view in mind: the theory of functional localization, or *Zentrenlehre*. Neuropsychiatrists in the nineteenth century had asserted the existence of sensory and motor "centers" in the brain, and they sought to map out the parts of the cortex dedicated to certain functions such as language.[2] For Jaspers, however, this approach was based on a deeply problematic rationale: "These anatomical constructions . . . became quite fantastic (e.g. Meynert, Wernicke) and have rightly been called 'Brain Mythologies.' Unrelated things were forcibly related, e.g. cortical cells were related to memory [*Erinnerungsbild*], nerve fibres to association of ideas. Such somatic constructions have no real basis. Not one specific cerebral process is known which parallels a specific psychic phenomenon."[3]

Jaspers criticized here the apparently direct translation process between psychological and anatomical categories. It was not that he denied the interdependence of mind and body; he maintained that bodily phenomena, such as digestion, were influenced by mental phenomena and, conversely, that mental phenomena "originate[d] in part from somatic sources." But in no case did he think that "some specific psychic event [was] directly associated with some specific somatic event." The simple one-to-one translation was, to Jaspers, an uncritical, "unreal" one.[4]

Jaspers's critique was marked by his larger project. In line with his later existential philosophy, Jaspers denied that there was a "uniform theoretical framework" of psychopathology and, by extension, subjectivity.[5] Unlike the sciences, which "rest[ed] on comprehensive and well-founded theories," such as atomic theory or cell theory, no theory for psychology or psychopathology had such a sure foundation. To Jaspers, then, not only was there no solution to the translation problem; the very attempt to provide psychology with a coherent theory (based on its relationship with physiology, or any theory at all) was misguided. Psychologists should strive rather for a pre-theoretical understanding.

In questioning the process of translation between psychological and anatomical categories that was constitutive of the *Zentrenlehre* project, Jaspers identified its central difficulty and limitation. In fact, as we will see in this chapter and the next, the process of translation was inherent to the practice upon which the *Zentrenlehre* was based, and the proponents of the very tradition that Jaspers attacked recognized it as a problem. But because it was a problem that they confronted directly, we should not see the translation between psychological and physiological categories as an uncritical one. Neuropsychiatrists did not simply map psychological functions onto brain anatomy. Rather, this mapping was conditioned by the institutional, practical, and social context of hospital medicine, especially in Vienna in the mid-nineteenth century. In navigating this context, the neuropsychiatrists developed complex and innovative strategies for bridging the apparent divide between the psychological and the somatic aspects of their work. Further, while Jaspers saw the difficulty as sufficient reason to reject the somatic approach wholesale—a critical attitude that was further developed in the interwar neurological "holism" of Kurt Goldstein and Henry Head—the neuropsychiatrists attempted to address its problems.[6] Indeed the most important criticisms, institutionally and intellectually, of the neuropsychiatric tradition came from within. As we will see over the course of this book, the developing disciplines of neurology, neurosurgery, and even psychoanalysis are the results of such engagement; rather than external criticisms like Jaspers's, they are better understood as internal reformulations.

Somatic Medicine in Vienna

The neuropsychiatric attempt to correlate mental pathology with lesions in different parts of the brain participated in a broader turn to the somatic in modern medicine. As Michel Foucault has prominently (if somewhat schematically) argued, the nineteenth century saw a dramatic shift in medical practice, what he calls a transformation of the medical "gaze" (*regard*). Gone was the old classificatory system that provided a hierarchy and family tree of diseases, where each was defined in relationship to the others. Location in the body, rather than location within a classificatory system became the structuring principle of the new nosology. In the earlier model, a disease could migrate, manifest itself across a range of symptoms, and metamorphize over time, while symptoms allowed a disease to "show through." But such migrations and metamorphoses were no longer possible in the new medicine, where pathology was related to a lesion located in a particular organ. In the new system, the disease was, in Foucault's words, "entirely exhausted in the intelligible syntax of the signifier."[7] That is to say, the lesion *was* the pathology, rather than being merely an effect of it. Connected to this new form of medicine was the conviction that, given the appropriate tools, all pathologies could be found in the body, that one could see a disease fully, and that it could not elude the eye of the doctor forever. Pathological anatomy practiced on bodies in the mortuary became the privileged way to access disease and learn about it.[8]

In *The Birth of the Clinic*, Foucault structured his narrative around a transformation in Paris at the turn of the nineteenth century.[9] In this period, the French capital attracted medical students and doctors from all over the world. But by 1840, Vienna came to rival Paris as a magnet for aspiring young doctors.[10] The "Second Vienna Medical School" followed in many ways its French counterpart. From midcentury until about 1890, Viennese medicine was dominated by a new somaticist paradigm promoted by Carl von Rokitansky (1804–1878), the first *Ordinarius* of pathological anatomy in Vienna (and later rector of Vienna University), and his collaborator, the internist Josef von Skoda (1805–1881).

Rokitansky presented the new somatic method as a scientific response to both a vitalistic *Naturphilosophie* and an unrigorous clinical method.[11] It was the second that was the most pressing concern at the beginning of his career. Rokitansky identified its central problem: "The same symptom complex was caused by very different material changes." He insisted that diagnoses would find a surer foundation through a study of "the changes of material substrates perceptible to the senses."[12] This could be achieved through the pathological anatomical method, which, as he pointed out in the preface to his three-volume *Handbuch der pathologischen Anatomie* of 1846, had the potential to become the new basis for medical theory and practice.

Rokitansky's aspirations for pathological anatomy did not immediately translate into institutional centrality. Indeed, at the time he wrote his *Handbuch*, pathological anatomy was relatively marginal in the Vienna medical landscape. True, since the 1820s it had been an independent discipline, having wrestled itself free from internal medicine, of which it had, under the auspices of Johann Peter Frank (with Aloys Vetter as prosector), been an ancillary specialty.[13] But independence was not tantamount to prominence. The old pathological anatomical building (*Alte Prosektur*) was a simple structure in the northern part of the hospital area (see fig. 1.1, top); a visitor in 1847 described it as rather "pathetic barracks."[14] Apart from a storage space for dead bodies, the institute consisted of merely two rooms, "a larger one for the dissections of those who died in the asylums and a smaller one for the forensic dissections." Rokitansky complained about the equipment and lack of heating. The *Prosektur*, he grumbled, was "well suited to ruin permanently the health of everyone working there."[15]

Rokitansky's early career was marked by steady progress through the ranks of a relatively marginal field. In 1827, when still a medical student, he accepted a post as unsalaried intern (*Praktikant*) at the Prosektur, where he became assistant in 1830.[16] Not long thereafter, he was promoted to *Extraordinarius* in pathological anatomy (1834), then a decade later to *Ordinarius* (1844).[17] By the middle years of the 1840s, Rokitansky had become a big fish in a small pond, destined, it seemed, for an honorable if not particularly illustrious career.

That all changed after the abortive revolutions of 1848. In quick succession Rokitansky gained a number of prestigious administrative and public posts (without, however, giving up his work in pathological anatomy). The following year, he was made dean of the medical school. In 1853 he became the first freely elected rector of the University of Vienna. Outside of the university, too, he was seen as a man on the rise: he became president of various learned societies, including the Imperial Academy of Science in July of 1848 and the Anthropological Society in 1870. In 1863 he became medical adviser to the Ministry of the Interior and was nominated by the emperor to the High Chamber of the Royal Council.[18]

Rokitansky's ascendence after 1848 can in part be explained by the political and cultural changes in postrevolutionary Vienna. Before the revolution he had often been considered apolitical, and this reputation

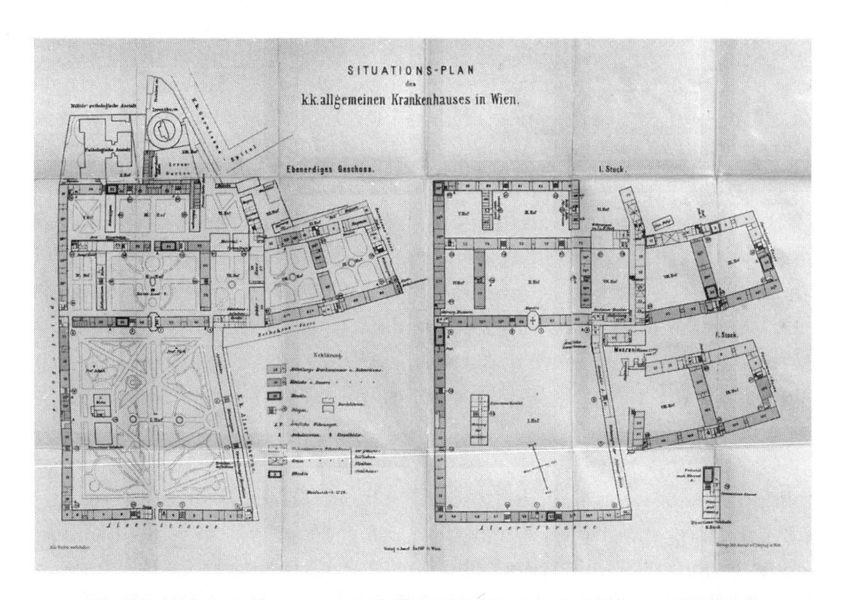

FIGURE 1.1. *Top*, map of the Vienna General Hospital at the time Meynert worked there (1886). The former "graveyard" (X. Hof) with the Pathological Anatomical Institute—also the site of the *Alte Prosektur*—is located at the very top of the drawing on the left. *Bottom*, the newly opened Pathological Anatomical Institute. (*Top*, aus dem Bildarchiv des Instituts für Geschichte der Medizin Wien. Courtesy of Collection of Pictures, Collections and History of Medicine, MedUni Vienna. *Bottom*, reprinted from Richter, "Eine Stätte," 748.)

served him well in the immediate wake of the revolution's suppression, when change at the political level remained elusive.[19] Rokitansky's particular brand of liberalism—which is evident in his speeches from the period—placed faith in science as the motor of human progress, valuable both for the production of wealth and for overcoming class differences.[20] In this way, it conformed to the norms of the moment. Much of the liberal energy from the Vormärz period was redirected after 1848 to cultural and social pursuits. Among other things, the Academy of Fine Arts (Akademie der bildenden Künste) was reorganized by Count Franz Thun, new construction projects were undertaken, and the period saw a great rise in associational activity, with the formation of numerous *Vereine* and artistic groups.[21]

Of all the post-1848 projects, the one that perhaps left the greatest mark on the city was the Ringstrasse.[22] After 1848, the Hofbaurat was dissolved, and proposals were invited for a new development in the location of the former city wall. Although none of the winning proposals were in the end implemented fully—instead the emperor chose a modified version drawing on several, in order to respond better to military concerns—the plan was accepted in 1859 and work began shortly after. The planning compromise was fitting; combining entrepreneurial spirit with authoritarian rule, it showed the Ringstrasse to be a child of its time.[23]

Rokitansky's rising prominence was mirrored by the status of pathological anatomy. As a sign of its importance, the old "barracks" were replaced by a "splendid building," in the words of a contemporary, to house Rokitansky's new Pathological Anatomical Institute (fig. 1.1, bottom).[24] It is significant that plans for the institute participated in the broader enthusiasm for renovation that came with the building of the Ringstrasse. Indeed, the new institute was built at the same time by Ludwig Zettl, the author of one of the winning proposals for the large urban renewal project. In both designs Zettl projected bourgeois self-confidence. In his plan for the Ringstrasse, he had lined the main boulevard with privately owned houses, bringing the city's wealthiest families into the center of the newly designed city.[25] Further, the grand design of the Pathological Anatomical Institute reflected his faith in science and material progress.

The Ringstrasse project raised the profile of Rokitansky's institute in another way. Whereas before, the General Hospital lay outside the city walls, now it was merely a block away from the new center of town. At the same time, Rokitansky's specialty of pathological anatomy had come to assume a central position within the hospital. Although in the same location as the former barracks (in the north part of the "corpse yard" [*Leichenhof*], directly adjacent to the clinics at the General Hospital), the new institute was substantially larger and could serve as a meeting place for physicians from the hospital's other departments. In addition to rooms for the storage of cadavers; rooms for scientific and legal dissections, pathological chemistry, and microscopy; and exhibition space for rare pathological specimens, the new building also furnished "working premises" (*Arbeitslocale*) for pathological anatomists, forensic anatomists, and clinicians from the internal, surgical, ophthalmological, and obstetric departments. In this way, physicians like the dermatologist Ferdinand Hebra, the forensic pathologist Jakob Kolletschka, and the surgeon Franz Schuh, as well as gynecologists, ophthalmologists, and pediatricians, came to the institute to cooperate with Rokitansky, following the pathological anatomical method.[26] The new pathological institute, a true "model institution" in the words of one commentator, served, then, to connect the various medical specialties. This unifying role was all the more important given the size and sprawling nature of the hospital. As one observer suggested, the "complex of buildings, . . . deserve[d] to be named a small city given its dimensions and number of inhabitants."[27]

Somaticizing Psychiatry

Psychiatry participated in the general trend toward somaticization, both in Vienna and in the German-speaking world more broadly.[28] The Prussian psychiatrist Wilhelm Griesinger (1817–1868) formulated the battle cry of a new, somatically informed psychiatry when, in his 1845 opus magnum *Die Pathologie und Therapie der psychischen Krankheiten*, he asserted that "mental disease" was "brain disease."[29] Griesinger's program influenced a whole generation of students in the 1850s and 1860s across the German-speaking lands.[30] Figures like Theodor Meynert, Carl Westphal, and Bernhard von Gudden took Griesinger's dictum to heart. They thought they could differentiate between mental diseases more effectively through the study of pathological anatomy and thus provide, as they saw it, a scientifically grounded nosology.

In Vienna, Theodor Meynert (1833–1892) became one of the key proponents of the new somatic psychiatry. In 1867, along with his colleague Maximilian Leidesdorf, he founded the first specialist journal of psychiatry, which foregrounded their somaticist credentials. The journal's full and unwieldy title was *Vierteljahresschrift für Psychiatrie in ihren Beziehungen zur Morphologie und Pathologie des Centralnervensystems, der physiologischen Psychologie, Statistik und gerichtlichen Medizin* (Quarterly for psychiatry and its relations to the morphology and pathology of the central nervous system, physiological psychology, statistics, and forensic medicine). A year later, Meynert, Leidesdorf, and Joseph Gottfried Riedl founded the Verein für Psychiatrie und forensische Psychologie, which manifested a similar orientation.[31]

Like Rokitansky, Meynert built his career out of the dissection room. As a medical student in Vienna, he worked in the histopathology laboratory of Carl Wedel, producing pathological specimens. Although after his *Promotion* (doctoral dissertation) in 1861, Meynert held a number of clinical positions—as *Aspirant* in internal medicine and surgery at the Vienna General Hospital in 1861, and as *Sekundararzt* in Gumpendorf near Vienna in 1865—it was clear that his interests lay elsewhere. During this period, he dedicated his free time to studying the anatomy of the nervous system.[32] He received the necessary material through various sources: his brother-in-law Scheuthauer, assistant to Rokitansky, gave him access to corpses from the Institute of Pathological Anatomy; Theodor von Oppolzer and Maximilian Leidesdorf allowed Meynert to examine the brains of their psychiatric patients after they had died; in addition, Meynert had the opportunity to work with the brains of dead animals from the Vienna zoo.[33]

Meynert's private work resulted in a number of scientific publications, which allowed him in 1864 to apply for the *Habilitation* in *Bau und Leistung des Gehirns und Rückenmarkes* (structure and function of the brain and spinal cord). The *Habilitation* was granted to him a year later. On the basis of this emerging academic reputation, Meynert was able to return full-time to pathological anatomy. Soon after becoming *Sekundararzt* at the Niederösterreichische Landesirrenanstalt (Asylum for lower Austria) in 1865, he was offered the newly created post of prosector there, a post that did not include any clinical duties. He held that position for four years, until 1870.[34]

The impact of this training can be seen in Meynert's theoretical and practical commitments. Especially when he had little direct contact with patients, he placed great value on the physical characteristics of the brain specimens he studied. For instance, in an 1865 paper, Meynert offered a description of the step-by-step process for the anatomical "exposure of the course of bundles in the brain stem,"[35] specifying how to separate the brain stem from the hemispheres, how to remove the meninges, and how to cut the brain in order to obtain the right surfaces.[36] When he touched on psychological function, his discussion was always directed by anatomical research. Meynert found in one case that the "cross section of the pyramidal tract [was] in direct relationship to the development of the hemispheres, that is to the sum of *Vorstellungen* [ideas] of an animal." As a result, he recommended that in the future, the physician would have to "explore the anatomical behavior of the pyramids in cases of sensory disturbance [*Empfindungsstörungen*]."[37] Similarly, in two papers from 1867 and 1868, Meynert compared the weights of brain parts with the gender, age, and various psychiatric diseases of the patients.[38] The priority of the somatic in Meynert's work, and his allegiance to Rokitansky, put him on the fast track to academic promotion. In 1870 Meynert was appointed head of the newly founded Psychiatric Clinic in Vienna (later the First Psychiatric Clinic). By 1873 he had become *Ordinarius* of psychiatry.

Dissonant Voices

The ascendancy of the somaticists in Vienna did not meet with universal approval. In psychiatry in particular, which had hitherto been dominated by asylum psychiatrists, or alienists, there was rising discontent with the academic pretensions and somaticism of the Rokitansky school. Psychiatry had long been based in the asylums, often in the countryside, and it felt peculiarly distant, both geographically and institutionally, from the university context where Rokitansky's revolution was under way.[39] Moreover, alienists had traditionally focused their attention outside rather than within the body. They hoped to treat their patients by removing them from the environments in which mental disorders arose and placing them within the patriarchal structure of the asylum. If anything, alienists believed that their main public role lay in guiding the political debates surrounding the reform of the asylum; they considered the lure of the dissection room foreign to their interests.

The editor of the *Wiener medizinische Wochenschrift*, Leopold Wittelshöfer, spoke for many when he attacked Meynert's appointment as head of the Psychiatric Clinic in Vienna in 1870. The appointment was made based on Meynert's anatomical credentials, but Wittelshöfer suggested that Meynert's work in the anatomy of the nervous system was not sufficient preparation for psychiatric work:[40]

> All respect is due for Mr. Meynert's anatomical work and its results. But with what right can the anatomist wish to line himself up for a professorship in psychiatry? Surely not because he has dealt with the anatomy of the brain or because he happens to be prosector at the local asylum? So far, Mr. Meynert has achieved nothing in the realm of psychiatry which would give him the right to the rank of professor in this field, and if what his partisans claim were true, that he will one day "cause a landslide [*Umschwung*] in psychiatry [*Irrenheilkunde*]," we believe that we should first wait for this landslide and only then reward it."[41]

Similar sentiments were expressed in a debate over Meynert's promotion to *Ordinarius* in 1873. Meynert's patron Rokitansky had written a glowing recommendation to support the promotion, emphasizing Meynert's "splendid reputation in the entire medical world, which probably no psychiatrist had enjoyed previously." To Rokitansky, Meynert had "led the field out of its former stagnation and superficiality onto progressive paths and tapped into its depths."[42] Identifying Meynert as one of his own, someone who had extended the pathological anatomical approach to the nervous system, Rokitansky put the full weight of his influence behind him. This support seems to have been decisive. A day later, Meynert's promotion was accepted, with twelve votes approving the promotion and seven against.[43]

But as this split vote suggests, the promotion did not please everyone. The alienist Ludwig Schlager, newly minted clinical director of the Niederösterreichische Irrenanstalt am Brünnlfeld, who had to work alongside the clinical professor, wrote a long and angry "separate vote." In his dissenting opinion, Schlager attacked Rokitansky, who, "with all due respect to his authority in the field of pathological anatomy cannot claim the right to be recognized as an authority in the field of psychiatry as well."[44] This lack of knowledge made Rokitansky unable to acknowledge sufficiently the achievements of other psychiatrists in the areas of forensic psychiatry, clinical observation, or psychic semiotics. Meynert, moreover, with "only two years of clinical work" and a complete lack of administrative experience, was utterly unprepared to represent the field as a whole. Schlager also attacked Meynert's pathological anatomy, which was, he argued, either old news or wrong, as recent criticism of some of Meynert's anatomical findings had suggested. More importantly, expertise in pathological anatomy—and Meynert had "so far almost exclusively published anatomical research"—was of no use for the clinic. Questioning Rokitansky's approach more generally, Schlager asked sarcastically: "Why doesn't one then appoint pathological anatomists as the heads of medical clinics at all universities?"[45]

The animosity between Meynert and Schlager deepened to the extent that it was eventually decided that the two men could not work together in the same institution. As a solution, the clinic was split in two. In 1875 the Second Psychiatric Clinic was founded in the rooms of the former psychiatric "observation ward" (*Beobachtungszimmer*) at the General Hospital. The existence of two psychiatric clinics was unprecedented in Vienna or indeed anywhere else.[46]

The Social Structure of Pathological Anatomy

While, as his detractors never failed to mention, Meynert had very little firsthand experience with patients, he had not been without contact with clinical medicine. This contact, however, took a peculiar form and was typical of the Rokitansky system of pathological anatomy in Vienna. In contrast to Paris, where, in the words of historian of medicine Erna Lesky, "every clinician was his own prosector," Vienna was both more centralized and more specialized.[47] Already under Frank, Vienna medicine had broken with the Paris model. Frank recognized that the rich and diverse pathological anatomical material would exceed the capacity of the clinician.[48] Specialized prosectors would deal with material sent from all corners of the hospital. The Viennese system separated the clinic from the mortuary.

This structure helped raise the status of Rokitansky's method, but it also meant that pathological anatomists depended in important ways on clinicians. Rokitansky was clear about the importance of clinical observation: "Just as pathology can henceforth not dispense with an anatomical basis (an anatomical element), so must pathological anatomy . . . be executed with constant consideration of clinical observation. In fact it has to take this practical direction if it . . . wants to raise itself to the most comprehensive anatomical description possible of disease processes."[49] But, because the pathological anatomist did not perform the clinical observation himself—by the time the bodies arrived at Rokitansky's institute, the moment for describing the symptoms was long past—for the system to work, a form of social interaction was needed. The pathological anatomist carried out the dissection, but in order to correlate his findings with symptoms he had never seen, he also relied on the diagnosis and clinical report that arrived in his morgue along with the corpse.[50]

The interaction between the clinicians and the prosectors was structured in a number of ways. First, all clinical *Chefärzte* (senior physicians) had the right, along with their assistants if they wished, to attend the dissection of their former patients. In this way, as one commentator noted in 1863, "the Pathological Anatomical Institute always remains closely linked to active processes at the bedside."[51] Second, Rokitansky considered it useful for the physicians at the General Hospital to keep diaries (*Tagebücher*) for all their patients, which he could consult. Third, clinicians examined the living patients following Rokitansky's suggestions, using a new battery of tests, such as auscultation and percussion. Rokitansky created a model for this kind of cooperation with the clinician Skoda, who demonstrated its fruits in his influential *Abhandlung über Perkussion und Auskultation* of 1839.[52] The Vienna system thus furthered "a fruitful back and forth between bedside and dissection table."[53] Pathological anatomy in Vienna was an intensely social operation. The same was true for Meynert. His work marked an explicit application of the principles of pathological anatomy to the nervous system.[54] When Meynert looked for changes in the brain to explain the symptoms that had been recorded during the patient's lifetime, he translated an often psychological language, written by colleagues in other clinics, into the idiom of a new somatic psychiatry.

Meynert's 1866 paper "A Case of Speech Disturbance, Explained Anatomically" is a good example.[55] A patient from Skoda's clinic, a twenty-three-year old domestic maid, had suffered from heart problems.[56] Two weeks before her death, she had become aphasic.[57] As the report stated, she was no longer able to "find words necessary for communication," such as "head" or "hand," and she drew on other, less appropriate words instead (e.g., "yellow"). After she died, her corpse was sent to Meynert, who carried out the autopsy. Meynert found two sets of lesions that, after characterizing them both macroscopically and microscopically, he correlated with the aphasic symptoms outlined in the report.[58] He wrote: "Functionally, local death had arrived in an area previously assigned to representations [*Vorstellungen*] . . . which will manifest itself as loss within representational life [*Vorstellungsleben*]." Although Meynert remarked that it was difficult to decide whether he was dealing with a lesion of the connections between the optical *Erinnerungsbild* (memory image) and the *Klangbild* (acoustic image) or a destruction of the nerve cells where the *Klangbild* was localized, from his anatomical findings he concluded that "the latter at least was part of what caused the phenomenon."[59] That is, Meynert assumed a direct connection between a damaged part of the brain and the psychological symptoms presented in the patient file. Further he used this connection to identify more generally those parts of the brain responsible for speech and memory.

As this example shows, Jaspers was right to see in the *Zentrenlehre* the elision of psychological and anatomical characteristics. As its name already indicates, the pathological anatomical method was a hybrid enterprise, and its hybridity can be traced back to the social practice on which, in Vienna, it was founded. For his correlations, Meynert relied on the patient files written by physicians from various departments at the General Hospital—internists, surgeons, obstetricians—alongside those provided by psychiatrists.[60] He correlated nervous damage with such disease categories as aphasia, epilepsy, and "paralytic idiocy" (*paralytischer Blödsinn*) or more traditionally psychiatric categories such as "primary insanity" (*primärer Irrsinn*) that he found in his patient files.[61]

But it would be wrong, following Jaspers, to see this as an uncritical translation. Confronted with the reports written by other doctors, who often worked with very different nosologies and who, in the case of the alienists, were often overtly hostile to his project, Meynert recognized the translation between psychology and anatomy as a problem. He wrote in 1876, "as the traditional psychiatrists still feel today, and the clinician who cannot do the same kind of work, just as little does traditional psychiatric theory in its psychological formation match up with the . . . special clinical theory of the diseases of the forebrain and their complications."[62] Meynert worried that the psychology underlying alienist practice was inexact and thus could not serve as a secure reference for pathological anatomical findings.[63] In Meynert's earlier work, he refused to accept any but the broadest of psychiatric disease categories, such as "primary forms of depression" (*primäre Depressionsformen*), "primary forms of excitement" (*primäre Aufregungsformen*), alcoholism, and epilepsy, because he considered that the more specific the presentation of the symptoms, the more unreliable it became.[64]

There were two possible solutions to the problem. Meynert could try to reform the language of psychology to make it more susceptible to a translation into physiological language. Or, he could construct a vision of the nervous system that would be primed to explain psychological functions.

Focusing on a psychological reform posed the greater challenge to Meynert, especially since before 1875 (the year when he moved to his own clinic) he struggled with Schlager for control over patients. Despite these difficulties, Meynert tried to influence the psychological side of the equation in two ways: first, he encouraged the use of those nosological schemes that he considered most free from ideology. For instance, he embraced the French psychiatrist Jean Etienne Esquirol's turn-of-the-nineteenth-century nosology, which he liked precisely for the "lack of systematicity [*Systemlosigkeit*] of his presentation." According to Meynert, the unsystematic nature of his account had allowed Esquirol to "present, in an unrestricted manner, the diseases that he [was] able to clearly define."[65] Second, Meynert attempted as far as possible to perform his own clinical investigations. In the exchange with Schlager over his 1873 promotion, Meynert assured that he left "not a single examination to the assistants, but instead individualize[d] every single case in the pursuit, already somewhat successful, of grouping and separating the diseases of insanity naturally." Only on days when he had to do dissections, he pointed out, would he miss the morning medical round.[66] Whenever he had to rely on secondhand accounts, Meynert made sure the clinical descriptions had been performed by his own assistants who were trained in pathological anatomy.[67]

Reforming the Language of Physiology: Association Psychology and the Reflex

In his earlier career, however, when Meynert's work and influence were mostly restricted to the dissecting room, he was not in a position to mold psychological practice and description to his purposes. He thus worked on the other part of the equation, reforming the language of *physiology* to equip it to explain psychology. Brain physiology was open to such a reformation, because while there was growing consensus that the mind had an ultimate material foundation, there was no such agreement about what that foundation might be. In contrast to other medical specialties, psychiatry lacked a clear route to somaticization; there was no current and accepted somatic model for brain function. In many ways, psychiatry suffered—and continues to suffer—from what Charles Rosenberg has called "procedure envy, or organic inferiority."[68] For Meynert, the protean nature of nervous physiology at the time furnished him the opportunity to craft it for his purposes.

Among the most promising approaches to the nervous system at the time was an essentially connective principle: reflex physiology, which considered the brain as the interworking of sensory and motor parts.[69] Reflex physiology found its roots in the Bell-Magendie law of sensory-motor separation. Physiologists Charles Bell and François Magendie had shown in 1811 that the posterior spinal root nerves contained only sensory fibers, and in 1822 that the anterior spinal root nerves contained only motor fibers.[70] Although Bell did not describe the separation of anterior and posterior spinal root nerves in this way, according to Franklin Fearing, he provided an "adequate conception of the neural structures which must serve as a basis for the reflex."[71] The prominence and academic respectability that reflex physiology had gained by midcentury made this approach an obvious choice for the neuropsychiatrists. As Meynert himself put it in 1890, the Bell-Magendie law of sensory and motor separation was the "first axiom of brain physiology."[72]

The impact of the connective principle of sensory-motor physiology is apparent in the reframing of localization. Localization thus far had been linked to the work of the Viennese physician Franz Joseph Gall and his phrenological followers. The phrenologists argued that the sites of psychological faculties could be located on the surface of the brain, and psychological characteristics could thus be read in the skull's bumps and hollows. But although Gall's phrenological ideas had a wide popular impact and continued to find support through the middle of the nineteenth century, the work was always treated with skepticism, both by the medical establishment and by political and church authorities who resented the materialism and secularism of the doctrine.[73] Localization was given a new lease on life in 1861. That year Paul Broca famously presented his patient Monsieur Leborgne to the Paris Anthropological Society. Leborgne had difficulty producing speech, a condition that Broca called "aphémie." This was correlated postmortem with damage to the third frontal convolution.[74]

Broca's institutional authority and appeal to pathological anatomy was crucial for returning localization to the medical mainstream, but its position was not fully secured until thirteen years later. In 1874 the young Carl Wernicke discovered another speech area. Though Wernicke's discovery came after Broca's, it was at least as significant in aphasia research, because it confirmed the principles of the *Zentrenlehre* within a sensory-motor structure.[75] For Broca, the function of "articulated language" depended on the integrity of both a motor and a sensory system, and he had not presented the center as a specifically "motor area."[76] Wernicke's most important contribution was not simply that he "discovered" the location for sensory speech on the cortex.[77] Rather it was that he turned Broca's one-to-one lesion model of speech into a sensory-motor system. Broca's area, as much as the area on the first temporal gyrus Wernicke had identified himself, was no longer an isolated center of speech; it was part of a reflex system, a circuit, along which damage could occur. Wernicke was thus able to rename Broca's "aphémie" as "motor aphasia."[78]

This new way of categorizing higher functions is also visible in the mapping projects of the period. In the 1870s, Eduard Hitzig, working as an electrotherapist in private practice in Berlin, had noticed that he could cause a contraction of the eye muscles if he applied his electrodes on the area behind the ear. He concluded that the eye movements had been caused by the stimulation of brain structures. In collaboration with the anatomist Gustav Fritsch, he began to explore systematically the functional anatomy of a dog's brain through direct electrical stimulation, noting down their findings on a brain map.[79] The two researchers identified various motor "centers" on the surface of the dog's brain, such as the "center for the neck muscles" or the "center for the extensor and adductor muscles of the foreleg."[80] Their work was later extended to describe somato-sensory areas as well.[81] Basic sensory and motor function, as well as higher functions such as language, could all be localized on the surface of the brain.

The application of sensory-motor principles to psychiatric pathological anatomy, and by extension to the *Zentrenlehre*, was not, however, without its difficulties. Academically respectable studies of the nervous system based upon such principles had in the early part of the century been confined to lower functions. Further, the reflex, one of the key neurological principles, had traditionally been restricted to the spinal cord, not the brain. An early proponent of Bell and Magendie's work was the Englishman Marshall Hall. Hall's reflex was a purely spinal principle, and he explicitly detached it from higher functions. To Hall, the reflex consisted in the reflection (modeled on optics) of an incoming nerve excitation into an outgoing nerve excitation, taking place in what he called the excito-motor, or excito-reflector system.[82] Hall chose the term *excito-reflector* to describe a reflex arc, rather than *sensory-motor*, in order to detach the reflex from sensation, which to him was conscious and thus only a function of the brain.[83]

To justify the extension of these neurological principles to higher functions, then, they would have to be radically revised. Indeed, in the mid-nineteenth century, though Hall's mechanistic model remained an inevitable reference point, many physiologists found it too restricted. The reflex could no longer be the simple stimulus-response/excito-reflector model that Hall proposed. Most reflex physiologists working after Hall criticized the narrow scope of his concept and work and challenged his strict distinction between lower and higher functions.[84] German anatomist and physiologist Johannes Müller styled himself as a successor to the early reflex physiologists. He is credited with experimentally proving the Bell-Magendie law in frogs in 1831, twenty years after its first formulation.[85] Further, he extended Hall's reflex beyond the "reflector" spinal cord and medulla oblongata to all elements of the "central organ" (*Centralorgan*), including the brain.[86] Müller thus broke down the distinction between higher and lower neural structures that had dominated Hall's model. As a consequence of extending reflection to the brain, Müller no longer excluded sensation from spinal action. In fact, he suggested that Hall was wrong to assume that the spinal cord and medulla oblongata only "excited," that is, initiated the motor action. Supposedly, lower automatic reflex function involved "real sensations [*wirkliche Empfindungen*]" too.[87] The (reflex) movements of coughing and sneezing proved this, because they comprised sensation even if the response appeared automatic.[88]

Müller also aimed to complicate Hall's simple "reflection," and his work marks the first in a long line of interventions that served to emphasize and elaborate on the connective properties of the reflex.[89] Müller suggested a "principle of proximity" that determined the process of reflection, explaining the "flow or oscillation" (*Strömung oder Schwingung*) from incoming sensory to outgoing motor fibers.[90] Often, one incoming sensory flow would affect more than one motor fiber, resulting in the action of several muscles. If the reflex was to be extended to higher functions, and thus to functions of a higher order of complexity than the automatic action of the spinal-cord reflex, the simple one-to-one relationship between stimulus and response was no longer sufficient.

The neuropsychiatrists followed Müller's example. Griesinger, when drawing on the reflex in his somaticization of the psychiatric project, also attempted to introduce a new degree of complexity to the intersection between sensory and motor arcs, corresponding to the complexity of higher function. Parallel to Müller's rather vague principle of proximity, Griesinger proposed a process of *Zerstreuung* in the *Centralorgan. Zerstreuung*, or dissipation, was made possible through a state of the *Centralorgan* that resembled the tension of muscles, what Griesinger called

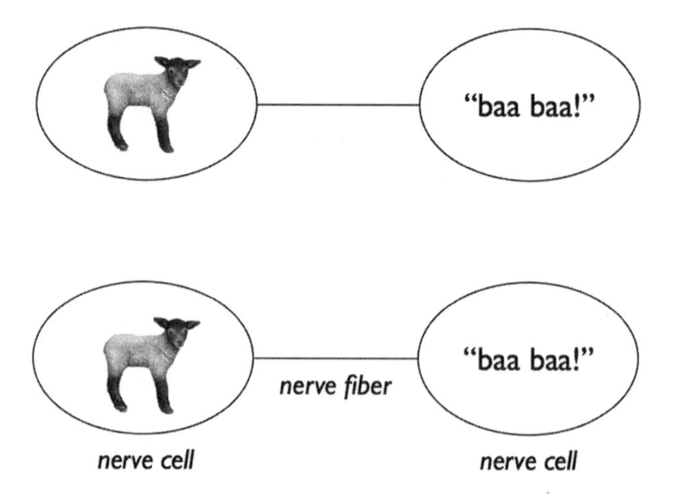

FIGURE 1.2. Schematic model of Meynert's physiologization of association psychology. *Top*, visual *Vorstellung* of lamb is psychologically associated with acoustic *Vorstellung* of bleating. *Bottom*, visual *Vorstellung* of lamb "located" in nerve cell is physiologically associated (through association fiber) with acoustic *Vorstellung* of bleating (also located in nerve cell). (Figure prepared by author.)

"tonus."[91] Although the tonus was produced by separate incoming sensory excitations, these excitations were not individually preserved but contributed to an "average level of excitation" (*mittleres Fazit der Erregung*) in the *Centralorgan*. While Griesinger did not propose an exact mechanism for *Zerstreuung*, he developed a speculative anatomical model to explain it. He suggested a complex connection between centripetal impressions and already-existing brain states: a combinatory process, in which different sensory impressions were redirected by *Strebungen* (volitional impulses) toward appropriate *Bewegungen* (movements).[92]

Such considerations lead us back to Meynert. In his 1870 article "Contributions to the theory of maniacal phenomena of movement," Meynert, too, sketched a supercharged reflex in order to develop a physiological model of the mind, foregrounding once again its connective properties.[93] But Meynert differed from his predecessors in that he found resources not in physiology textbooks, but rather in the work of association psychology. Like the association psychologists, Meynert asserted that two sensations might be associated if they occurred simultaneously in consciousness. Meynert differed from most of them, however, when he presented the process in physiological terms.[94] An example he used throughout his writings is that of a bleating lamb (fig. 1.2).[95]

A bleating lamb excited two separate cortical cells. In the first cell, which was connected to the eye through a system of so-called projection fibers, it produced the visual *Vorstellung* (representation) of a lamb.[96] In the second cell, connected to the ear, it produced the *Vorstellung* of bleating. The acoustic or visual *Vorstellungen* of the bleating lamb thus could be traced to the particular physiological modifications of individual nerve cell bodies. This simultaneous excitation caused the two cells to be connected through an association fiber. After the first coincidence of the two sensations, which created the association, both *Vorstellungen* became obscured (*verdunkelt*), drifting out of conscious life.[97] But in the future, when the sound of bleating was heard, exciting the auditory nerve, the nerve cell containing the lamb image would also be excited, and thus the complex audiovisual idea of a lamb would be produced in the mind, even if the lamb could not be seen.[98]

After accounting for the association of the two sensations physiologically, Meynert broadened the model to include motor function as well: if association could explain the connection between two sensations, it could also provide a mechanism for understanding how a sensory impulse could be connected to a motor response.[99] Like the physiologized connection between two sensations, sensory and motor images were physically linked through fibers of association. Take the example of a child touching a flame.[100] On first feeling the lit candle, a child would respond through an automated (reflex) response: withdrawing his hand from the source of pain. During this process, the visual image of the flame (A), the sensation of pain (B), and the *Innervationsgefühl* of the reflex movement (C) would be produced on the cortex, via the projection system. According to the principle of association, these three *Vorstellungen* would be connected by association fibers, forming the association bundles CB, CA, and BA. Thus in the future, the mere sight of the flame (A) would activate C, and the child would withdraw his hand before touching it (see fig. 1.3).

The Paradoxes of Localization

The project of physiologizing association psychology had a double pay-off for Meynert. It simultaneously provided a model of reflex function that would justify its application to the brain, and it facilitated his communication with the physicians whose reports accompanied the corpses he studied. Meynert primed his physiology with association psychology to allow

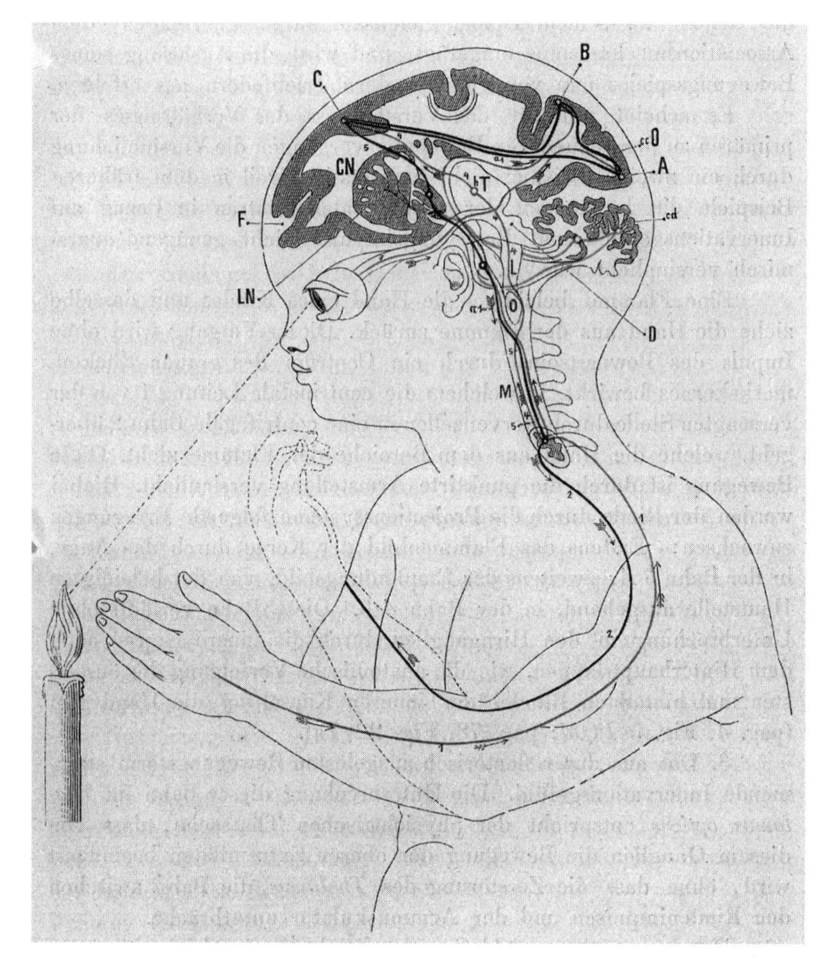

FIGURE 1.3. Schema for the conscious movement of the arm. (Reprinted from Meynert, *Psychiatrie*, 147.)

the application of pathological anatomy to psychiatric problems. He remained faithful to localization and pathological anatomy throughout his working life, as is evident from his programmatic statements. In 1876 he insisted that the pathological anatomical approach was justified and useful: "Exact disease theory [*exacte Krankheitslehre*] has gained its ground in Germany through Rokitansky, Skoda and Virchow alone, and for Germany there is no essential pathologist or adequate method to be found outside of this school." Criticizing the alienist emphasis on external factors such as building structures, asylum location, and occupational work for patients, Meynert stressed instead the importance of his own project of exploring the "physiological valence of localities" of the brain.[101]

Meynert remained confident even though he realized that compelling cases, like that of the aphasic patient described in the 1866 landmark paper, were very rare.[102] Only a minority of diseases manifested *Ausfallsymptome* (symptoms of loss) correlated with a clear lesion.[103] To respond to this problem while remaining within the somatic paradigm, in his later work Meynert expanded the scope of his research beyond the lesion proper, including less dramatic pathological processes in his explanations. As he pointed out in the section "Nutrition of the Brain" of his 1884 textbook and in his 1890 *Klinische Vorlesungen*, metabolic changes played an important role in the workings of the brain. Apart from participating in association, cortical cells had the ability to control vasoconstriction (*Gefässverengung*).[104] Dysfunction in this process could help explain mental pathology without recourse to a lesion. Similarly, Meynert developed the notion of *reizbare Schwäche* (excitable weakness), which described the "capacity of the forebrain . . . to inhibit excitatory states of other segments of the nervous system."[105] Diseases such as hallucinations or epileptic seizures could be explained this way.[106]

And yet, the impact of his revised reflex model was by no means clear-cut. While it legitimated the pathological anatomical method in psychiatry, it also provided resources for undermining the localization paradigm for which that method was deployed. The example of the *reizbare Schwäche* is instructive on this point. In drawing on the model of inhibition, Meynert placed emphasis on the connective elements of his theory, the way in which different reflex elements worked together: the *reizbare Schwäche* represented a "physiological intertwining (*Ineinandergreifen*) of the constituent parts of the brain."[107] To the extent that elements worked together, it became difficult to attribute function to any one individually. Such intertwining was most apparent in the association system that Meynert had invoked to justify his system but which, for this reason, remained profoundly corrosive of the localization project.

On the one hand, the localization of function depended most proximately on the *projection* system. As we saw in the example of the bleating lamb, nerves were connected to the sensory organs through the system of projection fibers. As implied in the term *projection*, Meynert assumed a direct, one-to-one connection between points of the body surface and points of the cortex. Even though the fibers, on their way from the periphery to the cortex, traveled through various areas of gray matter (e.g., the optical nerve, on its way from the eye to the brain surface, passed through the subcortical original cells [*Ursprungszellen*]), the fibers did not undergo any change at this midway point; one might say that the projection fiber preserved its identity across gray matter, what Meynert called the "principle of isolated conduction."[108] This fiber-based mapping thus provided a homology between sensation (and, by extension, motor processes) and specific centers on the brain surface.[109]

The association system, on the other hand, could not be considered a one-to-one mapping. It was precisely the complexity of the connections in the association system that had justified the application of the reflex to the brain.[110] Meynert was acutely aware that his embrace of the connective model of associationism troubled the appeal to simple localization, and he did not think it was possible to localize higher intellectual function. Associations linked areas of the brain together, and so the "intellectual processes in the cortex" that they described could not be restricted to sharply defined areas. Rather, the "uniform structure of the forebrain" that everywhere contained "nerve [cell] bodies able to create *Erinnerungsbilder*" made "every part of the brain the center [*Herd*] of induction processes."[111] For this reason, Meynert opposed accounts of localization such as the one by Eduard Hitzig, who suggested an area of abstract thought (*begriffliches Denken*) as a functional area localizable on the cortex. Meynert noted that "the unified anatomical structure of the cortex rendered absurd a lining up of thought processes with specific cortical parts such as the frontal lobe."[112] He generalized his critique by pointing out that the "very direct, crude mode of argumentation in psychiatric treatises" was problematic because the "localized separation" of simple *Vorstellungen* was "instantly reversed [*aufgehoben*] insofar as the use of sensory impressions in the process of thought happens through the associations, which join together the most distant parts of the cortex and thus nullifies [*beheben*] the localization."[113] For this reason, in his later writings, he distanced himself from the *Zentrenlehre*, referring to the "*so-called* cortical 'centers.'"[114]

The associative model of the nervous system was a double-edged sword in a second sense. Its connective properties helped to mitigate the fragmentation of the mind implicit in the *Zentrenlehre* paradigm, because they helped explain how anatomically separated functions nonetheless worked together. And yet, even as the model helped to reconstitute a unified self, it could also promote its later dissolution. These implications of

Meynert's system are clearest in his late synthetic writings. Take, for example, a lecture given at the tenth International Medical Congress in Berlin in 1890: "Das Zusammenwirken der Gehirntheile" (The collaboration of the parts of the brain).[115] "A word like brain," Meynert began, "signifies in our language a composition of individual parts." Although these parts consisted "at the level of the smallest constituents" of the same elements, namely cells, they assumed different functions. Such a presentation concurred with Meynert's previous presentation of the *Zentrenlehre*. But he also raised a new question: if the nervous system was composite, how did individual parts work together?[116] Meynert provided the answer: they did so in a social way. Meynert argued that "the gray and white matter of the brain can only be held together with a social grouping of living *beseelt* [animated] beings." The "living *beseelt* beings" were animated by different *Vorstellungen* and participated in a larger process of association.[117]

The social relations in the nervous system helped produce a quasi-unified organism. Drawing on military vocabulary, Meynert pointed out how the actions of the "social colony of cells in the cortex" formed a "tactical formation for the mastery, for the capture of external nature." They did so through the body, the "undeniable collective property of the colony that is conjoined with it," that could act upon the "limitless region of the so-called external world, even to the stars."[118] The "cavity-like form of each cortical hemisphere" was a "formation of its living elements that surrounded nature," seized nature through the sensory-motor arsenal of "feelers" (*Fühlfäden*) and "tentacles" (*Fangarme*).[119] The result of this collaborative action was the production of higher psychological functions: "The triumph of this attack of over a billion entities armed with their feelers and tentacles is the sparking of consciousness through images that are radiant, sounding, fragrant, and tactile."[120]

Meynert's social metaphors corresponded to his conception of the subject. To Meynert, the construction of individuality referred to "the anatomical structure of the cortex, and the simple physiological process which enters into our present discussion."[121] It was created through the "firmest associations" connecting "the aggregate of 'memories' [*Erinnerungsbilder*]," which "under ordinary circumstances are well-nigh inseparable."[122] Meynert distinguished between primary and secondary individuality. The primary individuality was bodily and was based on the child's distinction between the *Kennzeichen* (signs) of its own body and the *Kennzeichen* of the world. If the touch of a foreign finger elicited only one sensation, the touch of the child's finger on its own body elicited two: from the finger and from the body part it touched.[123] Through what was essentially a process of *Bahnung* (facilitation), in which "memories conveyed by those nerve fibres, which enter into definitely associated groups and constitute tracts along which a motor impulse is conducted, lend a helping hand," fixed connections were established, making up the child's primary individuality. Crucially, as Meynert pointed out, this primary individuality "will never again be able to dissolve."[124] In other words, association, the mechanism by which individual elements of the nervous system were linked together, acted as a unifying force, establishing a sense of the self that was, to a certain extent, unified.

But while the process of unification counteracted the fragmentation inherent in the localization paradigm, it did not necessarily stop at the boundaries of the individual. What Meynert called "secondary individuality" extended the process beyond the self and thereby threatened it. *Erinnerungsbilder* of the world, if repeated often enough, "will enter into firm associations," forming the "nucleus of a secondary individuality." The primary individuality, or ego (*Ich*), thus extended itself constantly "by association" through repeated *Vorstellungen*, to include loved ones and personal possessions, but also "skills obtained by constant practice in any art, science, a fondly cherished aim in life, convictions, patriotism, and honor."[125] This was not to say that the individual did not appear to herself unified. In fact, Meynert pointed out that the cortex "through its association processes appears to itself as a single entity." But Meynert emphasized that secondary individuality became "de-centralized and . . . made to include much of the external world."[126]

Extending this process of association beyond the self had an important function; Meynert pointed out that it could explain the emergence of feelings such as allegiance to a social group or a nation. But he suggested that it could also go too far. Secondary individuality might even endanger the corporeal "primary individuality" and, by extension, itself. In an arresting passage, Meynert acknowledged the possibility that the "component factors of the secondary ego [might attain] such a psycho-motor intensity in the play of association" that they become more powerful than the motive of the preservation of the primary ego (that is, self-preservation): "In fact, the very person who sacrifices his life believes, in so doing, that he preserves his own individuality, which now includes so much that lies beyond his physical self." It is in this explanation of an action as "complicated, problematic, or incomprehensible" as suicide that the self-corrosive tendencies of Meynert's model became most evident.[127] Not only did association present a challenge to the localization of psychological function; it also challenged the unity of the self.

Conclusion

The extension of the pathological-anatomical method to psychiatry was complex and contested. It was perhaps Rokitansky's toughest sell: although his somaticism held intellectual as well as professional promise, the dual power structure of psychiatry (asylums and university clinics) in the middle decades of the nineteenth century created problems that other specialties did not have to confront. That Theodor Meynert was able to rise to his position was thus simultaneously a testament to his resourcefulness and a sign that pathological anatomy had achieved hegemony in Vienna medicine in the third quarter of the nineteenth century.

But already at its height, pathological anatomy in psychiatry and the *Zentrenlehre* for which it provided empirical support showed internal cracks. Meynert had foregrounded the connective properties of the reflex in order to render a theory that had previously been restricted to lower, automatic function suitable to explain the workings of the mind. The integration of the association system into the reflex arc, however, made questionable the claim that functions could be localized in one place. Meynert felt the force of these difficulties only for the highest level of mental activity. While simple psychological functions could be localized, thinking relied on complex associative webs that spanned the brain.

Meynert was not the only neuropsychiatrist in Vienna. His growing reputation attracted numerous visitors and students from Austria and beyond in the 1870s and 1880s: neuropsychiatrist Gabriel Anton was Meynert's assistant in 1887; the American neurologist Bernard Sachs, who later translated Meynert's *Psychiatrie* into English, worked there in the early 1880s; the Russian neurologist Liverij Osipovich Darkshevich joined Meynert's lab in 1884; and the Swiss psychiatrist Auguste Forel, later director of the Burghölzli, worked there in 1871–72.

Two students in particular will interest us here, because in their work we can see the extent to which Meynert's theory of nervous actions challenged the *Zentrenlehre* at all levels. Sigmund Freud, who in 1883 worked in Meynert's laboratory and was *Sekundararzt* at his clinic, radicalized this challenge to shake the very foundations of neuropsychiatry.[128] I discuss his engagement with Meynert in chapter 3.

That Freud challenged Meynert's *Zentrenlehre* should come as no surprise. The orientation of Freud's later work makes it hard sometimes to recall that he ever worked within the framework of somatic medicine. It is perhaps more surprising that we can see a similar challenge, albeit less pronounced, in the work of the man who is regarded by many as Meynert's legitimate heir: Carl Wernicke. Wernicke worked in Meynert's laboratory in Vienna for six months in 1871, and his development of Meynert's work brought him fame and position. And yet, as Wernicke developed his own clinical practice, the fraught relationship between the connective model of the reflex and the localization project became increasingly clear. In chapter 2 we will see how, faced with a steady stream of patients, Wernicke came to sideline the pathological anatomical method and move beyond the localization project that had in the early part of his career made his name.

CHAPTER TWO

In the Lecture Theater

Reflex and Diagnosis in Carl Wernicke's Krankenvorstellungen

Vienna, one of the great cities of Europe, capital of a continental empire and home to a booming middle class, might seem an appropriate location for a medical revolution. Its hospitals and university clinics set trends, and doctors elsewhere, especially in the German-speaking world, took note. It is not surprising that Theodor Meynert and, as we shall see in chapter 3, Sigmund Freud made their careers in Vienna and used it as the base to promote their medical programs. The second city in the European part of our story is, in contrast, unassuming. Breslau, where Carl Wernicke (1848–1905) spent most of his life, figures less prominently in the European imagination. Today even the name, Breslau, is erased from our maps, the result of shifting borders in postwar Central Europe.[1] Today it is part of Poland, but in the second half of the nineteenth century, Breslau was located in the far east of the German Empire. Two hundred miles from the Prussian capital, it was for many out in the sticks; when economist Lujo Brentano accepted a position at the university in Breslau just after unification, he considered himself to be moving into a kind of "exile."[2]

On Brentano's arrival in Breslau, however, he found a city on the rise. It was the third-largest city in the newly founded Reich and still growing.[3] The city made the most of the new possibilities of the empire; it benefited from the growth of the railway system, which telescoped the distance separating Breslau from the rest of the empire and created a market for its manufacturing sector. In this way, it participated fully in the economic expansion in Germany following 1871. Especially important for Breslau was the growth of coal mining and iron production, woodworking, brewing, and the tobacco and textile industries.[4]

For an aspiring doctor, Breslau offered many advantages as well. For Breslau was also known as a university city. Apart from Munich and Berlin, Breslau was the only German city that had a university *and* a polytechnic, both with good reputations.[5] In particular, the medical faculty was highly regarded.[6] With its excellent laboratories, Breslau attracted doctors from abroad with interest in the basic sciences: William Welch, later a powerhouse in the medical faculty at Johns Hopkins, worked there in the late 1870s; a decade later, Ivan Pavlov worked in Breslau with the physiologist Heidenhain, studying gastric secretion in dogs, a project that laid the foundations for his later work on conditioned reflexes.[7] The university could claim a number of famous alumni, especially in the life sciences, including Paul Ehrlich, Albert Neisser, and Ferdinand Cohn. For the young Wernicke, Breslau was a place where one could make a career.

Wernicke did not have to wait long. As we saw in chapter 1, he was propelled to fame when he published a monograph on aphasia in 1874. He was only twenty-six years old. In the book, Wernicke described the brain area for sensory speech, an area that is still named after him. According to Wernicke, we could explain sensory aphasia by the existence of physical damage to this part of the brain. That is, he had shown how a "mental disease" (*Geisteskrankheit*) was in fact a "brain disease" (*Gehirnkrankheit*).[8] Wernicke's research was received as one of the most sophisticated confirmations of Griesinger's dictum: he became a poster child for somatic psychiatry.

Wernicke's position as Meynert's heir and the culmination of the neuropsychiatric tradition has profoundly marked his reputation and the scholarly writing on him. The familiar Wernicke is the one who in 1880 was the first psychiatrist to address the prestigious Gesellschaft deutscher Naturforscher und Ärzte. In his talk "On the scientific standpoint in psychiatry," Wernicke drew on his aphasia research.[9] In particular, he described the so-called sensory and motor "centers" on the surface of the brain, containing *Erinnerungsbilder* (lit., memory images) and *Bewegungsvorstellungen* (lit., representations of movement), which were the basis for understanding all mental processes. The discovery and description of such centers was, for Wernicke, the ultimate evidence of psychiatry's participation in modern scientific medicine. It demonstrated how neuropsychiatrists could recast psychiatry in the somatic mold of the era. In the 1880s Wernicke saw a golden future ahead for such a psychiatry. At the same time, he criticized those who rejected neuropsychiatry's somatic claims. Playing with the opposition of science and superstition (and the double meaning of the German word "Geist"), he pointed out: "Despite the name *Geisteskrankheiten*, nobody thinks of regarding them as diseases of the *Geist* ["mind" and/or "ghost"], at least nobody who does not believe in ghosts and specters otherwise. Rather all are agreed that *Geisteskrankheiten* are based on diseases of the brain and thus are, if still less well known, but diseases of a known organ."[10]

Less familiar is the more circumspect Wernicke of the 1890s. According to Karl Bonhoeffer, who was Wernicke's assistant from 1893 to 1898, at that time Wernicke showed only little interest in brain pathology: "It was painful for me at first, that what I had expected from the discoverer of sensory aphasia and after his two-volume work about brain pathology, the analysis of brain pathological organic cases, was then almost entirely sidelined by him. It went so far that in the clinic one could be fairly certain that he would pass on a case, if one told him that it was something neurological or organic-brain pathological. At the time he only wanted to see psychiatric [cases] in the clinic."[11] Over the course of his career, Wernicke had shifted interests. By the end of the century, he was concerned more with those diseases that seemed to elude pathological anatomy than with those that confirmed its claims about the correlation of lesion and symptom.[12]

In the 1890s Wernicke was not alone in his de-emphasis of the Griesinger model. Within neuropsychiatry, many were moving away from pathological anatomy and refocusing their attention on the clinic.[13] But Wernicke's early reputation as a "brain psychiatrist" (*Hirnpsychiater*) makes his trajectory particularly valuable for tracking the broader development in the field. Wernicke's original research—much like Meynert's, discussed in chapter 1—was based upon autopsies, and this provided the initial theoretical framework for his clinical practice. He drew on a modification of Griesinger's psychic reflex, asking his patients carefully chosen questions (stimuli) and examining their responses. The result of this analysis was a clear classification of mental disorder that could be related to postmortem findings. The patients he encountered in the clinic, however, did not always allow such a simple correlation between mental and brain pathology. An "anatomy of the living" did not map so easily onto the anatomy of the dead. As Wernicke's practice continued, he drifted away from the *Zentrenlehre* until, toward the end of his career, he only loosely grounded his clinical work on the localization of function. In its place, Wernicke had developed a subtle and nuanced strategy for reading clinical symptoms. As we shall see, his reflex transformed into an interpretive principle.

Aphasia in an Age of Lesions

When Wernicke published his monograph *Der aphasische Symptomen-complex*, he had just completed his medical studies and was working as an assistant in psychiatry with Heinrich Neumann in Breslau.[14] In his book, Wernicke described a center of speech that was the sensory complement to Broca's language area, connected to it by a system of transcortical fibers. Wernicke's book sparked a wave of interest in aphasia research. Several shorter papers on aphasia had already appeared,[15] but it was only after his book came out that the tempo of publication really picked up. By the end of the century, aphasia research was a vibrant field of investigation in the German-speaking world (with the work of such figures as Arnold Pick, Ludwig Lichtheim, Paul Flechsig, and Freud).[16]

What is perhaps most striking about the *Aphasischer Symptomencomplex* is its slender evidentiary base. Strictly speaking, the anatomical examples Wernicke discussed there should have been insufficient for drawing the far-reaching conclusions that permanently reconceptualized aphasia.[17] Of the ten cases that Wernicke presented to demonstrate the anatomical correlate of sensory aphasia, only three involved dissection, and of those, one did not even describe sensory aphasia.[18] Wernicke's thesis of a sensory speech area was thus dependent upon the examination of only two patients.

The fact that such meager evidence was sufficient for the book to become a success is testament to the hegemony of a *Zentrenlehre* based on sensory-motor principles. That is, Wernicke's thesis was well received predominantly because it adhered to the broader theoretical norms of neuropsychiatry, as described in chapter 1, not because it provided a thorough investigation of clinical cases. The priority of the theory is reflected in the structure of the book. As Gertrude Eggert has noted, unlike other neurological publications at the time, which usually presented a clinical picture with pathological anatomical evidence first and then drew their conclusions, in Wernicke's book this order was reversed. Wernicke outlined his theory of aphasia at the beginning of the book and only then provided a description of the pathological anatomical cases that seemed to confirm it.[19]

That Wernicke's theory of aphasia was reliant on the neuropsychiatric tradition is evident from his account; he drew explicitly on the ideas of both Griesinger and Meynert.[20] Griesinger, as we have seen in chapter 1, had argued that there was a parallel between the workings of the spinal cord and the brain. He formulated these ideas in the 1843 article

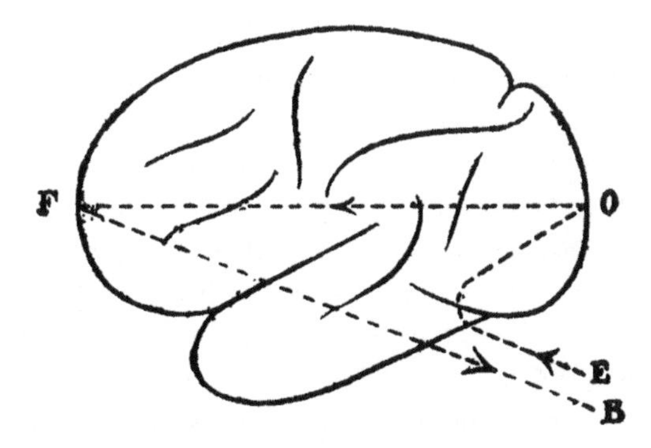

FIGURE 2.1. Schema for higher reflex. (Wernicke, *Symptomencomplex*, 8.)

"On psychic reflex actions."[21] Wernicke referred to this article in his book and placed a version of Griesinger's "psychic reflex" at the center of his theory. All human activities, including language, according to Wernicke, could be considered to result from such a reflex: "My interpretation of the speech process is merely the specific application of the process involved in spontaneous movement, the main features of which have already been well-established, to the movements necessary for speech."[22]

According to Wernicke's schema, higher functions could be explained in the following way (fig. 2.1). Sensation E traveled along the pathway EO toward place O in the parieto-occipital lobe (*Hinterhauptsschläfe-hirn*), leaving a memory image. If, later, another external stimulus arrived (not represented in figure 2.1), the "latent stimulus" at O was activated and moved toward F in the frontal lobe, along fiber tract OF. From there, it caused movement along the fiber tract FB. As Wernicke summed up the result: "Thus, pathway (E-O-F-B) adequately explains spontaneous movement in the mode of a reflex process."[23]

As this account shows, while Wernicke appealed to Griesinger for the general principle of psychic reflexes, he took the details from Meynert.[24] Though Wernicke had worked with Meynert only for a relatively short time, he acknowledged his debt to his senior colleague throughout his writings.[25] In particular, Wernicke's aphasia monograph was published not long after his stint in Vienna and bears the mark of Meynert's training.[26] In the opening paragraph of the book, Wernicke affirmed his adherence to Meynert's model of brain function: "The work here submitted is an attempt to provide . . . a practical application of Meynert's teachings of brain anatomy. . . . Whatever merit may be found in this work ultimately reverts to Meynert, for the conclusions here submitted issue naturally from a review of his writings and pathological studies."[27] In his version of the psychic reflex, Wernicke adopted Meynert's division of the cerebral cortex into an anterior motor part (F), where *Bewegungsvorstellungen* (movement images) found their place, and a posterior sensory part (O), the site of *Erinnerungsbilder* (memory images). These two parts were linked to the periphery (sense organs and muscles) by "projection" fibers (EO, FB), as in Meynert's model, and they were connected together through an "association" system (OF).

Wernicke's theory of the psychic reflex provided the rationale for his taxonomy of pathologies. First, pathologies were distinguished by the location of the lesion; lesions could occur in the projection system, leading to *Gehirnkrankheiten* (i.e., EO or FB in fig. 2.1); or they could affect the association system (OF), causing *Geisteskrankheiten*. Within the association system, a further distinction could be made. Wernicke identified two forms of *Herdsymptome* (symptoms caused by lesions).[28] Either the lesion could be located at a center (O or F), which would cause "symptoms of the stimulation or loss of function of circumscribed groups of psychic elements"—such as "maniacal movements" (if limited to certain muscle groups and relatively constant over time), asymbolia, and psychoses that followed on "auditory and visual hallucinations." Or the lesion might be located on an association fiber (OF), interrupting the connection between the two centers, what Wernicke called "sejunction" (*Sejunktion*).[29] Such lesions explained the symptoms of several types of aphasia.[30]

Second, lesions could have two kinds of effects: *Ausfallsymptome* (symptoms of loss, that is, *Herdsymptome*) or *Allgemeinsymptome* (general symptoms)—a distinction originally made by Griesinger.[31] Whereas *Ausfallsymptome* were defects that were permanent or at least persistent for some time and would correspond in their character to the function of the destroyed brain tissue, *Allgemeinsymptome* were transitory and included symptoms of "increased intra-cranial pressure" such as "papilloedema, apoplectic insult and the impairments of the sensorium known under the names of coma, somnolence and stupor," as well as dizziness, nausea, vomiting, and headaches.[32]

As this typology of disease suggests, Wernicke's characterization of nervous disorder was essentially a thought experiment based upon his theory of the psychic reflex. Imagining the effects of different types of lesions located at different points along the psychic reflex arc, Wernicke produced a powerful tool for distinguishing mental from purely nervous disorder and for explaining the differences between various types of aphasia. And yet it is clear that this taxonomy was dependent first and foremost on Wernicke's theoretical model and only secondarily on clinical evidence. The a priori model did not withstand the test of the clinic. While the examination of psychiatric patients played only a small role in Wernicke's early work, that quickly changed. As his career developed, he was confronted more and more by a range of mental disorders that caused him to call earlier certainties into question.

Neuropsychiatry and the Clinical Turn

Throughout his professional life, Wernicke was exposed to the clinic. As a young physician, he took part in the Franco-Prussian War of 1870, where he assisted surgeons treating the wounded on the battlefield.[33] Return from the war brought Wernicke psychiatric experience. Wernicke was an assistant to the Breslau psychiatrist Neumann and later worked under Carl Westphal at the Psychiatrische und Nervenklinik at the Charité in Berlin. After 1878, he went into private practice in Berlin. During this period, Wernicke made use of a wide range of diagnostic and therapeutic resources. Much like Sigmund Freud in Vienna only a few years later, he practiced electrodiagnosis and electrotherapy.[34]

Wernicke was thus always aware of the demands of the clinic, but as he rose up in the ranks, he felt them ever more insistently. In 1885 Wernicke moved back to Breslau to succeed Neumann. As a professor of psychiatry (*Extraordinarius*), Wernicke had responsibility for a growing number of patients.[35] The centrality of his clinical work is most evident in the deliberations surrounding his promotion, four years later. In 1889 the medical faculty of the University of Breslau recommended Wernicke's promotion, emphasizing the "close connection (*innige Verbindung*) between most detailed anatomical examinations and thorough practical experience" in Wernicke's work. This connection, they argued, was demonstrated most notably by his three-volume *Lehrbuch der Gehirnkrankheiten*.[36] Wernicke's reputation for aphasia work was surely still important, but for his employers, his clinical expertise was his most important selling point.

Wernicke's increased exposure to psychiatric patients coincided with a broader shift in Wilhelmine medicine. The 1880s saw an explosion in the numbers of patients. This was in part due to the newly instituted social insurance programs. Of the social reforms of the 1880s that Otto von Bismarck initiated, the sickness insurance (*Krankenversicherung*) was the first (1883) and most successful. The programs were part of the chancellor's attempt to undermine the growing strengths of social democracy. In Breslau the first deputy to the Reichstag of the Sozialistische Arbeiterpartei, the precursor party to the Social Democratic Party, had been elected in 1878, and in the ensuing decades, the city became a stronghold of social democracy.[37] But whatever their political goals, the social insurance programs had a profound effect on society more broadly. Hospitals were no longer sites of last resort, where people would go to die; they became institutions dedicated to returning patients to their working lives.[38] Most importantly, the sickness insurance scheme vastly expanded the population served by the empire's hospitals.

In Breslau, in the decade 1875–85, the population of psychiatric patients rose by a factor of three, and admission numbers nearly doubled.[39] Wernicke's Abtheilung für Geisteskranke was accommodated in the Allerheiligenhospital, Breslau's largest hospital and one of the largest in Germany at the time.[40] It consisted of eight wards and 209 beds, but since the beginning of the decade, it had been hopelessly overcrowded.[41] The move of a considerable segment of the patient population to a newly opened branch at the Wenzel-Hancke Hospital alleviated but did not solve the problem.

To accommodate the patients in Breslau, the city funded a neurological polyclinic and constructed a new psychiatric clinic, and Wernicke moved in in 1888. The clinic, located on Göppertstraße in the Odervorstadt, some 1.5 miles north of the city center, was a new and attractive building designed by the architects Johann Robert Mende and Richard Plüddemann (fig. 2.2).[42] As one commentator noted, the hospital was "equipped in the most opulent way."[43] By the end of the nineteenth century, further additions were made: the clinic now possessed gardens where patients could while away their time, but also a morgue, service rooms, and a residence for doctors. The agreeable space of the clinic did not fail to attract middle-class patients. Among the patients presented in Wernicke's *Krankenvorstellungen* (which I will discuss shortly) were, for example, a chemist, a student of theology, two engineers, a civil servant's widow, and wives of businessmen. Similarly, some of Wernicke's patients in Halle, where he moved in 1904, presented themselves as distinctly middle-class, as the photos taken at their admission suggest (fig. 2.3).

In the first flush of enthusiasm for his new clinical responsibilities, Wernicke did not see them as an impediment to a research program based

FIGURE 2.2. Psychiatric clinic in Breslau, around 1900. (Photograph by Eduard van Delden. Kiejna and Wójtowicz, *Z dziejów Kliniki Psychiatrycznej*, 38.)

upon pathological anatomy. After all, Griesinger himself had asserted the importance of the clinic for research and in the 1860s had developed a reform program for German-speaking psychiatry to optimize their relationship. In his program, Griesinger had proposed institutional changes to maximize the number of patients available for study. He had suggested

FIGURE 2.3. Two of Wernicke's patients at Halle. (Courtesy of Carl Wernicke Patient Files, Universitätsklinik und Poliklinik für Psychiatrie, Psychotherapie und Psychosomatik, Halle. File number 12372, admission date 3/30/1905; and file number 12488, admission date 5/15/1905.)

the division of asylums into two types of clinics, a rural asylum for the care of long-term patients and an urban asylum to accommodate short-term patients. Working in the urban asylum, research psychiatrists would have access to a large and constantly changing patient population. Moreover, as a result of Griesinger's emphasis on early admission, the patient material would shift from mostly chronic to more acute cases, which displayed the most vivid symptoms.[44] As Griesinger made clear, access to such patients was crucial, because psychiatrists could not understand the development of psychiatric symptoms entirely through pathological changes in the brain.[45]

Wernicke's new psychiatric clinic matched Griesinger's model of the urban asylum. Indeed, in a programmatic article from 1890, "City Asylums and Psychiatric Clinics," Wernicke claimed that his clinic in Breslau was "the only example where Griesinger's idea has been implemented."[46] The proximity of large provincial *Irrenanstalten*, such as Leubus, assured the "discharge into the provincial asylums," which was required for a high patient turnover.[47] Further, since the pathological anatomical method required postmortem dissection, Wernicke was enthusiastic about the "unusually high number of deaths," more than twice as many as in the *Provinzialanstalt*. As Wernicke pointed out, "it is the dissections after all, which provide the working material so essential to the scientific goals of the clinic."[48] His clinic thus promised to provide rich material for study.

The Structure of Wernicke's *Krankenvorstellungen* in Breslau

Griesinger's system also privileged the kind of patient demonstrations that Wernicke was going to perform in his clinic. Consonant with the broader research goals of his psychiatry, Wernicke's *Krankenvorstellungen*, where patients were presented to medical students and assistants, came to assume a central place in his work. Such clinical demonstrations of neurological or psychiatric patients had become increasingly common over the latter half of the nineteenth century: Jean-Martin Charcot's Tuesday lessons at the Salpêtrière, which were translated into German by Sigmund Freud in 1892, are only the most famous example.[49] Earlier concerns about their effectiveness were largely put aside.[50] Some doctors even began to argue that clinical demonstrations were beneficial to the patients' health.[51] Other concerns, for example about the symptom-altering nature of the demonstrations, were still common: anxious patients might gain confidence in front of an audience, or maniac patients could become quiet and shy.[52] One of Wernicke's patients who on the ward had shown "profound unrest" showed "only hints of this" in the demonstration.[53] On the other hand, the demonstrations could bring to light symptoms that were otherwise hidden: Wernicke wrote about one case, "With greater confidence than before, in today's examinations we can note the presence of two symptoms that are characteristic for affective melancholia: the blunting (*Abstumpfung*) of emotions and the inability to make decisions. In the asylums, we have never been able to observe these symptoms."[54]

Wernicke considered his clinical work to be in the service of his research, and so his *Krankenvorstellungen* were organized according to reflex principles: *Geisteskrankheiten*, like aphasia, were caused by a lesion on the speech reflex arc, and for this reason the examination of psychiatric patients would be analogous to the reflex exam: "What is the main approach to mental illness? No doubt the clinical exam, in which we address our questions to the patient and infer from his answers the state of his thought activities. In this sense we think of the patient as receiving our questions in *s* [sensory pathway] and answering them in *m* [motor pathway], and regard his answer as a kind of reflex movement."[55]

As Wernicke imagined the process, an incoming sensory stimulus was carried along sensory pathways (named *s* in the quotation) via the spinal cord to the brain and, after passing various projection and association centers, traveled down motor pathways (*m*) to cause the patient's response (instead of producing a twitch of the target muscle in the spinal reflex exam). Depending on the examiner's question, the reflex would follow different paths through the brain. For this reason, by carefully choosing his words, Wernicke could target distinct cerebro-anatomical structures, like a psychic sonar that could determine the seat of a lesion along the "mental reflex arc [*psychischer Reflexbogen*]."[56] The patient's answer to the question might point the examiner toward the existence of a lesion on that particular mental reflex arc.

Wernicke's model of conscious life fits into this model of localizable damage. Wernicke divided human consciousness into three units, or spheres: the *allopsyche* (consciousness of the world), the *autopsyche* (consciousness of the self), and the *somatopsyche* (consciousness of the body). Drawing on evidence from comparative anatomy and from physiological experiments by Gustav Fritsch, Eduard Hitzig, and Hermann Munk, Wernicke suggested in his 1879 paper on consciousness that "the seat of consciousness is to be located in the great hemispheres, more specifically in their cortex and medulla."[57] Each sphere had a separate anatomical location, and for this reason pathological damage could occur in each one independently (resulting in allopsychosis, autopsychosis, or somatopsychosis).[58]

If we look at the *Krankenvorstellungen*, we see that the tripartite division was a major structuring principle in Wernicke's clinical practice. In a standard psychiatric exam, Wernicke would test one or more of the patient's spheres of consciousness. These did not necessarily follow a particular order, nor did all examinations cover all three, but Wernicke's questions were formulated in order to examine one of these fields:

Auguste T., seamstress, age nineteen.
[Allopsyche]
[WERNICKE:] "Who am I?"
[PATIENT:] "The Herr Professor."
[W.:] "Who are these gentlemen[59] here?"
[P.:] "Doctors." (laughs wildly.) . . .
[W.:] "What kind of institution is this?"

[P.:] "Institution for the mad."
[Autopsyche]
[W.:] "Do you belong here?"
[P.:] "Of course, I was mad after all."
[W.:] "Not any more? Do you still have all your marbles?"
[P.:] "Sometimes."
[Somatopsyche]
[W.:] "Listen, your body is healthy, isn't it?"
[P.:] "Yes, yours is, but not mine!"
[W.:] "What is wrong with your body?"
[P.:] "I don't have to tell you that."
[W.:] "What is wrong with your body?"
[P.:] "I will kick you out next time."
[W.:] "Did something change?"
[P.:] "Very much."
[W.:] "What?"
[P.:] "The poison maker has changed."
[W.:] "You mean, poison has changed your body? How did you notice this?"
[P.:] "I can't tell you that."[60]

Wernicke's understanding of the topography of the three parts of the psyche also allowed him to assess the severity of a particular complaint. In anatomical terms, the somatopsyche was the "first relay of the cortex . . . , which ha[d] to be passed before the stimulus enter[ed] consciousness of the world" (the allopsyche) and before it finally reached the autopsyche, which was the top layer.[61] Wernicke considered the consciousness of the self (autopsyche) to be informed by the other two spheres; it was a "function of the consciousness of the world and the body."[62] The consciousness of the body (somatopsyche) was thus considered to be primary, the most fundamental of the three spheres, from which the other two spheres, most notably the consciousness of the self, were constructed.[63] Not only was the consciousness of the body the most basic; it was also the most robust.[64] Wernicke compared the cortex to a photographic plate that collected incoming stimuli. While experience of the world constantly changed, because the somatopsyche was founded on the fixity of the arrangements of bodily organs, it was, once established, relatively stable:[65] "Because the signals that the body sends into consciousness in similar circumstances are always the same, the memories of the organ sensations [*Erinnerungsbilder der Organempfindungen*] are associated in a very robust way, which would never be the case for the memories [*Erinnerungsbilder*] of the external world."[66]

Exceptions to this rule were life stages when the body underwent notable change, as, for example, in puberty, menopause, pregnancy, and old age. Consequently, these were the times when patients were most vulnerable to mental disease.[67] At other times, if the somatopsyche did, in fact, become "disoriented"—the term Wernicke used to refer to a disturbance within one of the three spheres—this was generally considered a serious condition. In such disorientations, patients might believe that they had "no head, no tongue, or no ears, no stomach, or three legs, the head turned to the back, or interchanged arms," or that they were "made of ivory, or glass, and therefore [would] be numb."[68] Disorientations of the autopsyche occurred much more regularly but on the whole were less serious. Easier to disorient, diseases of the autopsyche were also easier to realign; autopsychosis was generally more susceptible to therapy.

The Carnivalesque Character of the *Krankenvorstellungen*

As we have seen, Wernicke organized his clinical practice to promote his pathological anatomical research. Adopting Griesinger's program of asylum reform, he had assured himself a constant flow of patients, with a relatively high death rate for postmortem dissections. Wernicke's working conditions seemed ideal for the future development of his work. Moreover, the centrality of the reflex in Wernicke's theory seemed to bode well for its application to clinical practice; it offered a clear and obvious frame for the clinical test.

In practice, however, his high hopes turned out to be overly optimistic. In the same article in which he declared the value of the new Breslau psychiatric clinic for a Griesinger research project, he admitted that "crude reality" had obstructed the "fulfillment of the ideal requirement." A large number of nonpsychiatric patients and patients who required significant time and attention—dangerous patients and paralytics constituted up to 68 percent of the patient population, alcoholics and epileptics another 13 percent—made the character of the urban asylum less like a site for research and more like a *Pflegeanstalt*.[69] Patient care got in the way of Wernicke's research.

The single greatest obstacle to Wernicke's research program was, however, internal to his reflex practice. As I suggested in the introduction, when reflex physiology was used to inform clinical work in psychiatry, it functioned in unpredictable ways. The reflex test aimed to pinpoint nervous damage that distorted the usual functioning of reflex arcs, that is, damage that disrupted the *combined* functioning of a number of nervous elements. Because no single nervous element could be tested on its own, it was difficult if not impossible to infer specific nervous damage from particular responses. Instead, reflex testing encouraged the doctor to read the battery of reflex stimuli and responses as a whole, according to a shifting and complex interpretive practice. And as Wernicke's practice brought the interpretive aspects of the reflex to the fore, his clinical material turned out to be less conducive to the lesion model than he had hoped.

Wernicke's *Krankenvorstellungen* in Breslau took place in the psychiatric clinic's lecture hall, which simultaneously served as a teaching space and the clinic's laboratory.[70] Clinics were held twice a week, on Wednesday and Saturday mornings, and in each session usually two, sometimes three, patients were presented. When one of the approximately two hundred patients from the hospital had been selected for the *Krankenvorstellungen*,[71] he was guided from the distant patient wings—separated for men and women—to the central block of the building where the auditorium was located.[72] An attendant nurse brought the patient into the room, where Wernicke sat in his armchair, facing an audience of assistants and medical students; the chair beside Wernicke was empty, waiting to be occupied by the patient. Over the course of the presentation, Wernicke would ask the patient questions, and a team of assistants would record every response.

Reading the *Krankenvorstellungen*, it becomes evident very quickly that the patients played an active role in affecting the course of the exam. The German noun *Vorstellung* (as in *Krankenvorstellung*) has multiple meanings, including "introduction" (as in introducing someone or something) and "presentation," but also "show," used in compounds such as *Theatervorstellung* and *Zirkusvorstellung*. This latter aspect strongly marked the presentations.[73] Many patients acted like performing ballerinas. The note-takers remarked at the end of one exam that Bertha P. "[walked] out with grand airs, and by means of a soubrette [left] the stage."[74] Asked why she was there, the patient responded that she "believed she was [there] to debut."[75] It is remarkable how often the *Liebigsche Lokal*, a local pub and informal theater in Breslau, was mentioned in the *Krankenvorstellungen*.[76] Patients referred to it during their dances, stating their desire to be there on stage.[77]

Much like the better-known Tuesday lessons of Jean-Martin Charcot at the Salpêtrière, the *Krankenvorstellung* not only showed a close relation to the theatrical at the level of patients; the spatial arrangements of the *Krankenvorstellungen* resembled those of a theatrical *Vorstellung*.[78] Wernicke sat like a director in the director's chair, governing the clinical exam. Doctor-director and patient-performer faced and spoke to an audience of students and assistants. The doctors themselves frequently made the link to the realm of amusement and theater. They would often characterize the patient's movements as clownlike: for example, referring to "the slow and composed motion of a pantomimic presentation, but wrong, or carried to extremes." They would also encourage the patients in their theatrical performances, asking a patient to dance: "Upon request he waltzes on the tip of his toes, in an extremely graceful way, all the while whistling a melody and saying: 'That goes, as if I was a fly.'"[79]

But if the *Krankenvorstellungen* were like theater, they were closer to a form of improvisation or burlesque, and the patients often went off script. Thus, while Wernicke's model of the three spheres provided a structuring element, this structure was challenged by the patients' actions. As we have seen, in the original structure of the exam, Wernicke's questions constituted a stimulus, which would elicit a response pointing to damage within the patient's mental reflex system. But in the exam, Wernicke could not so easily control the stimuli; any element in the room might elicit a response—people, objects, and even the process of note-taking. Once the exam started, it would rarely continue according to plan; despite the overall similarity of the cases in the *Krankenvorstellungen*, no two were ever the same.

Some patients did not even wait for the doctor to question them, but rather opened up the conversation by asking questions themselves. "[Patient:] 'Why are you always wearing a pince-nez?' [Wernicke:] 'It makes me see better.'" Similarly, a patient, asked by Wernicke about his age, turned the question back on his interrogator: "How old do *you* think I am?" When Wernicke took a guess, the patient said: "Yes, I mean no, I don't know, I won't tell you, I will be very careful not to tell you."[80]

Some patients reacted angrily to the students and assistants taking notes during the *Krankenvorstellungen*: "[The patient] repeatedly moves toward the note-taking assistants, especially when they fixate on him, [and] tries to take away their notebooks from them." Others became curious: "You have all got those notebooks there. What are you writing?" Still others tried to make sense of the note-taking process by relating it to themselves: "Wernicke: What are these gentlemen here for?—Patient: For note-taking. I told them to come here."[81]

In one case, the patient, instead of waiting to be examined herself, made it her task to examine the professor:

> [PATIENT] jumps up, takes the professor's hands, makes dancing movements and shakes her upper body....
>
> [WERNICKE:] "Why so cheerful?"
>
> [P.:] "I am happy because I can be close to you, sit next to you, have been given the pleasure to be close to you, Herr Professor![82]

The same patient grabbed for Wernicke's cuff links and commented on them ("Those are fine cuff links. You are a well dressed man...."), touched objects in the room, went to the open window, looked outside and waved.[83]

Other patients undermined the examination by refusing to give answers to Wernicke's questions or by parodying them. One patient, responding to the question "Are you being treated well?" said "I cannot answer this, I am keeping the result to myself."[84] Another patient did not answer Wernicke's repeated questions, continuing to talk about other things: "'Who am I?' The professor calls her energetically and repeatedly by her name. 'Alas, you are the man' ... 'Who am I?' Alas, you are the same one who was in the *neue Weltgasse*—then you will get the money, and the man gave me the watch." One patient simply answered a question from Wernicke with "Silence is golden."[85] Yet another patient, asked to memorize the name of the Assyrian king Ashurbanipal, ridiculed the question by saying what sounded like "ashurbanipality" (*Assurbanipalität*)."[86]

Wernicke was very aware that the complex behavior of his patients undermined the neat structure of his mental reflex exam and its goal of tracing mental disorder to discrete nervous damage. In his work on Duchenne, Wernicke had addressed this very issue. Guillaume-Benjamin Duchenne (1806–1875) is best known today for his study of facial expressions. He electrically stimulated single muscles of the face and documented the results in series of photographs. Following the practice of physiologists such as Gustav Fritsch and Eduard Hitzig, Duchenne mapped out individual muscles through localized electrical stimulations, which showed strong parallels with the methods of pathological anatomy. It was another part of Duchenne's work, however, that interested Wernicke. He translated Duchenne's *Physiology of Motion* into German in 1885—almost two decades after publication of the French original—thus indicating the importance and currency he believed the work still had. In this book Duchenne had laid out his criticisms of pathological anatomy: he thought of his work as an important corrective to conventional anatomy, because his was an "anatomy of the living" that worked "without pricking or cutting the skin."[87]

The error of the pathological anatomists was that they mistook an experimental artifact, the independent activity of different muscles, for an example of normal functioning.[88] The isolated stimulation of the deltoid (shoulder) muscle, for example, did bring about the lifting of the arm, as they had thought, but at the same time it also caused the scapula to assume a "faulty position" (*attitude vicieuse*), suggesting that it did not operate this way in real life.[89] The functioning of the deltoid muscle could not be determined by the experiment alone. The experiment showed that "the movements caused by the partial muscular contractions produce[d] deformations that were nonphysiological," and this further suggested that "there *was* no isolated muscular action in nature." It was thus necessary to complement the electrophysiological experiment by clinical studies on living subjects, to "illuminate the electrophysiology of muscles through the control of clinical observation."[90]

When Wernicke introduced Duchenne's *Physiology of Motion* to the German-speaking medical world, juxtaposing Duchenne's experiments on living bodies with the corpse-based pathological anatomical method, he too recognized that the workings of the body were always more complicated than was suggested by the simple correlation between lesion and symptom. Lesions might still be responsible for disease, but their effects were as complex as the nervous system they affected. The first step was thus a careful examination and description of any symptoms the patient presented.

Notation, Text, Interpretation

Because, as we have seen, the responses to individual questions could be considerably more complex than initially expected, Wernicke placed great emphasis on the recording of the patient's actions and words.[91] During the lectures, Wernicke's assistants took copious notes according to an elaborate system, detailing the patient's behavior in its full range. Owing to this careful and laborious work, in reading the *Krankenvorstellungen*, we get a sense that we are witnessing the event ourselves. Wernicke's assistant

Hugo Liepmann recalled the procedure: "Storch and I . . . were instructed to produce a photograph-like picture of each patient who was presented. The roles were split: Storch wrote down the patient's speech in shorthand: I registered his facial expression, his gestures and actions. . . . I believe that this way the maximum truth to nature, as far as it can be attained through words, was achieved."[92] The notation system was able to capture even subtle motor expressions in the patients, in particular their facial expressions and their way of speaking. Throughout the *Krankenvorstellungen*, we find passages like the following: "The by now intensely reddened face assumes an anxious expression; trickles of sweat become visible on the nose. Despite the very energetic and urgent questioning, the patient answers very hesitantly now, with a hoarse, anxious-sounding voice and not always meaningfully."[93]

Once the *Vorstellung* had been completed, the two accounts—of the speech and of the patient's movements—had to be synthesized to provide a full report. For this purpose, Wernicke and his assistants developed a notation system to allow the synchronization (see fig. 2.4). In the system, the underlined accent sign above a syllable indicated that the syllable was to be synchronized with the movement described within the ensuing parentheses; for example, when the patient said "I was too well *behaved*," he took a bow. This way, the reader could fully picture the moving and speaking patient.[94] As one of Wernicke's assistants noted, the emphasis on clinical observation in his exams allowed Wernicke to "really learn much more about the patient than one previously thought," so that one found "in Wernicke's descriptions of illness a descriptive acuity of individual symptoms [*Einzelsymptomschilderung*] that we so far have hardly seen in other authors."[95]

The result of these recording techniques was thus a long and detailed script of the patient-doctor encounter. No longer a simple list of stimuli and responses, where an individual unexpected answer would indicate on its own the presence of a lesion, the report had become a complex interwoven text.[96] First, nonquestions could act as stimuli as well. We have seen how any element within the *Krankenvorstellungen* could incite the patient's responses: notebooks, cuff links, Wernicke, the situation itself. Second, even when the patient's actions responded to a question, there was not always a clear and direct relationship between the two: questions aimed at one sphere could lead to responses indicating disturbance in another. For example, one patient, when asked if he lived with his parents, qualified: "Yes, what one might call parents." He added: "You know that

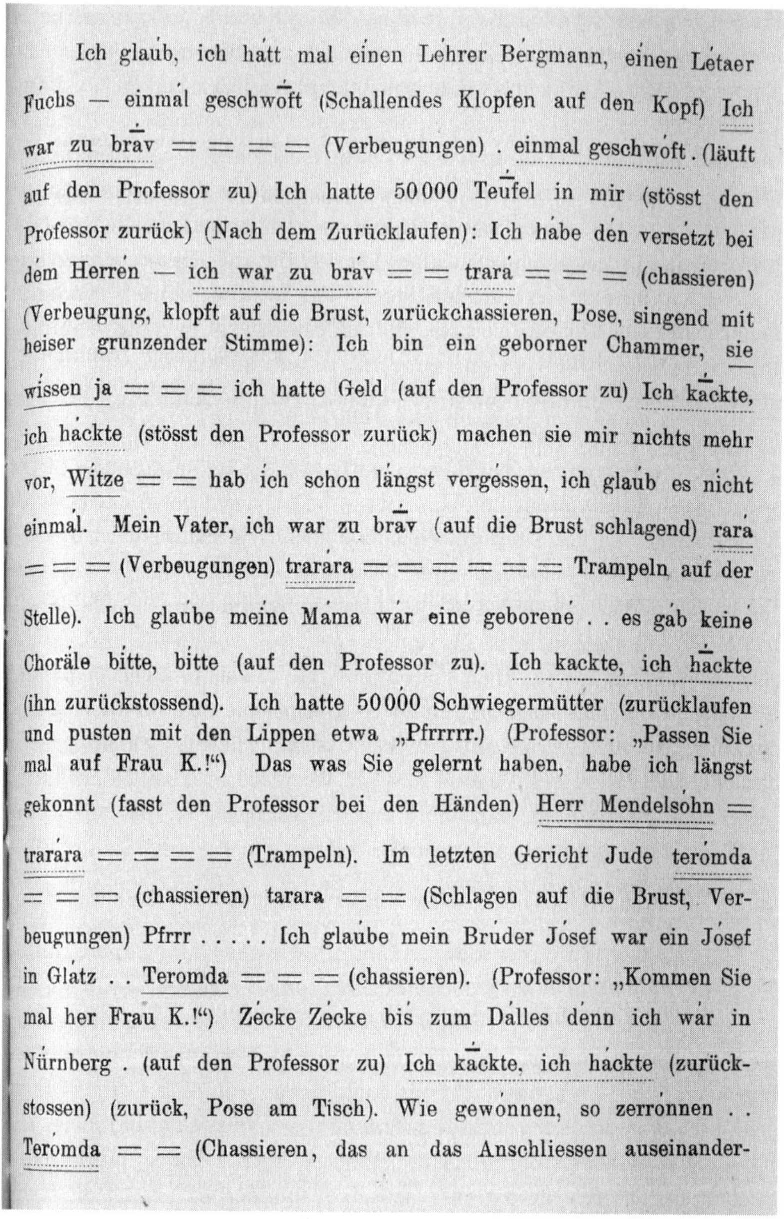

Ich glaub, ich hatt mal einen Lehrer Bergmann, einen Letaer Fuchs — einmal geschwoft (Schallendes Klopfen auf den Kopf) Ich war zu brav = = = = (Verbeugungen) . einmal geschwoft. (läuft auf den Professor zu) Ich hatte 50000 Teufel in mir (stösst den Professor zurück) (Nach dem Zurücklaufen): Ich habe den versetzt bei dem Herren — ich war zu brav = = trara = = = (chassieren) (Verbeugung, klopft auf die Brust, zurückchassieren, Pose, singend mit heiser grunzender Stimme): Ich bin ein geborner Chammer, sie wissen ja = = = ich hatte Geld (auf den Professor zu) Ich kackte, ich hackte (stösst den Professor zurück) machen sie mir nichts mehr vor, Witze = = hab ich schon längst vergessen, ich glaub es nicht einmal. Mein Vater, ich war zu brav (auf die Brust schlagend) rara = = = (Verbeugungen) trarara = = = = = = Trampeln auf der Stelle). Ich glaube meine Mama war eine geborene . . es gab keine Choräle bitte, bitte (auf den Professor zu). Ich kackte, ich hackte (ihn zurückstossend). Ich hatte 50000 Schwiegermütter (zurücklaufen und pusten mit den Lippen etwa „Pfrrrrr.) (Professor: „Passen Sie mal auf Frau K.!") Das was Sie gelernt haben, habe ich längst gekonnt (fasst den Professor bei den Händen) Herr Mendelsohn = trarara = = = = (Trampeln). Im letzten Gericht Jude teromda = = = (chassieren) tarara = = (Schlagen auf die Brust, Verbeugungen) Pfrrr . . . . . Ich glaube mein Bruder Josef war ein Josef in Glatz . . Teromda = = = (chassieren). (Professor: „Kommen Sie mal her Frau K.!") Zecke Zecke bis zum Dalles denn ich war in Nürnberg . (auf den Professor zu) Ich kackte, ich hackte (zurückstossen) (zurück, Pose am Tisch). Wie gewonnen, so zerronnen . . Teromda = = (Chassieren, das an das Anschliessen auseinander-

FIGURE 2.4. Notation system in Wernicke's *Krankenvorstellungen*. (Wernicke, *Krankenvorstellungen*, 3:14.)

I'm the Son of God."[97] Originally tested for disorientation within the allopsyche, the patient gave insight into his autopsychic world. Similarly, a patient displayed allopsychic disorientation in his belief that the Breslau psychiatric clinic was a military hospital. When asked about current politics, he answered promptly, albeit temporally disoriented; he believed he lived a quarter of a century earlier, in the 1870s. Wernicke probed this disorientation in his questioning:

[WERNICKE:] "What is the Emperor's name?"
[PATIENT:] "Wilhelm I."
[W.:] "The chancellor's?"
[P.:] "Bismarck."
[W.:] "Leader of the Center Party?"
[P.:] "Windthorst."
[W.:] "Of the National Liberals?"
[P.:] "Lasker."[98]

The continuation of the questioning—all directed at the allopsyche—pointed toward a disturbance in the autopsyche as well. When asked about the program of the social democrats, with whom he seemed to sympathize, the patient squirmed and answered only with marked reservation.[99] After a pause, he said carefully: "The social democrats want to dictate for themselves the conditions under which they want to work." Imagining himself to be a soldier with social democratic leanings in the 1870s, the patient had reason to be hesitant in sharing his political opinions. The German chancellor Otto von Bismarck had launched a campaign against the Social Democratic Party and, after an attempt on the Kaiser's life, banned it in the 1878 Anti-Socialist Laws. Wernicke noted, "This topic is apparently too delicate for him as a putative soldier."[100]

The example points toward a third difference, the need for context to understand a patient's disturbance. The patient's autobiographical delusions made him suspicious of the questioning and thus were pertinent to understanding what was wrong with him. From this example, we can see how this encounter differed from the simple investigation of a patient with *Herdsymptome*. The examination was not a simple synchronic investigation of nervous pathology with right or wrong answers. It was instead transformed into a model for accessing the diachronic mode of patient biography. It made the patient's history, and the political history of Germany, for that matter, relevant for the understanding of the pathology.

In these three ways—questions blurring together, the difficulty of targeting particular spheres of the mind separately, and the requirement of broader context for understanding symptoms—the text of the *Krankenvorstellungen* no longer resembled a set of discrete reflex tests with clearly delimited stimuli and responses. Rather, they provided a rich account of the patient, which needed to be interpreted as a whole. The development had a significant effect on the understanding of the case. Wernicke dealt with a whole texture of symptoms from which he would have to draw its meaning; it required interpretation.

The compilation of the *Krankenvorstellungen* corresponded to this changed understanding. Once the diagnosis had been decided, the notes were "fixed in written form, and, supplied with epicritical remarks, handed out to each student at the end of the semester."[101] Similarly, the clinic owned a "Golden Book," into which each case of psychosis had to be written.[102] The book presented a loose arrangement of cases, each unique in its own way and deserving full documentation and interpretation. This book was used as the basis for discussions during the regular staff conferences of the clinic.[103] Wernicke's former colleagues liked to refer to Wernicke's favorite anecdote: when asked by a visitor where his clinic's library was, Wernicke pointed at a pile of case records and said, "*This* is the library."[104]

The original patient files from Breslau no longer exist, but we can gain a sense of how they were used by looking at those produced after Wernicke moved to Halle in 1904. A substantial number of the nearly nine hundred patient files at Halle bear the marks of intense study: telling phrases are underlined, and notes are written in the margin (fig. 2.5).[105] These annotations, mostly written in pencil, were usually in a different hand from the one seen in the case history. It is likely that Wernicke's assistants sat down and studied the patient files after their completion: the work of diagnosis took as its object of study the case notes in addition to the patient. Often, the notes drew on Wernicke's system, which in Halle was new to his assistants.[106] In the example, the patient's protocol was interpreted thus: the patient thought he had "sinned," having "tempted God and Jesus," and thus he "belonged in hell." Further, he was "oriented within the environment" but had "ideas of impairment" (*Beeinträchtigungsvorstellungen*) in the form of refusing food because he thought it was poisoned. Using Wernicke's system, these symptoms were interpreted as "secondary autopsychic anxiety ideas (physical explanation delusion)," a diagnosis that was written in the margins.[107]

The development of this interpretive system had an important effect on Wernicke's diagnostic system. Each case ended in a diagnosis of the patient, allo-, auto- or somatopsychosis (and others such as motility psychosis,

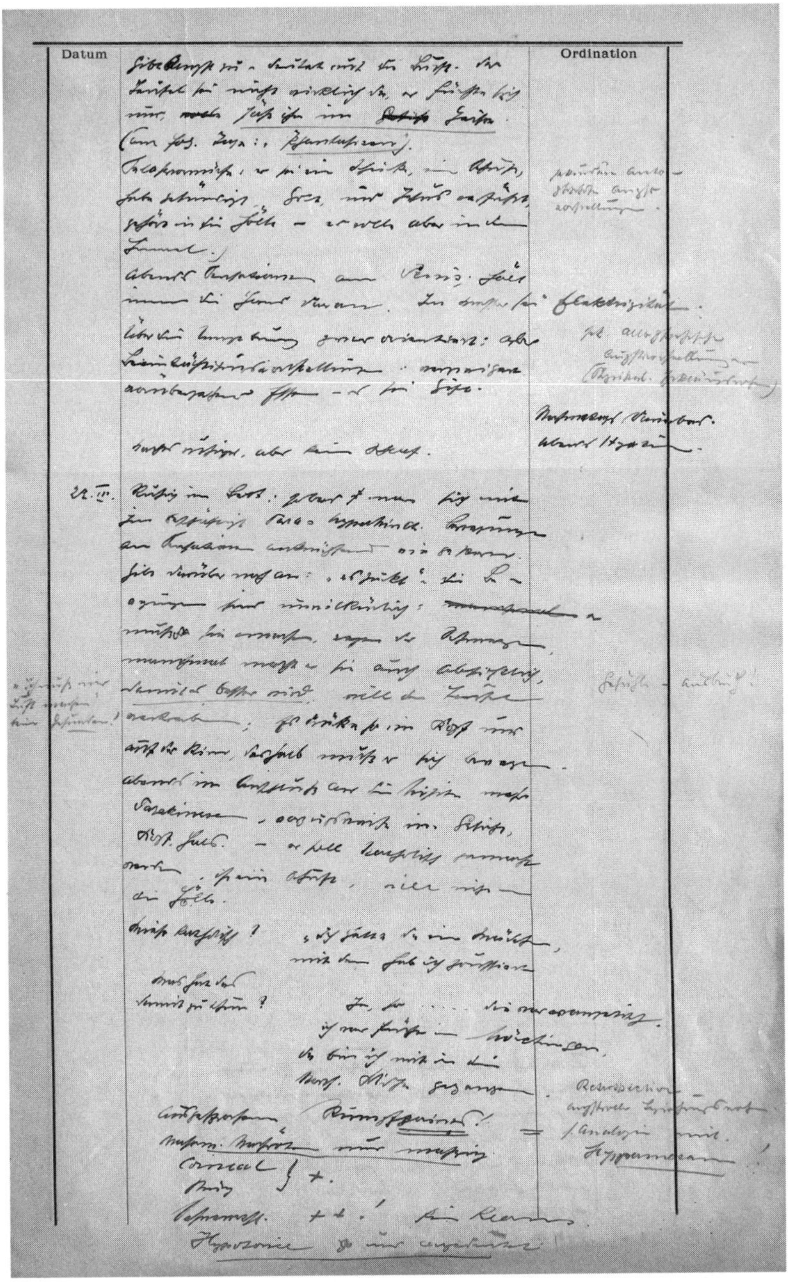

Datum

Ordination

FIGURE 2.5. Patient file with marginalia. (Courtesy of Carl Wernicke Patient Files, Universitätsklinik und Poliklinik für Psychiatrie, Psychotherapie und Psychosomatik, Halle. File number 12348, admission date 3/19/1905.)

a state of "motor helplessness" [*motorische Ratlosigkeit*)], which could take the form of hyper-, a-, or parakinesis). But as we have seen, the primary product of the system was the full description of the patient's behavior during the presentation, not the one-word diagnosis. The three areas of consciousness that Wernicke had first understood as anatomically localizable areas of the brain, areas that could serve to confirm and reinforce an existing nosological canon based upon the location of lesions, were now primarily used as a way of classifying and ordering that clinical observation.[108] As Karl Bonhoeffer emphasized, Wernicke "frequently pointed out that the principal value of his nosology was not that it revealed completely the inner logic [*innere Wesen*] of the psychoses. . . . From the beginning he was [more] concerned with a viable examination technique."[109]

The priority of interpretation over diagnosis is visible in Wernicke's treatment of the cases. Several cases in volumes 1 and 2 of the *Krankenvorstellungen* were continued in the next volume.[110] In contrast to the *Herdsymptome*, which could be conclusively diagnosed only through dissection, the pathology of his patients was subject to evolution; his system was never closed. Each diagnosis was merely a provisional reading of an ongoing text that could always be revised. Indeed, the diagnoses themselves were often just a redescription of the clinical symptoms. Throughout the *Krankenvorstellungen*, Wernicke often moved from an adjective construction, such as "allopsychic disorientation," to a diagnostic term, such as "allopsychosis."

The Move away from the Lesion Model and the Foundation of a New Psychiatry

The Griesinger-Meynert model remained active in Wernicke's work throughout his life. He never explicitly rejected the idea that symptoms corresponded to the damage of clearly delimited brain areas. But given the results in the *Krankenvorstellungen*, it became clear that lesions caused only a tiny minority of all mental disorders: four out of a total of ninety-eight patients in volumes 1 to 3 of Wernicke's *Krankenvorstellungen*.[111] Instead, Wernicke had to draw on other explanations of mental disorder. In volume 1 of his 1894 *Grundriss*, Wernicke drew the distinction between *Herderkrankungen* (or *Herdkrankheit*) and *Allgemeinerkrankungen* and suggested that "mental disease will certainly not be subsumed under the former." Perhaps, he suggested, it might be included "under the latter."[112]

Because *Geisteskrankheiten* were no longer understood to be caused by lesions, Wernicke increasingly lost interest in pathological anatomy. In his later life, he rarely performed dissections himself; that work was carried out by his assistants. Often, he did not even supervise them. Wernicke had hired a separate assistant, Heinrich Sachs, to oversee the microscopic works in the clinic's laboratory.[113] As Bonhoeffer remarked: "We passed the day at that time doing patient examinations and in the laboratory," where "one became acquainted with the newer methods of von Marchi, Nissl, Weigert and Pahl and lived in the hope that one would find the anatomical basis of the psychoses via the study of histopathology."[114] But the assistants had to teach themselves ("autodidactically"), and Wernicke focused his attention elsewhere.

The radicalness of the transformation in Wernicke's work and interests is most clearly visible in his changing relationship to aphasia. His reconceptualization of mental disease called into question the work that had made his name. Because in the 1890s Wernicke asserted that *Geisteskrankheiten* could not be *Herderkrankungen*, his earlier work on aphasia now became an embarrassing wrinkle in his system. At this later time, he noted an overlap, a "common trait" shared by *Geisteskrankheiten* and certain cases of aphasia. More specifically, transcortical aphasia affected the fiber tracts connecting cortical areas with each other, that is, the association system—the same fibers targeted by *Geisteskrankheiten*. But because aphasia was caused by lesions, now Wernicke needed to differentiate between transcortical aphasia and lesion-free *Geisteskrankheiten*. He decided that the difference between the two could be presented thus: "Mental disease affects these tracts [of the association system] in a scattered way, with individual selection[;] the *Herdkrankheit* on the other hand destroys a compact mass."[115] According to his later understanding, Wernicke's 1874 description of aphasia did not confirm Griesinger's dictum, because aphasia could no longer be seen as a *Geisteskrankheit*.

* * *

We have seen in the previous sections how Wernicke's work in the clinic made him lose interest in his original project of pathological anatomy and put in its place the detailed recording of the patient. The change in emphasis required a reorientation in the field. Pathological anatomy had previously served as a legitimation for psychiatry by aligning it with somatic medicine. When the lesion model no longer seemed to play this role, Wernicke needed to find a substitute. As he wrote in the first issues of the *Monatsschrift für Psychiatrie and Neurologie*, which he launched in 1897 together with his colleague Theodor Ziehen at Jena, clinical observations rather than pathological anatomy would be the new foundation of psychiatry.[116]

To give this new clinical science an institutional grounding, Wernicke tied it to pedagogy. As he wrote in a programmatic article in 1889, psychiatric clinics "had to address a dual task, of which one presented the precondition for the other. They should first offer as good as possible a working environment for the teacher, including patient material that encompasses the whole subject area as well as equipment with all necessary means of examination. The clinician is thereby enabled to follow his second task and now in turn teach the student and introduce him to the field."[117] Clinical research and teaching were thus combined.

As Wernicke saw it, this combination was necessary because psychiatry was not a settled discipline. It was rather a system in flux, for which research was ongoing. Wernicke did not use a textbook in his teachings until his own *Grundriss der Psychiatrie* appeared in its first edition in 1894, even though there would have been a wide variety of options for him to choose from. From the late 1870s, spurred by a new optimism that psychiatry would soon become a mandatory part of the medical curriculum, a range of textbooks in psychiatry appeared in Germany, authored by Emminghaus (1878), Schüle (1878), Dittmar (1878), and Krafft-Ebing (1879)—probably the most widely used at the time, along with Griesinger's classic—as well as translations into German of Blandfort (1878) and Weiss (1881).[118] But to Wernicke, none of the proposed "systems" were adequate. He preferred to work with what was basically a loose collection of patient files—even his textbook, the most comprehensive published account of his nosology, was based on roughly five thousand case records (*Krankengeschichten*) collected at his clinic.

The dissatisfaction Wernicke felt with respect to the neuropathological paradigm was not uncommon. From the 1880s onward, the affirmed somaticism of the discipline was ever more under attack, even from those who had previously espoused it. Criticism was voiced at a number of levels: the anatomical work by Theodor Meynert and Paul Flechsig was attacked for making stronger claims about brain function than their evidence allowed; after the turn of the century, such work was called "brain mythology"[119]; the histopathological study of the nervous system was criticized from within (e.g., by neuropathologist Franz Nissl) and without (e.g., by the Würzburg psychiatrist Konrad Rieger) for being too labor-intensive and producing relatively few and minor scientific results; and finally psychology gained a new self-confidence as a nonsomatic discipline.[120] Moreover, internal battles, such as the so-called neuron controversy, divided neuropathologists and prevented any unified resistance to the onslaught.[121] The hegemony of a neuropsychiatric approach was a thing of the past. If anything, as Friedrich Jolly, Westphal's successor at the Berlin Charité hospital, asserted, anatomy and physiology should from henceforth have only auxiliary status within psychiatry.[122] Clinical work would now have precedence.

The most influential proponent of this turn to the clinic, and the best-known example of the new psychological orientation in particular, was Emil Kraepelin. Trained under the Leipzig psychologist Wilhelm Wundt and one of the major proponents of Wundtian experimental psychology, Kraepelin rejected the unification of psychiatry and neurology outright and instead advocated the use of experimental psychological methods.[123] This, of course, did not imply a turn away from scientific medicine. Quite the contrary. As Volker Roelcke has shown, Kraepelin drew on the principles of causality and disease specificity—established principles in somatic medicine since the successes in bacteriology in the 1870s and 1880s—in addition to his psychological experimentation.[124] Most importantly, in Kraepelin's view, Wundtian psychology was able to mediate between clinical observation and the established somatic sciences.[125]

Like Wernicke and other psychiatrists, Kraepelin responded to the clinical demands placed upon him. Pathological anatomy remained insufficiently supple to describe the range of pathologies psychiatrists confronted in clinical work, and psychiatrists were in urgent need of a classification system, a nosology, that would be up to the task. This is what Kraepelin hoped to provide, and his nosology became the dominant system in early-twentieth-century psychiatry; it was even revived toward the end of the twentieth century and beyond.

Kraepelin's system was successful for a number of reasons. First, it could present itself as scientific because it drew on an enormous number of patient files. As Eric Engstrom has shown, the complex disciplinary economy in Kraepelin's clinic went hand in hand with the construction of his nosological system, which convinced by its sweep and scale. The reputation of Kraepelin's system was built more on its solid evidentiary base than on its somaticist beliefs. Second, the system was eminently practicable. Kraepelin's emphasis on etiology and prognosis, which was related to his newly introduced distinction between dementia praecox (incurable) and manic-depressive illness (curable), helped the everyday organization of psychiatric care by informing directly the treatment of the insane. Finally, Kraepelin got the timing right. By developing a psychiatric nosology, he provided a good defense against newly developing antipsychiatry sentiments in the 1890s. Further, editions of his textbook appeared in 1896 and 1899, a time when the introduction of psychiatry as a field in the state examination was being discussed.

If these factors help explain Kraepelin's success, they were valid for Wernicke as well: he, too, distanced himself from an excessive reliance on neurological methods and developed a new "scientific" approach to characterizing and classifying diseases of the mind; he even published a psychiatric textbook around the same time as Kraepelin (the three volumes of the first edition of the *Grundriss* were published in 1894, 1896, and 1900). Also, his system seemed practicable. As we can see from the patient files at Halle, his students universally adopted his system. Some of Wernicke's students continued to use it after his death.[126]

But overall, the impact of Wernicke's nosology on German-speaking psychiatry was marginal. Several smaller problems conspired against him. Wernicke's *Krankenvorstellungen*, the most important exposition of his clinical practice besides his textbook, was published by the local Schletter'scher Buchhandlung, which had a limited reach.[127] His textbook as well, although published with the prominent press Thieme in Leipzig, never managed to compete with the more popular manuals by Krafft-Ebing and, in the 1890s and early 1900s, Kraepelin. Moreover, although Breslau was famous for its research university, within Prussia it still remained in the shadow of Berlin. Kraepelin's Heidelberg, on the other hand, was the major university in the growing and progressive southern state of Baden.[128]

Wernicke's impact was also weakened by an ongoing conflict with the city authorities in Breslau. Relations had been bad from the beginning. Until 1898, Wernicke held the dual position of clinical director and *Primararzt*; he was thus employed both by the university and by the city, and felt constantly stymied in his work by the city authorities.[129] He demanded the "right to intervene autonomously in urgent cases and make commands also in administrative issues," as was the case "in all public asylums in Germany."[130] The city, however, refused to give Wernicke greater administrative powers; he had no more rights than *Primarärzte* in other fields.[131] Then in 1898, Wernicke's contract as *Primararzt* was not renewed, and the position was separated from that of clinical director.[132] Consequently, Wernicke's opportunities for research were drastically reduced. Although as director of the clinic, he retained the right to examine and select patients for the purpose of teaching, he had to be accompanied by the new municipal *Primararzt*, Dr. Ernst Hahn, or his assistants, on his rounds, and this was a nuisance to both parties. To make matters worse, the city claimed ownership of the patient files, disputing Wernicke's assertion that they were his "scientific" and thus "intellectual property."[133] This meant that he could take the files home for study "only with a written receipt from the clinic." Further, because Wernicke's role was purely pedagogical, "after a completed *Krankenvorstellung*" the files had to "be returned immediately."[134] The increasing hostility faced by Wernicke in Breslau explains why the prospect of a position at Halle would have been so attractive to him. In Halle he was offered the directorship of a university clinic, and a modern and progressive clinic at that, devised and built by Eduard Hitzig in 1891.[135] But whatever potential the position held, Wernicke was unable to exploit it. He died in a bicycle accident just fifteen months after his arrival, at age fifty-five.

There is, I think, another reason why Wernicke's new nosology lost out to Kraepelin's. Although, as we have seen, Wernicke increasingly turned toward the clinic and came to emphasize the psychological symptom over somatic pathology, he never fully gave up on the pathological anatomical model. His tripartite model of the psyche remained—if loosely—tied to the localization paradigm, even if this no longer played an explicit role in his clinical practice. Wernicke remained, both by his reputation and by his own account, a member of the old guard of neuropsychiatrists, and around 1900 the future seemed to be psychological. Kraepelin's clean break with that past made him a better leader for the developing field.

It would be wrong, however, to suggest that the only path to psychology was the one that led away from neuropsychiatry. As we shall see in chapter 3, a reassertion of psychological factors in the emergence of mental disorder could also arise from a close engagement with the neuropsychiatric tradition. And this led to an even more momentous shift in the mind sciences. Kraepelin rejected the physiological pretensions of neuropsychiatry and came to dominate German psychiatry. Sigmund Freud held neuropsychiatry to its physiological promise and founded a new discipline: psychoanalysis.

CHAPTER THREE

On the Couch

Sigmund Freud, Reflex Therapy, and the Beginnings of Psychoanalysis

From the periphery of the German Empire, we return to the Austrian metropolis. But we are not back exactly where we started. Rather, our ongoing story takes us to a residential area a few blocks away from the Allgemeine Krankenhaus: Berggasse 19. The address evokes cigar smoke and Persian rugs, bourgeois angst and cultural sophistication, worlds apart from the clean Spartan spaces, whiff of disinfectant, and white coats of the hospitals we have considered so far; this was private rather than university medicine.

From 1891 to 1938, Berggasse 19 was both Sigmund Freud's home, where he raised his six children, and the site of his practice. Freud would welcome patients into his house five or six days a week and normally saw each patient individually for fifty-five minutes.[1] Only Sundays were kept regularly free. In between patients, he would allow himself a five-minute break. At the peak of his practice, Freud saw up to eight patients a day.[2] He engaged with his patients in two rooms: a study and an adjacent consultation room, which were located in the back of the apartment, overlooking the chestnut trees in the backyard. The very first meeting with each patient would be at Freud's desk in the study. Afterward, doctor and patient moved through an open door to the consultation room. It was here, in the furthest recesses of the apartment, shielded from the noises of the street, that the analysis would take place.

We are able to gain a sense of the spaces in which Freud practiced his therapy through the work of photographer Edmund Engelman (fig. 3.1). In 1938, just before Freud fled the country, the psychoanalyst August

FIGURE 3.1. Freud's consultation room. (Courtesy of Sigmund Freud Museum.)

Aichhorn commissioned Engelman to take a series of about a hundred pictures of Freud's premises.[3] The photographs give a sense of a well-furnished and warm if slightly cluttered environment, filled with the assortment of artifacts that Freud had gathered over a lifetime of collecting. For the analysis, the patient lay on the famous couch, an ottoman draped with cushions and an oriental rug, leaning against a wall covered with pictures. Freud took his seat, "tucked into the little corner," and remained quiet "like an old owl in a tree," his feet up on a stool placed just behind the patient's head.[4] The seating arrangement meant that the patient could not see the therapist. During consultation she faced, instead, a porcelain stove, which kept the room warm, and if she turned to her left, she would see on the wall a heavy Persian rug that matched the rug on the couch. Freud's vantage point offered more possibilities. His gaze could wander around the consulting room, from the cabinets filled with antiquities, past the open door, the window, perhaps to the copy of André Brouillet's famous picture *Une leçon clinique à la Salpêtrière* that hung on the wall.[5]

An examination of Freud's consultation room, its furnishings and ambiance, directs our attention to the world of ideas, of inherited cultures, of human relations. It lends credence to those accounts of psychoanalysis that place it among the *Geisteswissenschaften*, at odds with the somatic orientation of Meynert or Wernicke. True, Freud had begun his career in the neuropsychiatric and neuropathological fields.[6] After laboratory work with the physiologist Ernst Brücke from 1876 to 1882, Freud worked in Meynert's laboratory in 1883 and was his *Sekundararzt* at the Allgemeine Krankenhaus.[7] But according to the standard story, this earlier period should be cordoned off; at best it was irrelevant, at worst a hindrance, to the emergence of psychoanalysis.[8]

The story of a clean break between the preanalytic and analytic periods of Freud's career is compelling because it seems to do justice to the vast differences between neuropsychiatry and psychoanalysis: Meynert sought the causes of mental disorder in lesions, Freud looked for trauma in past experience; Meynert explained higher functions as a form of reflex action; Freud saw in them the result of unconscious activity; and while Meynert remained content with diagnosis, Freud looked for ways to relieve his patients' symptoms. Yet, significant as these differences are, I will argue here that they are best understood as the result of an internal criticism and reformulation of the neuropsychiatric tradition.[9] More specifically, I propose to read the development of Freud's psychoanalysis as the patient and careful working out of those tensions within Meynert's neuropsychiatry that I discussed in chapter 1. It was this process, throughout which Freud tried to remain loyal to the somatic aspects of Meynert's work, that best explains the ways in which he broke with it.[10] By radicalizing the associative elements of Meynert's reflex, Freud came to reject the lesion model of mental disorder, posit the existence of the unconscious, and open up new therapeutic possibilities.[11]

Freud and Neuropsychiatry

Historians and critics who posit a break between the early and the late Freud often locate it in the 1891 text *On Aphasia*.[12] It is in this text, written eight years before the pathbreaking *Interpretation of Dreams*, that Freud delineated his critique of the dominant localization discourse in the brain science of his time and first turned to questions of language. And yet it is an oft-neglected aspect of this work that Freud's criticism of contemporaneous brain sciences was not that they were insufficiently psychological and too mired in a restrictive somaticism. Freud identified in localizationists like Carl Wernicke and Theodor Meynert a diametrically opposed error. In his criticism he asked rhetorically whether it was "justified to immerse a nerve fibre, which over the whole length of its course has been only a physiological structure subject to physiological modifications, with its end in the psyche."[13] That is, he suggested that the neuropsychiatric understanding of the nervous system was unjustifiably contaminated by psychology.

Freud's judgment of the neuropsychiatric tradition makes sense in light of the historical argument I presented in chapter 1. There I showed how a new generation of psychiatrists renewed the localization project by making the connective principle of the reflex productive for it. In psychiatry, the marriage of the pathological anatomical method with sensory-motor physiology (in lesion studies or experimental physiology) allowed for the possibility of the project—that higher functions (such as language) could be localized in circumscribed areas of the cortex—while integrating it into the somatic paradigm of the age. But as we saw, this project was not without difficulties. Sensory-motor physiology had traditionally been restricted to the study of the spinal cord, and researchers often explicitly denied that it could be applied to higher functions. To be raised successfully to the brain, therefore, the reflex model had to be reformulated and made adequate to the task. This motivated Theodor Meynert's appeal to association psychology, which allowed him to complicate the simple reflex by extending its connective properties. Meynert and his student Wernicke linked sensory and motor images through the physicality of nerve fibers, within what Meynert called the "association system."

According to Freud, however, the translation of association psychology into physiological terms was fraught with problems.[14] In particular, Freud thought Meynert and Wernicke had not been sufficiently critical in this translation, because they assumed that the building blocks of the new association physiology would correspond directly to the building blocks of association psychology. The associations for Meynert and Wernicke were now instantiated by nerve fibers, but according to Freud, at their ends those fibers were "immersed in the psyche," by which he meant that they still supposed that those fibers connected the direct physiological correlates of psychological ideas. The force of the argument in *On Aphasia* was to show that the project of localization (which in Freud's mind remained governed by the psychological concerns that had first motivated it) was incompatible with the connective properties of the reflex.

Thus, though Freud's criticism might simply be seen as a turn away from brain science to psychology, it is significant that he gave it a different interpretation: Freud criticized the localization of function because it was not physiological *enough*. He wanted to construct a purely physiological model of the nervous system, which could give meaningful insight into mental processes (and consequently psychology) in a way that was blocked for a simple and introspective psychological account. Indeed, Freud's *Aphasia* proposes a thoroughgoing somaticism in order to be able to explain aphasic symptoms: the book is an object lesson in the need for a detour through physiology.[15] Freud's critique of localization, therefore, was informed by insights from the very tradition he attacked, and consequently his formulation of psychoanalysis was not so much a break from earlier brain science, but rather can be more productively understood as a radicalization of its principles.[16]

Freud's Physiology

Freud's appeal to "physiology" in *On Aphasia* has often been overlooked. In part this stems from an ambiguity in his use of the term. Freud often connected physiology to a "functional" perspective. But in his early texts, and especially in the *Aphasia* book, Freud used the term "function" in two distinct ways. First, in the sense of "localization of function," functions were understood as psychological, the solidary function of large structures of the nervous system, controlling speech, movement, and other "psychic" functions that would be visible in a clinical setting. "Function" in Freud's work, therefore, could refer to functions of the nervous system as a whole—functions that could be related to the "psychic." Following this understanding, scholars have suggested that, in his physiology, Freud left behind the nitty-gritty of an anatomical understanding to be able to account for a wide range of clinical (especially hysterical) symptoms. For them, Freud's "functional" and consequently his "physiological" account privileged a nonmaterialist understanding of mental processes.[17] And yet, function had a second, and for our purposes crucial, meaning: it could refer to the functions of the basic building blocks of the nervous system, which did not have a direct or obvious relationship to clinical symptoms. Freud drew on the English physiologist Charlton Bastian's notion of functional changes to the nervous system, in particular changes to the nerves' excitability. In Freud's usage, "functional" here referred to biological processes at the level of the nerves themselves.[18] This biological meaning of "function" was central to Freud's *Project for a Scientific Psychology* (1895), where he distinguished primary and secondary function.

The first notion of "functional" appears almost exclusively in the discussion—and criticism—of the localization-of-function paradigm.[19] When Freud uses the term on his own account, the notion of function as biological processes at the level of the nerves predominates.[20] For this reason, we should not regard Freud's use of the term "functional" to infer that he had adopted a "functional*ist*" perspective, if we mean by this a methodological indifference to the physical processes that produced "function." Freud's understanding of physiology corresponds to "function" in its second sense; it considered nerve function (in particular, excitability) in addition to brain anatomy.[21] Freud's emphasis on physiology, even a "functional physiology," showed that he was concerned to give a thoroughgoing materialist account of brain function that would not be distorted by the uncritical importation of psychological ideas.

Take, for example, Freud's unpublished 1887 manuscript "Kritische Einleitung in die Nervenpathologie" (Critical introduction to neuropathology). Here Freud elaborated his *Gehirnarchitektur* (brain architecture), which, to him, was the "complete knowledge of fiber pathways [*Faserverlauf*]."[22] He realized that on occasion anatomical methods might be inadequate to the task of gaining this knowledge: When tracing the path of fiber bundle 1 into gray matter out of which three other bundles emerged, anatomical methods "ha[d] no means of deciding into which of the bundles 2, 3 and 4, that originate from the same gray matter, bundle 1 continues."[23] Freud suggested that a "physiological" method could help. Because the nerves directly connected to a damaged nerve would show "secondary degeneration," if a brain scientist deliberately damaged a nerve, he could visualize its connections to other nerves across gray matter. Thus an appreciation of physiological processes could help "fill the gap in the tracing" of fiber systems left open by anatomical methods.[24] A consideration of physiological processes did not mark an attempt to circumvent the intricacies of brain anatomy in Freud's particular formulation of *Gehirnarchitektur*; rather, it helped Freud gain a fuller understanding of the structure of the nervous system, with respect to which anatomical approaches showed themselves to be essentially limited.

This understanding of physiology was the basis of Freud's criticism of Meynert. From Freud's perspective, Meynert had not carried his project of providing a physiological account of associationism to its end.[25] True, Meynert's model had given association psychology a physiological basis: nerve cells were hardwired to points on the body surface and gained their content through the projection of sensory stimuli. But the elementary units that were associated were still *Vorstellungen*; that is, Meynert had simply and uncritically transposed into physiological language the structures of association *psychology*. His project of providing a truly physiological associationism was thus incomplete; while Meynert had given a physiological explanation of the "associations," the fact that he considered that nerves associated *Vorstellungen* showed that he had not escaped the influence of the psychological model. According to Freud, the "elliptic phrase: an idea is localized in the nerve cell" led to a confusion of things "which need have nothing in common with each other."[26] It was invalid, Freud argued, to assume that the simplicity of psychological elements (the basic *Vorstellungen*) corresponded to a similar simplicity at the physiological level: "In psychology the simple idea is to us something elementary which we can clearly differentiate from its connection with other ideas. This is why we are tempted to assume that its physiological correlate, i.e., the modification of the nerve cells which originates from the stimulation of the nerve fibres, be also something simple and localizable. Such an inference is, of course, entirely unwarranted; the qualities of this modification have to be established for themselves and independently of their psychological concomitants."[27] By taking elementary *Vorstellungen*, the basic building blocks of psychology, and placing them in individual cells, Meynert had (falsely, for Freud) concluded that the basic building blocks of *physiology* were organized in the same way, that the elements of physiology corresponded directly to the elements of psychology.[28]

But what would a fully physiologized associationism look like? If we could not simply transfer our understanding of psychology's most elementary particles to the physiological realm, what, then, was the physiological correlate of a *Vorstellung*? Freud suggested that a *Vorstellung* was "nothing static, but something in the nature of a process," a process that spread over the cortex along specific pathways, forming specific routes of excitation in the brain. Once established, the routes persisted, leaving behind "a modification, with the possibility of a memory, in the part of the cortex affected."[29] *Vorstellungen* were really associations, or rather, the two were "terms by which we describe different aspects of the same process."[30] Rather than associations tying different *Vorstellungen* together, those *Vorstellungen* were constructed from associations themselves.

Of course, this was not completely alien to the Meynert-Wernicke system. For the two older neuropsychiatrists, association was crucial to the formation of concepts (*Begriffe*) where different *Vorstellungen* were gathered together into complex groups: for instance, the unified concept of the lamb comprised its visual and acoustic elements. Freud wanted to say that the basic *Vorstellungen* were already complex, too. He merely transferred to the elements the associative structure that Wernicke and Meynert had applied to complex ideas.[31] Such a change, however, together with Freud's adoption of the neuron theory, which ramped up the possible complexity of associative systems, radically altered Meynert's model. Now rather than qualitatively different *Vorstellungen* being associated by a nervous connection, those nervous connections were ubiquitous. Structures could no longer then be considered as the groupings of heterogeneous elements. Variety was structural rather than substantial; *Vorstellungen* were distinguished because they were made up of different patterns of (essentially indistinguishable) nerve cells.

As a consequence of his pushing associationism further, replacing what he saw as the psychological vestiges in Meynert and Wernicke's system by physiological excitation patterns, Freud extended the site in which associations took place in two ways, which had profoundly disruptive effects for the project of localization.

First, whereas for Meynert *Vorstellungen* were localized in the gray matter of the cortex and were associated by white-matter fiber bundles, for Freud nervous connections also occurred within the gray matter: "We have no need to call on white fibre tracts for the association of ideas within the cortex. There is [even] a post-mortem finding which proves that the association of ideas takes place through the fibres situated *in the cortex itself*."[32] Though this meant that localization broadly understood still held—the "localization of a perception means nothing else but localization of its correlate"[33]—since this correlate was now a physiological modification or pattern, *Vorstellungen* could no longer be localized in one single cortical point. Different *Vorstellungen* did not correlate to the differing locations of cells; they were determined rather by differing structures of nervous connections, spread across the brain surface.

Second, Freud argued that the process of association might extend to what Meynert had called the "projection system," the very projection system that in Meynert's model determined the location of cortical cells containing basic *Vorstellungen*. Freud did not agree with Meynert's "principle of isolated conduction" that was essential to the projection model. In fact, to him, there was compelling histological evidence suggesting that it must be false. Results from contemporary brain science, including research on fiber reduction by Jacob Henle and Benedikt Stilling, Paul Flechsig's research on myelination, and Freud's own anatomical work, suggested that Meynert's view on "isolated conduction"—the simple and one-to-one connection between center and periphery—was unjustified.[34]

Freud's rejection of the "principle of isolated conduction" had consequences for his view of the projection system. Although the fibers arriving in the cortex still were in a certain relation to those at the periphery, this relation was no longer one of similarity: "They [the fiber tracts] contain the body periphery in the same way as—to borrow an example from the subject with which we are concerned here—a poem contains the alphabet, i.e. in a completely different arrangement serving other purposes, in manifold associations of the individual elements, whereby some may be represented several times, others not at all."[35] While it was justified to speak of projection in the *spinal cord*, because there a fiber would in fact travel from the periphery to the spinal cord without interruption, the term that best characterized the relationship between *cortex* and periphery was "representation" (*Repräsentation*).[36]

Because a point on the periphery was no longer "projected" onto a point on the cortex, a simple localization of function was no longer plausible. As we have seen for Meynert, direct projection privileged specific locations on the brain surface, determining their functions by their connections with the periphery. But if fiber pathways between periphery and center were more complicated, then the location of specific cortical points became less important. Because some transformation of the stimulus occurred on the way to the center, the whole system of fiber connections including the subcortex could no longer be sidelined and ignored.

While Freud's theory provided a powerful rebuke to the localizationists, he also had to contend with the vast and increasing evidence gleaned by the pathological anatomical method that lent authority to the localization-of-function paradigm. We have seen how the impressive results provided by pathological anatomical investigations had helped secure the localization project in the 1860s and 1870s. Freud considered himself well equipped to counter any problems that the method might present. In fact, through his radicalization of association physiology, he was able to reinterpret, and perhaps better explain, the practice that provided such strong evidence for the theory of localization. In brief, he suggested that the destruction of parts of the brain through a lesion caused symptoms not because it resulted in the anatomical alteration of a center, but rather because it cut off nervous connections or associations within a broader "speech territory" (*Sprachfeld*).[37]

For Meynert, the centers were connected through numerous associations. Thus in between the centers there was an area of the brain consisting purely of association fibers. But in Freud's model, the processes of associa-

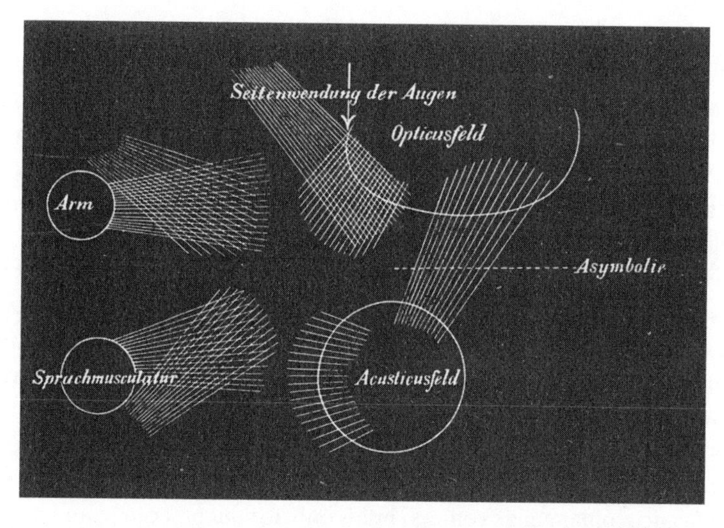

FIGURE 3.2. Anatomical schema to explain the appearance of language centers. (Sigmund Freud, *Zur Auffassung der Aphasien: Eine kritische Studie* [Leipzig: Franz Deuticke, 1891], 83.)

tion extended into those centers themselves, and thus there was no essential difference between the "centers" and the association system between them. Recast thus, what Wernicke and others had labeled "centers" looked now like far-flung regions of a larger area that Freud called an "association area," or "speech territory."[38] Freud's speech territory included both the different "centers" of language—the sensory (Wernicke), the motor (Broca), and the optical "centers" (for reading)—and the space between them. Broca's and Wernicke's "centers" then appear at the periphery, as "corners of the speech territory" (*Ecken des Sprachfeldes*) (fig. 3.2).[39]

This new mapping allowed Freud to reinterpret the results of the pathological anatomical method: a lesion at the periphery (the old "centers") would be more likely to cut off the majority of connections from a particular brain area. For example, if a lesion was located close to the acoustic field, it would cut off its connection to the speech territory and thus disturb the acoustic elements of speech.[40] This would result in sensory aphasia. It was not that acoustic *Vorstellungen* inhabited this field; rather it was the thoroughfare between the auditory and speech areas.[41]

If, on the other hand, the lesion was located in the center of the speech territory—an area that the localizationists had ignored, because lesions here provided no clear and consistent correlation with clinical symptoms—its symptoms would be far more diffuse. In the center of the speech territory, a lesion would not sever a sufficient number of fibers from any one particular area to have a clearly defined effect, even if the same number of connections were cut.[42] Thus, while in Meynert's (and his student Wernicke's) model, most pathological damage was caused by lesions in the "centers," and their theory led them to downplay the effect of a lesion located between those "centers," Freud's system allowed him to understand how this type of lesion, too, could be debilitating.[43]

Freud's model of the speech territory also shows that he did not abandon localization altogether. He still believed that the nervous system was the anatomical substratum of mental functions and that damage to those anatomical structures led to a loss of function. But because there were no longer any centers that could be pinpointed, but rather organizations or webs of nerves that spanned a larger area, one could not draw a direct correlation between nervous damage and the location of function: "This significance [of language centers] holds only for the pathology, and not for the physiology of the speech apparatus, because it cannot be maintained that in these parts other, or more important, processes take place than in those parts of the speech area the damage of which is better tolerated."[44] Just because a lesion in one area had a less clear effect than one in another did not mean that this area played a smaller role in speech function; its damage might just cause a more diffuse symptom. In a system where connections and processes were all-important, an emphasis on precise location was out of place.

The Disappearing Lesion

Freud's model of the ubiquity of associations and his concept of the speech territory were the first steps in his move away from the concept of the lesion. As we have seen, the idea of the lesion and its use in nineteenth-century physiology were strongly connected to the idea of localization through the pathological anatomical method. As Freud moved away from localization, the idea that all nervous disease could be traced to a lesion was set aside, making way for new possibilities. But this development did not happen all at once, and it is instructive to follow the line of Freud's thought.[45]

At first, Freud continued to use the word *lesion*, even as its meaning changed.[46] For a short period, Freud appealed to a modified idea of the lesion to explain hysteria.[47] Already in *Aphasia*, Freud adopted a more functional concept.[48] Because connections were cut rather than centers excised, a lesion no longer performed a precise role, nor did it usually entail complete destruction. As we have seen, Freud was sympathetic to Bastian's distinction between three types of lesion, corresponding to three levels of reduced excitability of a "center."[49] In Bastian's view, a lesion was not always absolute but led to a more generalized functional reaction.

Two years later, Freud developed the concept of "functional lesion," moving yet further away from anatomical understandings. In his 1893 article "Some Points for a Comparative Study of Organic and Hysterical Motor Paralyses," which he was asked by Charcot to write after his visit in 1885–86 but did not complete until eight years later, Freud discussed the differences between organic and hysterical paralyses, comparing them to each other.[50] The problem with hysterical symptoms was that, in contrast to most organic paralyses, they did not correspond to observable lesions in autopsy. To account for this absence, Charcot and his followers insisted that the lesions simply could not be seen. As dynamic or functional lesions, they had dissipated by the time of the autopsy. In principle they could be detected, but only if technological methods could be made more refined.[51]

Freud, on the other hand, suggested that Charcot was wrong to try to look for hysterical lesions in the autopsy material. Hysterical paralyses could not be explained by reference to anatomy; one needed to consider the "nature of the lesion" rather than its "extent and localization."[52] Hysterical lesions were no longer understood purely as physical damage to the brain. To Freud, they were without "concomitant organic lesion—or at least without one that is grossly palpable."[53]

If there was no palpable damage, why did Freud continue to use the term *lesion* to describe this problem? In part, we can suggest that it was out of respect for Charcot, who himself used the term.[54] But more importantly, the answer lies in the revision of the concept of the lesion that had occurred in Freud's *Aphasia* book. We have seen how, in *Aphasia*, Freud developed a new understanding of the concept of lesion in his notion of the speech territory. A lesion consisted not in the damage of a *center*, but in the cutting off of connections, in the case of aphasia, from the auditory cortex, the motor cortex, and other areas, to the speech territory.

This notion of interrupted connectivity, isolating certain areas of the nervous system, underlay Freud's notion of hysterical lesions as well. A hysterical lesion, to Freud, was an "alteration of a functional property," for instance a "diminution in excitability."[55] As an example, Freud presented the case of a loyal subject who refused to wash his hand because it had been shaken by his king. By not washing his hand, he prevented its representation from entering into new associations with other, less valuable, objects, and he thereby rendered the idea "inaccessible to association."[56] The same mechanism applied to hysterical paralyses: "Considered psychologically, the paralysis of the arm consists in the fact that the conception of the arm cannot enter into association with the other ideas constituting the ego of which the subject's body forms an important part. The lesion would therefore be the abolition of the associative accessibility of the conception of the arm. The arm behaves as though it did not exist for the play of associations."[57] In "Organic and Hysterical Paralyses," as in *Aphasia*, pathology arose from the interruption of connections between areas of association, which meant that a network of nerves was no longer available for excitation.

If the similarities between the idea of a lesion in the two works justified the continued use of the term, the differences were still important. In *Aphasia* this exclusion from the "play of associations" was caused by a physical lesion, *cutting* connections, and yet in "Organic and Hysterical Paralyses" the lesion did not have to correspond to physical damage: hysteria occurred without organic lesions, much like, in Freud's example, the concept of the arm that was lost "without being destroyed and without its material substratum (the nervous tissue of the corresponding region of the cortex) being damaged."[58] Hysteria, thus, was not a case of pathological anatomy, that is, a case of damaged structure; it was a different kind of pathological process.

The developments in Freud's conception of the lesion explain, then, why he became increasingly wary of appeals to "anatomy"; in a world of ever-changing connections, clearly defined and localizable functions were no longer in the cards. But this increasing skepticism with respect to anatomical explanations did not mean that Freud rejected the possibility of a *physiological* and materialist correlate to hysteria. This is nowhere clearer than in his 1888 article "Hysterie." Having asserted that hysteria was a "neurosis in the strictest meaning of the term" because "no visible changes of the nervous system [have] been found in this disease,"[59] Freud continued to assert its physiological nature: "Hysteria consists entirely in the physiological modification of the nervous system, and its character [*Wesen*] would have to be expressed in a formula that takes into account the patterns of excitation [*Erregbarkeitsverhältnisse*] in the different parts of the nervous system."[60] Hysteria, to Freud, was not anatomical but physiological. As in the simile of the camera presented in the *Interpretation of Dreams*, nervous changes could not necessarily be seen and mapped, and yet they were still physically instantiated in the machine, somehow causing a picture to emerge.[61]

If physical damage was not necessary for mental disease, new possibilities for understanding hysteria emerged. With nervous organization at the core, perhaps the absence of association was not the only form of pathology; mental disorder might also arise because unhelpful associations had been formed, a situation where the word *lesion* would no longer have any traction. In Freud's *Project for a Scientific Psychology* (1895), where he developed a thoroughgoing physiology of association to explain normal and pathological states, lesions dropped out of the picture entirely.

The *Project for a Scientific Psychology* should be understood as the culmination of Freud's physiological investigations. Although historians, for various reasons, have often characterized it as an anomaly, in the context of the previous discussion, it is best understood as a continuation of the themes elaborated in Freud's earlier work.[62]

First, in the *Project*, Freud, like Meynert, makes use of an associationist model. As in Meynert's model, Freud explains how stimuli (creating an excitation, *Erregung*, in the nervous system) were transferred from the periphery to the central nervous system. Parallel to Meynert's projection system, Freud described a structure of "φ neurons" bringing quantity Q to a more complicated system of "ψ neurons" (the association system) at the nervous system's core.[63] The ψ system was a collection of neurons constituted such that new connections could be made.[64] It explained the process of association so central to Meynert's model.

The physiological mechanism that Freud relied on for this process was the notion of *Bahnung* (facilitation), a concept developed by Sigmund Exner. In his chapter on the "experience of satisfaction," Freud described how this process of *Bahnung* or facilitation explained the "basic law of *association by simultaneity*."[65] If two neurons α and β were cathected simultaneously (corresponding to two simultaneous stimulations at the periphery), Freud asserted that it was then easier for a quantity Q to pass from one to the other; simultaneous cathexis reduced the resistance of the barriers between cells, such that Q would be more likely to push through and carve a more permanent path.[66] In this way Freud's model explained how "facilitation comes about between two mnemic images."[67] In the particular example that Freud gave, the feeling of satisfaction a child experienced on being presented with (and thus perceiving) the mother's breast and the crying preceding that presentation were linked in the baby's mind, such that an association was created between two memory images (of the object wished for and of the crying). With this physiological connection established, the resurgence of hunger in the future would lead the baby to reenact the action (crying) that previously led to satisfaction.

But second, and as before, Freud's appeal to associationism required a thoroughgoing physiologization of the process. One reason why the *Project* was "scientific" in Freud's mind was that it did not appeal to qualitatively different psychological ideas that could be "contained" within individual nerve cells.[68] Freud's model was, instead, purely "quantitative": a single and undifferentiated quantity Q determined whether a particular cell was cathected or not. There were no substantial differences between cathected cells (a cell cathected with the idea of the breast, the cell cathected with the idea of crying). Thus, as in his *Aphasia* book, to account for the differentiation of *Vorstellungen*, Freud had to extend the process of association from the links *between Vorstellungen* to the construction of *Vorstellungen* themselves.[69] *Vorstellungen* were differentiated not by the content of the nerves that corresponded to them but by their organization. We are thus not surprised to find that Freud explicitly stated several times in the *Project* that mental states were never localized in just one cell: "For the time has come to remember that perceptual cathexes are never cathexes of single neurons but always of complexes."[70] Whenever he did associate the *Wahrnehmungsbesetzung* with an individual neuron, he pointed out that this was a simplification: "For the sake of simplicity, however, I must now replace the cathexis of the complex perception by that of a single neurone."[71]

The first chapters of Freud's *Project* explain this construction of *Vorstellungen* by showing how a quantity Q that was derived from an external stimulus could carve a particular path through a network of nerve cells. As a quantity Q, which arose from external stimuli, reached the ψ system, the *Kontaktschranken* (i.e., barriers between cells) impeded its way. Each nerve had *Kontaktschranken* with numerous other nerves, and as the Q cathecting the ψ neuron rose, greater pressure would be applied to each one. Eventually one *Kontaktschranke* would succumb to the pressure and allow a flow between the two cells it divided.[72] According to the mechanism that Freud described, the flow of Q across a *Kontaktschranke* would weaken it—"contact-barriers becoming more capable of conduction"—and thus the next time the first cell was cathected with Q, it would be more likely to pass through that particular barrier. Over time and with repetitions, a path between two ψ neurons (and, by extension, complex paths across a network of neurons) would become ever more deeply engraved; the brain would form memories. As Freud suggested, this provided a physiological explanation for the "psych[ological] knowledge [that] the memory of an experience (that is, its continuing operative power) depends on a factor which is called the magnitude of the impression and on the frequency with which the same impression is repeated."[73] In the terms used by Jacques Derrida in his influential 1966 paper "Freud and the Scene of Writing," in this process we see the first signs of an increasingly insistent appeal to writing and the trace, which marked Freud's work throughout his career, cutting across any division between the preanalytic and analytic periods.[74]

Returning to the process of association *between Vorstellungen*, it becomes clear that this merely extended the process of *Bahnungen*, and complex *Vorstellungen* were not essentially different from so-called simple ones, as Meynert previously had suggested. And here again, we see how Freud drew on the resources of Meynert's model while at the same time going beyond it. His model of association was more strictly physiological than Meynert's because it did not rely on the localization of basic psychological functions in individual cells; that is, it moved away from the elision of elementary *Vorstellungen* with the basic elements of the nervous system. *Vorstellungen* were not unified "things" that could be connected and localized at a single point; they were already complex, already associated, patterns of cathected cells.

Thus, third, showing that *Vorstellungen* were more complex than Meynert thought and offering a mechanism to explain their formation, Freud was able to push his critique of localization to what seemed its ultimate end. Even as the opposition between φ and ψ neurons seems to map onto Meynert's distinction between projection and association systems, in both the essential structure of the nerves was the same: φ and ψ neurons differed because of the situation they found themselves in, not because they were substantially different. Freud was adamant that "the nervous system consists of distinct but similarly constructed neurones" and that the *Kontaktschranken* were not *essentially* different in different neurons: "To assume that there is an ultimate difference between the valence of the contact-barriers of φ and of ψ has . . . an unfortunate tinge of arbitrariness."[75] Instead, Freud suggested that it was the difference in the level of Q cathecting the neurons that determined their character. The *Kontaktschranken* of ψ neurons, open to the extreme stimuli of the outside world, offered no real resistance to the flow of Q, while ψ neurons, safely ensconced in the center of the organism, only had to deal with smaller levels and thus were able to direct its flow. If we exchanged "locality and connections" (*Topik und Verbindungen*) of a φ and a ψ neuron, it would make no difference to the functioning of the system: "They retain their characteristics, however, because the φ neurone is linked only with the periphery and the ψ neurone only with the interior of the body. A difference in their essence is thus replaced by a difference in the environment to which they are destined."[76]

But if the basic structure of Freud's *Project* can be understood as a continuation and indeed culmination of his earlier physiological investigations (of 1891 and 1893), there was one crucial distinction. For though Freud's *Project* was in one sense thoroughly physiological—because it was even more careful to avoid the "qualitative" psychological distinctions that still inhabited Meynert's work and replace them with a purely "quantitative" and thus "scientific" model—the key term of his earlier physiology, the "lesion," is conspicuously absent here.

Freud's analyses in his earlier work on hysteria had shown that pathology could arise not only because brain areas had been damaged, or connections cut, but also through a pathological *organization* of nervous elements. The term *lesion* was retained in "Organic and Hysterical Paralyses" because Freud still considered pathology to be caused by the absence of associations. But in the *Project* Freud suggested that organizational pathology might derive not only from the *lack* of association or the *inability* to associate. Instead, some associations themselves, created in the mind through the process of *Bahnung* explained earlier, might turn out to be pathological. And since *Bahnung* between neurons arose through the application of different external stimuli, pathology might be caused by lived experiences that left memory traces.[77] In this Freud developed a possibility that was latent in Meynert's work, because he too asserted the malleability of the nervous system and described the way it changed in response to different experiences. But because Meynert's associationism was so closely tied to the localization project, and thus to the lesion model of mental pathology that had always accompanied it, the notion of pathological nervous *organization* was never developed in his work.

Freud's *Project* leads up to a discussion of hysteria that makes use of this new possibility. As Freud explained it, hysteria can be described symptomatically as an "excessively intense idea [A], which forces its way into consciousness" to an extent not justified by its manifest content (e.g., excessive anxiety at entering a shop alone).[78] But analysis reveals that A has this effect because it is associated with another idea B, the two having been connected by a particular event in the patient's personal history. Hysteria corresponded, then, to what Freud called a "pathological symbol-formation," that is, a pathological association.[79] A causes the emotional response that would be understandable for the idea B but remains incomprehensible when considered with respect to A alone. And crucially, because the "association A-B, and B itself plays no part at all in [the patient's] psychical life," the patient's behavior seems completely incomprehensible.[80] Mental disease no longer relied on physical damage; instead, experiences—and the associations they created—could cause difficulties in a patient's life. The de-psychologization of associationism that allowed Freud to move beyond the lesion paradigm thus led him, ironically, to make room for the etiology of nervous disorder in individual "psychological" experience. Indeed it is telling that when searching for a word to describe this pathological experience, Freud turned to a term that until then had mostly been tied to physical damage: *trauma*.[81]

The Collapse of Localization and the Birth of the Unconscious

There are two important consequences of Freud's reformulation of the Meynert-Wernicke model, and they mark most clearly his overturning of it. As I will argue, it allowed him both to posit the existence of the unconscious and to develop a new form of therapy: psychoanalysis. The argument that Freud's engagement with the brain sciences allows us to see how he developed something like the unconscious has been proposed before, most famously by Marcel Gauchet in his book *L'inconscient cérébral*. Gauchet's book provides a history of the cerebral reflexes. Before Freud, the distinction between higher and lower functions mapped directly onto that between conscious and unconscious activity, for only lower functions such as digestion, heart regulation, and simple reflexes were considered to be nonconscious. By integrating Freud into a tradition of "cerebral reflexes," Gauchet was able to suggest how Freud could identify the role played by nonconscious processes in higher functions.[82] The cerebral reflex explained mental acts of which the subject might not be aware. That is, Freud opened up a space for nonconscious *higher* functions; he broke the traditional identification of the conscious and the psychical.[83]

But as we have seen, Meynert had also developed a theory of cerebral reflexes, and yet the unconscious did not appear as a central part of his theory. In fact, Meynert's example suggests that merely using reflex action to describe mental activity was not in itself sufficient to explain the emergence of the unconscious. Meynert complicated the reflex by adding an association system between the afferent and efferent reflex arcs, thus extending its connective qualities. Only in this way could the reflex be adequate to the task of explaining thought; for Meynert, association became the distinctive mark of higher functions. Not only was association closely bound to higher functions; it also remained intimately tied to consciousness. First, associations were created only if two *Vorstellungen* were conscious simultaneously. Second, though the brain was, in Meynert's terms, always in a state of "partial sleep," and the majority of *Vorstellungen* "dimmed [*verdunkelt*]," that is, lying outside the realm of consciousness, the association between two elements entailed that if one was raised into consciousness, the other would be, too. In Meynert's example of the bleating lamb, if one heard a lamb bleating (thus raising the acoustic image of bleating [B] into consciousness), one could recall the visual image of a lamb [A]. Meynert's development of a cerebral reflex, then, did not sever the essential connection between higher functions and consciousness.

Freud's model of brain action was different, and he teased apart Meynert's triple identification of higher function, association, and consciousness. He denied Meynert's excessive topological restriction of the association system. As we saw, Freud extended association into the projection system and into the realm of gray matter.[84] For Freud, association was sufficiently pervasive in the nervous system that it could no longer mark the distinction between higher and lower functions. At the same time, and in part as a consequence, Freud also broke down the connection between association and consciousness. As we have seen, it was central to his theory of mental pathology that an unconscious element could be associated with one that could remain unconscious even when the other was experienced excessively intensely. The hysterical symptom was a "symbol" for an element that was repressed. The association between the manifest and the latent was both the reason for and the ultimate object of analysis. In this way the radicalization of the association model in Freud's work allowed him to break the relationship between it and consciousness and consequently to imagine the possibility of higher functions detached from consciousness. The unconscious was born.

Reflex Therapy

The thoroughgoing physiologization of association had another effect on Freud's work: it opened up the possibility for a new form of practice. As we saw, in the last fifteen years of the nineteenth century, Freud distanced himself increasingly from the idea that physical lesions were a primary cause of mental pathology. Now he regarded pathology as a problem of nervous organization, often caused by traumatic psychological experience. Freud's emphasis on psychological experience could cut both ways. Without the finality of nervous damage, which rendered the idea of therapy tenuous at best, Freud's new model offered the possibility of improvement. According to this logic, pathological associations could be rerouted, isolated elements brought back into the nervous economy, and networks of nerves reorganized. And the physician would be able to encourage this not by intervening on the brain with a scalpel, but by carefully directing questions at the patient. The development is most striking if psychoanalysis is placed within the tradition of psychic reflex exams along the lines of Wernicke; what had previously been a purely diagnostic process had been transformed into a therapy.

To justify this comparison between psychoanalysis and the psychic reflex exam, I will take two steps. First, I will draw out the parallels between Wernicke's practice and Freud's early "cathartic method." Second, I will suggest that Freud's rejection of the "cathartic method" and elaboration of psychoanalysis can be read as a further step away from the lesion paradigm, bringing to its most radical conclusion Freud's internal critique of neuropsychiatry. And though the "cathartic method" was therapeutic, it was only in Freud's later psychoanalysis that therapy was to become the central structuring element of the practice. Psychoanalysis would then be a psychic reflex exam, but one where the reflex had undergone a profound transformation.

Freud's most prominent discussion of the cathartic method can be found in his *Studies on Hysteria* from 1895, which he wrote with his senior colleague Joseph Breuer. In a hysterical patient, "pathological associations" would cause somatic symptoms, such as hysterical coughs or paralyses, because normal processes of abreaction or association were inhibited. As Freud wrote, "One can thus say that the ideas that had become pathogenic are kept so fresh and intense [*affektkräftig*] because the normal coping mechanisms [*Usur*] through abreaction or reproduction in states of uninhibited association are taken away from them."[85] The search for these pathological associations constituted the bulk of the practice, and Freud and Breuer would often use hypnosis in order to bypass the patients' repression and resistances and access the unconscious directly.[86] Once the doctor had identified the pathogenic idea, he could reveal it to the patient, reentering it into the "traffic of associations" [*Associationsverkehr*] and thus making the symptoms disappear; Freud called the process "associative correction."[87]

Like Wernicke, Freud understood his practice as a direct intervention on the nervous system. This fact is clearest when the cathartic method is placed in the context of Freud's broader therapeutic arsenal, especially electrotherapy. Freud had learned electrotherapy, alongside a number of other physical methods—hydrotherapy, massage, and rest cures—during various clinical internships before 1883.[88] In early 1884 he spent ten gulden to acquire the equipment to perform it in his private practice. As he wrote to his fiancée Martha Bernays in early 1884, the purchase "had made him poor," but it was a necessary investment.[89] Freud continued to use electrotherapy until at least 1895, several years after he had begun working with the cathartic method.[90]

And while the cathartic method used very different tools from electrotherapy—intervening with words rather than electricity—Freud understood the two practices as complementary. Because Freud considered hysteria to be the result of a physiological modification of the nervous system, based on an altered pattern of nervous excitations, any practice that altered the distribution of these excitations could be effective. The fact that the cathartic method came to eclipse electrotherapy was not, then, a shift in orientation from physiology to psychology. Rather, it followed the development of his theory. As Freud came to locate the etiology of hysteria in traumatic psychological experience, he came to regard electrotherapy as a relatively undifferentiated approach for reorganizing nervous excitations. His new cathartic method followed the paths of the original trauma more closely and thus targeted the problem more specifically. As Freud noted in 1888, the cathartic method was effective because it "exactly imitates the mechanism of origin and disappearance of such hysterical disturbances."[91]

Not only did Freud envision the cathartic method as a practice taking the nervous system as its primary object; he also continued to understand the nervous system in reflex terms. This is why he remained comfortable using reflex language far beyond any putative "break." In *Etiology of Hysteria* (1896), Freud offered an explanation for the seemingly inadequate response to a given stimulus in the hysterical patient, the "mismatch between psychic stimulus [*psychisch erregendem Reiz*] and psychic response."[92] Even as late as 1901, Freud reaffirmed his understanding of mental processes as reflexes. In chapter 7 of *The Interpretation of Dreams*, Freud was definitive in his assertion: "Reflex action remains the model . . . for all psychic action."[93]

This reflex context provides insight into the development of Freud's interpretive practice. As I showed in chapter 2, Wernicke's engagement with his patients encouraged him to see the reflex as a principle of interpretation. The multiple and systemic functioning of nervous elements prevented any simple and immediate diagnosis of mental pathology. But the form of reflex interpretation can be specified more closely. For, when applied to higher functions, the reflex tended to introduce a distinction between manifest (conscious) and latent (unconscious) content. In a normal (psychiatric) patient exam, where the patient was asked to identify pains and aches, to recall strange events in the past, the doctor depended on the patient's powers of introspection and his ability to relay a particular internal state (memory, sensation, thought). The manifest content of the patient's words, what he meant to say, took center stage. In the psychic reflex exam, however, it was how a patient deviated from this script that was most important: slurring, the inability to say particular words, the inadvertent replacement of those words by others, slips of the tongue, and the like. These were caused by nonconscious psychic mechanisms—which in turn resulted from a trauma, whether physical or psychological—working behind and often at cross-purposes to conscious thought.

Moreover, for both Freud and Wernicke, the goal of analysis was the identification of that trauma, an identification achieved through the careful reading of the distortion in the patient's speech. Freud's early interpretation in the cathartic method was parasitic of reflex interpretation. Though the cathartic method was figured as a therapy, it is striking how, like Wernicke's psychic reflex exam, it remained predominantly focused on etiology. Therapy was merely a happy side effect of a thorough patient examination. The discovery of the pathology's cause would itself redistribute the nervous *Erregungen*.[94]

The Case of the Rat Man

Treatment became an autonomous and central aspect of Freud's practice as he grew increasingly dissatisfied with the cathartic method. This change can be seen in the analysis of one of the most famous occupants of Freud's couch: Ernst Lanzer. Today, he is better known as "the rat man," the subject of one of Freud's five long case histories.[95] Lanzer came to see Freud in October 1907. "A youngish man of university education," Lanzer presented with a number of obsessional ideas and impulses: he feared that two persons of importance to him would be harmed, and he felt compelled to cut his throat with a razor.[96]

By the time Lanzer first came to visit Freud at Berggasse 19, Freud had already dispensed with hypnosis in his practice. He had confronted a substantial number of hysterical patients who could not be hypnotized, and so he had begun to look for alternatives, to "circumvent hypnosis while still gaining access to the pathological memories."[97] The revised practice dramatically transformed the relationship between doctor and patient. In opposition to the earlier cathartic method, in analyzing Lanzer, Freud assumed a predominantly passive role; the analyst, according to Freud had to "suppress his curiosity."[98] As Freud wrote later, the analyst should not direct his "notice to anything in particular," but maintain instead "the same 'evenly-suspended attention.'" The relative passivity of the doctor was expressed in Freud's metaphor of the telephone receiver: the analyst should "turn his own unconscious like a receptive organ towards the transmitting unconscious of the patient. He must adjust himself to the patient as a telephone receiver is adjusted to the transmitting microphone."[99]

The patient, in contrast, became comparatively active. At the beginning of the treatment, Freud urged Lanzer "to say everything that came into his head, even if it was *unpleasant* to him, or seemed *unimportant* or *irrelevant* or *senseless.*"[100] "Free association," the "fundamental rule" of psychoanalysis (*psychoanalytische Grundregel*), became the center of Freud's treatment.[101] It was up to the patient to decide about the "order in which topics shall succeed each other during the treatment."[102] From the very first session, Lanzer began to share details about his early sexual life. He remembered that early on (he believed that he must have been around four or five years old) he had the strong desire to see the female human body naked. This desire was accompanied by the seemingly unrelated obsessional idea that something terrible would happen to his father—that he would die, or that he would undergo what Lanzer would later describe as the "rat punishment," Lanzer's compulsive fantasy that gave the case its name.

Free association, however, had its limits. Repression placed certain memories out of reach, so while they might perhaps be hinted at in the patient's account, they remained absent from it. The patient might even strongly resist the attempt to uncover them. Indeed, when Lanzer began to touch on that "specially horrible punishment," he "broke off, got up from the sofa, and begged [Freud] to spare him the recital of the details."[103] And yet, resistance should not be seen simply as an obstacle to the analysis, as it had been for the earlier cathartic method. Then, as we have seen, resistance could be bypassed through hypnosis. Now, for Freud, it was one of the "pillars of psychoanalytic theory,"[104] and Lanzer's early resistance constitutes the most significant passage of the case.

First, the moment of resistance gave Freud further insight into Lanzer's pathology. From his free association, Freud had already concluded that Lanzer as a child "had been guilty of some sexual misdemeanour . . . and had been soundly castigated for it by his father." He suspected that this event had "established" the father "for all time in his role of an interferer with the patient's sexual enjoyment."[105] Lanzer confirmed this reading. The patient suggested that when he was a young child, his father had once beaten him. Although he himself had no direct memory of the incident, his mother had repeatedly mentioned it, explaining that "he had been given the punishment because he had *bitten* someone."[106] Lanzer's resistance suggested that the rat punishment provided the key for unlocking the case: its "nodal point," as Freud called it.[107] Freud pressed Lanzer for the details: "Was he perhaps thinking of impalement?" Freud asked. "No, not that; . . . the criminal was tied up," answered the patient, "'a pot was turned upside down on his buttocks . . . some *rats* were put into it . . . and they . . .'—he had again got up, and was showing every sign of horror and resistance—'*bored their way in* . . .'—'Into his anus,'" Freud suggested.[108] From Lanzer's account, Freud was able to draw together an intricate interpretation of his pathology, an interpretation involving unwanted marriages, debt, and family disruption.

But second, and more importantly, the moment of resistance provided Lanzer with an opportunity to work through his pathology. Freud, in his theoretical papers of the time, compared his role to that of an educator (*Erzieher*). Psychoanalytic treatment could be considered the "*after-education in overcoming internal resistances*."[109] Indeed, it is significant that although Freud had begun to develop theories about early childhood trauma in the second session, he held his revelation back until fairly late in the analysis. Unlike what would occur in the cathartic method, the doctor's insight into a case was in itself insufficient to effect the cure; the patient had to be ready to accept it.[110] Lanzer strongly resisted Freud's "elucidation of the matter," and it was only in the further course of the treatment that the patient "was forcibly brought to believe in the truth of [Freud's] suspicion."[111] Similarly, when Lanzer asked Freud how the interpretation could have a healing effect, Freud answered that it was not the "disclosure" of this interpretation but the "process of discovery [*Auffindung*]," that was therapeutic.[112] As Freud argued, the "overcoming of resistances was a law of the treatment, and on no account could it be dispensed with."[113]

The most important tool in the psychoanalyst's kit for helping the patient work through these resistances was transference, where the patient came to project upon the analyst emotions associated with important figures in his psychic life. Such transference could be valuable, because it tied the patient emotionally to the practitioner. The confrontation with traumatic events from the past was unpleasant, so it helped if the patient had "formed a sufficient attachment (transference) to the physician for his emotional relationship to him to make a fresh flight impossible."[114] In Lanzer's case, his transferential relationship with Freud, taking the place of his father, encouraged him to accept Freud's explanations.

The working-through of resistances came to take priority over and even eclipse the interpretive aspect of the treatment, the discovery of the original trauma. As the case developed, Freud uncovered a number of earlier "traumas" in addition to the first "misdemeanour." In particular, he referred to Lanzer's reactions to his parents' marriage, which had impacted upon his own marital affairs. As the traumas multiplied, they also became less real. In a long footnote, Freud remarked that he could actually not be sure if the sexual "misdemeanour" had really taken place. He wrote: "It is just as possible that the child was reproved by his nurse or by his mother herself for some commonplace piece of naughtiness of a non-sexual nature, and that his reaction was so violent that he was castigated by his father." In this situation, it was no longer possible "to unravel this tissue of phantasy thread by thread."[115] But crucially, for Freud this did not seem to matter. If the patient worked through the resistances, whether some ultimate cause for the pathology had been discovered or not became increasingly irrelevant. No longer the analyst's virtuosic reading of the patient's symptoms, psychoanalysis was more about the painstaking process of helping the patient confront them.[116]

As we saw, over the 1890s, Freud slowly came to give up on the idea that lesions were the primary cause of mental illness. And yet at the same time, in his early discussion of hysteria, his analysis was still structured by it. Freud's cathartic method resembles Wernicke's psychic reflex testing, because both aimed to identify a single discrete cause for mental disruption, whether in the lesion or in a mental trauma. The relative de-emphasis on the reality and discovery of the trauma in psychoanalysis thus marks Freud's definitive rejection of the lesion model and is the clearest sign of the distance he had traveled since his days in Meynert's clinic.

Instituting Psychoanalysis

The practice of psychoanalysis very quickly gained an array of institutions to promote and guide its development. The first of these institutions, founded in 1902, was the Wednesday Psychological Society (it became the Vienna Psychoanalytic Society in 1908).[117] Meeting in Freud's home on Berggasse, the Wednesday Psychological Society initially consisted of five members (in addition to Freud, it included the Viennese physicians Wilhelm Stekel, Max Kahane, Rudolf Reitler, and Alfred Adler). As the years passed, the group was joined by a number of other physicians, such as Adolf Deutsch, Paul Federn, and Eduard Hitschmann.

As psychoanalysis moved beyond the borders of Austria, regulation of the practice became a central concern. We see debates in the regional centers for psychoanalysis, especially in the Berlin group, which in the 1920s became increasingly important for the movement, and, after 1938, in London and New York. When Sándor Ferenczi proposed his idea for an International Psychoanalytic Association (IPA), he pointed out the need to defend psychoanalysis not only from academic psychiatrists but also, and especially, from enemies within. The IPA would "offer some guarantee that Freud's own psycho-analytic methods were being used, and not methods cooked up for the practitioner's own purposes."[118] The IPA thus became the site of numerous debates over the training and regulation of psychoanalysts. As psychoanalysis became increasingly Anglo-Americanized, control over entry and qualification of members became stricter, causing further controversy; some of this will be discussed in chapter 5.[119]

The psychological aspects of psychoanalytic practice left their imprint on the developing institutions. As the movement grew, it began to attract cultural critics, sexologists, and sexual reformers, in addition to the physicians that had made up its initial core membership. This tendency was most visible in the new organ of the movement, the *Jahrbuch für psychoanalytische und psychopathologische Forschungen*, founded in 1909. Carl

Jung, in his preface to the first edition, emphasized this new scope for the movement, arguing that psychoanalysis was of interest to scholars of the *Geisteswissenschaften* (an orientation that came to be expressed most clearly in the later journal *Imago*) as well as those interested in nervous disorder.[120]

The disputes that divided psychoanalysis from the beginning show that there was no unified plan, no grand vision, or at least none that was dominant. But because these debates often took as their major object psychoanalytic practice, and because, as I have suggested, that practice marks the culmination of a complex and thoroughgoing criticism of the neuropsychiatric tradition, the institutions of psychoanalysis came to be the clearest marker of the gulf that had opened up between the two disciplines. While retaining links to hospital and university medicine, psychoanalysis increasingly showed itself to be a different track, taking place in private environments, treating different kinds of patients, requiring specific training, developing its own methods, eventually providing its own qualifications, and so forth.

And yet an excessive emphasis on these institutional differences obscures the curious development that had brought Freud to this point. Though the practice, and then the institutions that were built upon it, seem to locate psychoanalysis on the psychological side of the soma-psyche divide, the development that produced the practice, the rejection of the lesion model, was motivated by an attempt to rid neuropsychiatry of its psychological elements.

Conclusion

Freud's development highlights the way in which an uncritical assumption of oppositions like physiology-psychology or soma-psyche can obstruct an understanding of psychoanalysis and its history. Freud's working through of the tensions within localization helps explain his move from a thoroughgoing materialism to psychology without assuming an inherent opposition between the two. When writing about the development in Freud's thought, scholars have often characterized his turn to psychology as a move away from his earlier—in modern terms we would say neuroscientific—work. Similarly, his *Project* has been understood as an inessential and thus dispensable neurologization of key psychological concepts that were already in place in Freud's thought. But as we saw, Freud's extension of the scope of association was intended to "de-psychologize" Meynert's association physiology and localization project. And it was this de-psychologization of physiology that allowed Freud first to move away from the notion of lesion as physical trauma, casting it in his article "Organic and Hysterical Paralyses" rather as the isolation of nervous elements; then in the *Project* he developed an idea of a "psychological" trauma; and finally, in his later psychoanalytic practice, he even de-emphasized the search for any trauma at all. Freud was able to provide a psychological etiology of mental disease, not by rejecting but rather by radically adhering to his physiological roots.

This trajectory helps explain the parallels between Freud's developing practice and another reformulation of the neuropsychiatric model: that of the Breslau neurologist and neurosurgeon Otfrid Foerster. For while at first glance it might appear that Foerster's neurological work on the spinal cord was worlds apart from Freud's investigations of the unconscious, we shall see that Freud's "talk therapy" found its counterpart in the neurological gymnastics that Foerster called *Übungstherapie*.

CHAPTER FOUR

In the Exercise Hall

Otfrid Foerster, Neurological Gymnastics, and the Surgery of Motor Function

On May 26, 1922, the leader of the Russian revolution suffered a stroke.[1] Less than five years after he and his Bolshevik allies had overthrown the Provisional Government, Vladimir Ilyich Lenin found himself hemiplegic, unable to make full use of his right arm and leg and suffering from a disturbance of speech.[2] Soon afterward, an international team of doctors arrived in Moscow, charged with Lenin's care.[3] Breslau neurologist Otfrid Foerster (1873–1941) was the chief physician of the group. He lived with the Russian leader between 1922 and Lenin's death in 1924, first in Moscow, then near Lenin's family dacha in Gorki. Supervising all aspects of Lenin's medical treatment—which included warm baths, the administering of drugs such as iodide and quinine, and the use of an orthopedic boot to support his paralyzed right ankle—Foerster spent an hour every day leading his patient through a number of exercises, what he called *Übungstherapie* (exercise therapy).[4]

The therapeutic aspect of Foerster's work should hold our interest. Neurology is not often considered a therapeutic specialty; according to a common trope, neurology is "therapeutically nihilistic." That is, it is concerned with the study and diagnosis of nervous diseases, rather than their treatment.[5] Similarly, the neuropsychiatry that was the prominent institutional home for neurological activities in the last decades of the nineteenth century had a poor reputation for actually curing patients. As we saw in chapters 1 and 2, neuropsychiatry was concerned primarily with the diagnosis and the exact localization of nervous damage, rather than with ways to improve the patient's health, and because the pathological anatomical method often required autopsies to gain its results, they were of little use to the patient. Why was it then that—of all German doctors—Otfrid Foerster should be sent to treat the father of the Soviet revolution? How was neurology able to play a role in the developing diplomatic relations between Germany and the Soviet Union?

In this chapter I will attempt to answer these questions by considering the history of neurology as a therapeutic specialty. I will show how Foerster's development of both surgical and nonsurgical therapies for a number of different diseases was instrumental in giving neurology the institutional capital to detach itself from both neuropsychiatry and internal medicine.[6] Yet, while Foerster's therapies were critical to the institutionalization of an independent neurology, I want to argue that they resulted from an engagement with the principles of the neuropsychiatric tradition, an engagement that in many ways mirrors Sigmund Freud's. Like Freud, Foerster refigured the reflex model in ways that distanced his work from the simple *Zentrenlehre*: he explained nervous pathology as a breakdown in nervous organization, not simply as nervous damage, and he placed the "plasticity" of the nervous system at the center of his theory and practice.[7] And as in Freud's practice, these adjustments provided the resources for developing his therapeutic techniques. Although in the first decades of the twentieth century, psychoanalysis and neurology became increasingly divergent both disciplinarily and institutionally, they effected their break from neuropsychiatry in similar ways.

Foerster and the Neuropsychiatric Tradition

Foerster is a major figure in the history of neurology: he was president of the prestigious Gesellschaft Deutscher Nervenärzte from 1924 to 1932, he attracted students from all over the world to work with him in Breslau, and, together with Oswald Bumke, he authored the monumental, seventeen-volume *Handbuch der Neurologie*.[8] In 1931 he received a prestigious grant from the Rockefeller Foundation to support the founding of a research institute in neurology, later named the Foerster Institute.

Like many other neurologists in this period, however, Foerster began his career embedded in the neuropsychiatric tradition. He received his early training from Wernicke, working with him from 1899 to 1904, first as his *Volontärarzt* and then as his assistant in the psychiatric clinic and the neurological polyclinic. Foerster's early work was much in the spirit of

Wernicke's psychiatry: his first papers touched on topics such as the compulsive reproduction of memory images, the cortical paralysis of touch (*corticale Tastlähmung*), and the somatopsychoses.[9] After writing his *Habilitation* under Wernicke in 1903, Foerster's inaugural address (*Antrittsvorlesung*) as *Privatdozent* was titled "Comparative Investigations of Motility Psychoses and Diseases of the System of Projection Fibers," a topic recognizably marked by Wernicke's system.[10]

Foerster worked in Wernicke's neuropathological laboratory, where he was in charge of producing the third volume of a photographic atlas of the brain.[11] Foerster also shared Wernicke's interest in the neurology of movement: his *Habilitation*, in the field of *Nervenheilkunde*, was titled "Beiträge zur Kenntnis der Mitbewegungen." As one commentator later suggested, "it is probable that nobody had a more complete knowledge of the innervation of the human body than Otfrid Foerster."[12] To complement his education, Foerster spent time abroad, receiving neurological training from Heinrich Frenkel (1860–1931) in Heiden, a small spa town (*Kurort*) in Switzerland, and from Joseph Dejerine (1849–1917) in Paris.[13]

But while Wernicke's neuropathological and movement research remained part of his broader psychiatric project—as we have seen, even his engagement with Duchenne's *Physiology of Movement* informed his neuropsychiatric work[14]—Foerster's work was of little obvious psychiatric import. In the diverse ecosystem of Wernicke's clinic, Foerster was able to find his niche, but when the older man left Breslau in 1904 to become director of the Psychiatrische und Nervenklinik in Halle, Foerster was left without a position. The newly built neurological and psychiatric clinic in Breslau did not open until 1907, and even then it remained an unlikely source of employment for Foerster. It was headed by the psychiatrist Karl Bonhoeffer, who, as we saw, had been Wernicke's assistant but who considered neurology subordinate to the needs of psychiatry. Foerster's focus on the body and movement rather than psychological disturbances made his work seem at best irrelevant to those upon whom his career depended. During this period, Foerster's position was precarious. Like so many young neurologists at the time, he found only unsalaried positions, first as a *Privatdozent*, and from 1909 as a *Titularprofessor*.[15] He lived off the insecure income he received from tuition fees for his lectures and tutorials.[16]

Foerster's personal difficulties in academic medicine reflected a broader struggle over the disciplinary status of neurology.[17] For much of its history, neurology in Germany was caught between psychiatry and internal medicine. The first neurological textbook, the *Lehrbuch der Nervenkrankheiten*, was published in 1840 by Moritz Heinrich Romberg, an internist at the medical polyclinic at the Berlin Charité hospital. At around the same time, Wilhelm Griesinger published his textbook *Pathologie und Therapie der psychischen Krankheiten*, which became the bible for the neuropsychiatrists. Of the two approaches, it was Griesinger's that won out. He succeeded Romberg in Berlin, simultaneously holding the chair of psychiatry and running a neurological department from 1865. It was the first combined neuropsychiatric chair in Germany.[18]

We saw in chapter 1 that many psychiatrists opposed the neuropsychiatric appropriation of neurology. But that appropriation also elicited resistance from other doctors. The neurological internists Wilhelm Erb and Adolf Strümpell expressed their opposition to the "increasing usurpation by psychiatrists of nervous disease."[19] Together with Friedrich Schultze and Ludwig Lichtheim, in 1891 they founded the *Zeitschrift für Nervenheilkunde*, in which they tried to promote a neurology that was oriented toward internal medicine. In 1904, at the 29th Congress of Southwest-German Neurologists and Psychiatrists (Wanderversammlung der Südwestdeutschen Neurologen und Irrenärzte) in Baden-Baden, the debate culminated in an open conflict between the neurological internists Schultze and Erb and the psychiatrists Karl Fürstner and Eduard Hitzig, with each side claiming neurology for itself.[20]

Nevertheless, in the short term at least, the institutional dominance of psychiatrists in neurology seemed unstoppable. In Berlin and Halle, two of the leading universities in Prussia, chairs went to professors who preached the unity between psychiatry and neurology: Friedrich Jolly and Eduard Hitzig. The trend for combined neuropsychiatric university clinics continued in the early twentieth century in Greifswald (1905), Göttingen (1906), Breslau (1907), and Bonn (1908). Even in institutions where neurology had already been established in independent form—Breslau, Hamburg, and Heidelberg—additional neuropsychiatric clinics were founded.[21]

Displaced Lesions

One of the major arguments made by psychiatrists for their control over neurology was its lack of a specialized treatment.[22] In 1915 Bonhoeffer, who had moved to Berlin to become *Ordinarius* (full professor) of psychiatry and neurology and was thus (institutionally, at least) the most powerful psychiatrist in Germany, argued that "the development of a specialized examination and therapeutic technique led to the separation of otology, laryngology and orthopedics from surgery." Similarly, the "practical importance of infant care (*Säuglingsfürsorge*)" had led to the separation of pediatrics from internal medicine.[23] Therapeutic application was a precondition for institutional autonomy. Neurology, in Bonhoeffer's eyes, could not furnish a therapy that would justify its independence. Of the treatments it did offer, few were specific to it: general surgeons who had "learned the ropes of the specialty" could perform neurosurgery satisfactorily. Moreover, neurology's second major form of treatment, electrotherapy, was, for Bonhoeffer, a "disputable field with respect to organic effectiveness."[24]

For Foerster, the psychiatric disdain for neurological treatment presented a challenge. For his own career success and for the establishment of neurology as an autonomous discipline, Foerster would have to present its therapeutic bona fides. His earliest breakthrough was a new therapy for patients suffering from tabes dorsalis, or third-stage syphilis.[25] These patients displayed a range of debilitating symptoms, including paralysis, tremor, the inability to walk, sensory disturbances, and pain. Some of the tabes patients manifested a condition that was particularly troublesome. They suffered from "organ crises," that is, "over-excitement [*Reizerscheinungen*] in the sensory and motor spheres of a certain organ [that occurred] in a seizure-like way."[26] Although the attacks could affect any organ of the body—the heart, the kidney, the lungs, the intestines, even the sense organs—crises of the stomach were the most frequent. They were characterized by periods of severe stomach pain combined with violent contractions of the abdominal wall and vomiting and hypersecretion of gastric juice. Often the symptoms were so pronounced that food intake became impossible. What made the condition particularly confusing was that despite the wide range of symptoms—sensory, secretory, and motor—the only lesion that could be found was physical damage to the "fibers of the sensory roots . . . of the affected organ" (tabes targeted the dorsal [posterior] roots and the fibers that emerged from them, e.g., to supply the internal organs).[27]

Foerster's major insight was that the symptoms were caused indirectly, through a reflex mechanism.[28] For example, a gastric crisis could be triggered by food in the stomach. A hypersensitivity of the stomach lining, caused by the irritation of the spinal sensory nerve affected by the disease, would cause a reflectory motor response: vomiting.[29] However, emptying of the stomach was rarely total, and food residues would trigger a second incident of vomiting. Because of the reflectory hypersecretion, which, in addition to the motor response, was caused by the irritation of the sensory nerve, further vomiting was produced, so that vomiting bouts could last for hours.[30] Foerster thus presented the symptoms as a pathological motor response caused by damage to the sensory structures of the nervous system.[31] The reflex mechanism explained the noncoincidence of symptom and underlying process.

A similar indirect relationship between lesion and symptom could be seen in a second group of patients, who suffered from Little's disease. The disease was seen at the time as one of the worst of all childhood afflictions. The young patients spent their lives in immobility and dependence, lying bent in their beds or sitting on their chairs in contorted positions. According to Foerster's co-worker, the Breslau surgeon Alexander Tietze, "with their crooked, spastically contracted limbs," they were "truly pitiful creatures."[32] One cause for spastic contractures was physical damage to the nervous system, more specifically the sensory nerves. This damage could happen through various processes, such as the constriction of nerves, bone or metal splinters, or aneurysms. Spastic contractures could also be caused by damage to the pyramidal tract of the nervous system, which normally exerted an inhibiting influence on the spinal reflex arc. If the inhibiting influence was lost, the spinal reflex arc would be continuously active; that is, incoming sensory excitations that constantly arrived from the "skin, the joints, the ligaments, and . . . the muscles themselves," would constantly "be reflected" into the muscles, causing a spastic contracture.[33] As in the tabes dorsalis cases, though the damage was either to the *sensory* nerves or to the pyramidal tract, the symptom appeared elsewhere, causing the contracture via the spinal *motor* nerves. In both cases, then, the lesion had only an indirect relationship to the symptom; the cause of the pathology was distant from the pathology itself.[34]

The Development of a Therapy

Foerster's reconceptualization of nervous pathology opened up new therapeutic possibilities.[35] In a simple lesion model, damage to the nervous system was necessarily pathological, removing one or another element of the nervous system and thus eliminating one function or another. The doctor's scalpel, as much as accidental physical trauma, could only have a negative effect. But if a lesion did not simply cause a loss of function, but rather provoked disorganization of a reflex system, causing pathology elsewhere, its disorganizing effects could be counteracted by the doctor. Much like Freud's bad associations, Foerster's disrupted nervous organization was amenable to therapy.

This was all the more remarkable because patients with tabetic organ crises had been notoriously difficult to treat. Only general cures, such as the application of mercury or iodine, or exercise therapy, electrotherapy, and sweating cures, existed, and although they could produce acceptable results, a powerful specific treatment was still a desideratum.[36] When developing his treatment, Foerster reasoned in the following way. Since the lesion caused disturbances indirectly, affecting the nervous organization, the doctor might be able to intervene surgically, causing another "lesion" that would limit the pathological effects of the first. In other words, a well-placed "lesion" could be beneficial to the patient by establishing a new order within the nervous system. Foerster initiated a cooperation with the Breslau surgeons Alexander Tietze at the Allerheiligen hospital in Breslau, as well as with surgeons at the university surgical clinic, including Johann von Mikulicz, Hermann Küttner, and K. H. Bauer. The surgeons would conduct the operation following Foerster's neurological guidance.

Take the example of Eduard G., one of Foerster's patients suffering from tabetic organ crises. He was "skinny to the bones" because, during his frequent crises, which usually lasted about two weeks, he consistently threw up any food he was given.[37] The seventy-two morphine shots that he received per day could not control his pains or allow him to eat. As we have seen, according to Foerster's understanding of the nervous system, the crises were caused by a breakdown in the organization within a reflex system: the damaged sensory part of the patient's reflex arc fired constantly, causing the patient to throw up any food that was given. If the reflex arc was cut, Foerster reasoned, the vomiting would stop. In the operation, Foerster told Küttner to intervene at the level of the patient's afferent reflex arc, resecting the 7th to 9th dorsal root responsible for the sensory supply for the stomach. Through this intervention, "the source of irritation [*Reizquelle*] [would be] separated from the central nervous system, and thus the sensory and consequently also the motor and secretory symptoms should disappear."[38] If the sensory *Reizquelle* was eliminated, its systemic effects would be eliminated as well; the organizational change would result in a cure.

As Foerster noted himself, the procedure was a great success. As a result of the operation, the patient lost sensibility in a beltlike zone on his trunk, but the frequency and severity of his crises were significantly reduced: "Pain and nausea had disappeared and . . . the feeling of strong dislike of food of all kinds was displaced by a healthy appetite." On the day after the operation, "the patient . . . ordered himself a meal," which he was able to keep down.[39] He gained weight continuously, so that two years after the operation, he was able to fully resume his work as an architect.[40]

Foerster's neurosurgical operations on patients with tabes received favorable reviews. The Viennese psychiatrist Wagner-Jauregg had been developing his own treatment for syphilitic patients, and in the first decade of the twentieth century he began experimenting with the induction of fever. His malaria fever therapy became widespread between the 1920s and 1940s and earned him the Nobel Prize in medicine in 1927, the first time the award was given to a psychiatrist. But as Wagner-Jauregg admitted, three years before receiving the Nobel Prize, fever therapy was not sufficiently specific to be successful in the treatment of syphilitic organ crises. Foerster's "ingenious idea," in contrast, was a "liberating feat" in their treatment.[41]

Foerster's operation was similarly effective in the treatment of his patients with Little's disease. There had been little therapeutic hope for these patients. In his 1897 monograph *Die infantile Cerebrallähmung*, Sigmund Freud had called the treatment of the disease a "pathetic and hopeless chapter" in the history of neurology, dedicating only three out of three hundred pages to the discussion of its therapy.[42] As we have seen, according to Foerster, the mechanism of the disease in his patients with Little's disease was the following: in these children, the afferent part of a spinal reflex arc was constantly firing, forcing the muscles into a permanent state of contraction. Whereas in a healthy individual, the motor cortex would inhibit the reflex via the pyramidal tract, for the affected patient the disease prevented the control of the spinal reflex arc. Given this understanding, the symptoms could be reduced if the firing of the reflex could be prevented. More specifically, in this case, Foerster realized that if the spinal dorsal roots were cut through, then the sensory stimulations from the body would no longer cause a motor response; the spastic contracture of the muscles would be alleviated.[43]

Again, the operation was very successful. Almost 80 per cent of the children operated on by Foerster showed clear signs of improvement. The efficacy of Foerster's operation becomes evident in his photographs, for example, the depiction of a boy before and after the operation in figure 4.1.

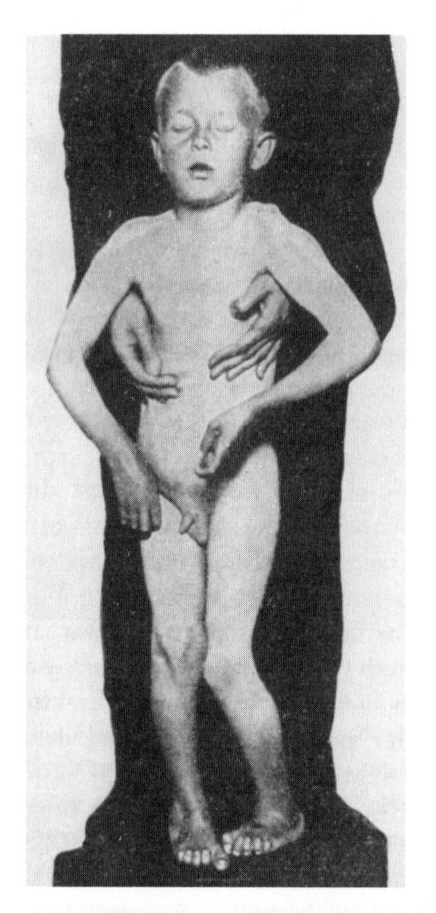

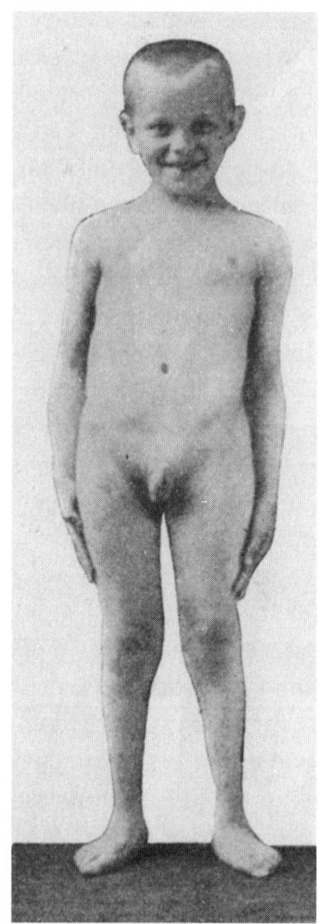

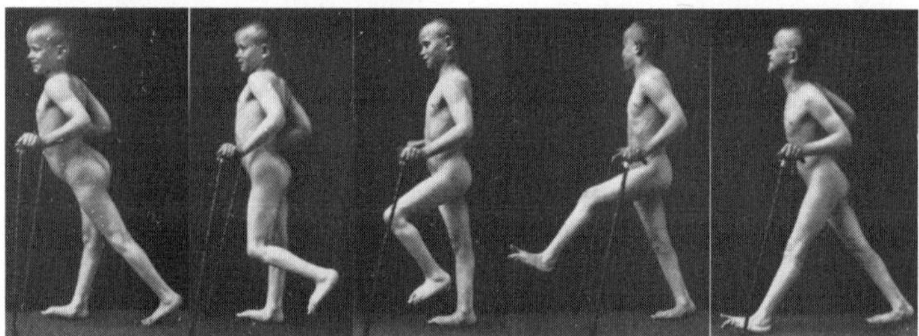

FIGURE 4.1. *Top left*, patient with Little's disease, held by Foerster. Note the boy's inability to extend his arms and legs. *Top right*, the same patient after the interruption of reflex arcs in Foerster's operation. *Bottom*, a sequence of photographs demonstrates the boy's regained ability to walk. (Modified from Foerster, "Beeinflussung spastischer Lähmungen," 149, 152, 161–164.)

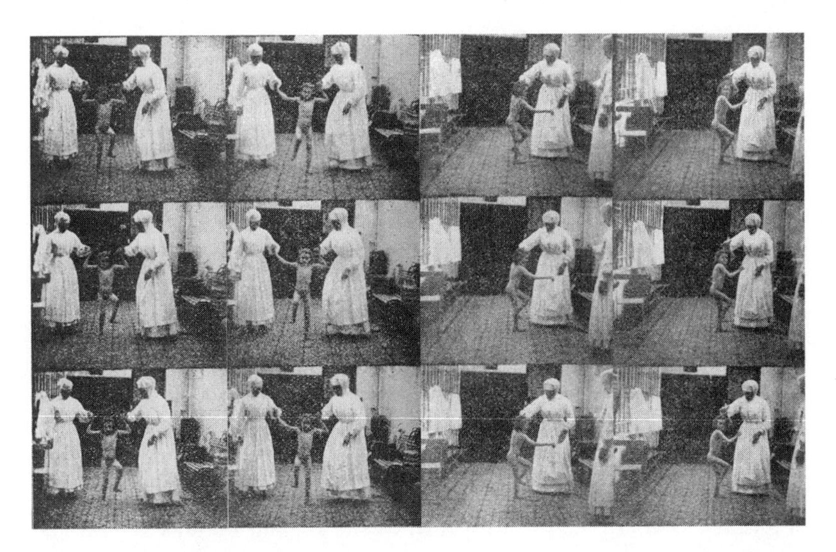

FIGURE 4.2. Walking attempts after a Foerster operation. (Van Gehuchten, "La radicotomie postérieure," 272, 273.)

In another example, Foerster compared this therapy to the alternatives. Before the operation, Foerster noted, the patient was "an unmoving brick, condemned to permanent bed rest." Therapeutic attempts based on "conventional orthopedic methods, such as myotomy of the adductors, or prolongation of the plantar flexors of the foot" had been entirely ineffective. But "through dorsal root resection, the spastic contractures were considerably reduced. . . . The child now walks around with crutches all by himself."[44] In the second decade of the 1900s, the operation gained wide acceptance within the medical community inside and outside Prussia. For example, the Belgian neurologist Arthur van Gehuchten documented the result of his own modification of Foerster's operation on film. A 1910 film shows a nine-year-old girl suffering from Little's disease; she, like Foerster's patients with the condition, was severely handicapped. Several days after the operation, the girl was able to walk when supported by a nurse, as depicted in the film sequence (see the screenshots in fig. 4.2). Although her gait was still not entirely normal, the operation resulted in a great improvement over her previous, entirely dependent situation.[45] By the 1930s, Foerster's procedure was well known even outside of medical circles.[46] In 1932, a man transported his paralyzed daughter all the way from Frankfurt am Main—a long and strenuous journey at the time—to receive treatment from Foerster.[47]

Foerster's Mapping Project: Brain Maps and Body Maps

Foerster's conceptualization of the relationship between lesion and symptom marked a significant step away from the localization tradition. There, as we have seen, a lesion corresponded to a symptom directly. In fact, based on the correlation method in pathological anatomy, if a lesion caused a sensory disturbance, this justified the identification of the lesion's location as a sensory center. Foerster, in focusing his attention on the way in which a lesion worked within a larger reflex (read: connective) system, was able to break the identification between lesion and function. In the two examples above, Foerster understood pathology not as an *Ausfallsymptom*, the *loss* of a localized function, but rather as a breakdown of a reflex system, a *disturbance* of the function of the reflex working as a whole. Because of the relationship between sensory and motor elements of a reflex system (what in Meynert, Wernicke, and Freud's system had been labeled the association system), a lesion on one part of the reflex arc could cause dysfunction in another. Here we have the development of a systemic model of reflex function, concentrating less on the identification of functional centers than on the interrelation of nervous elements. This was the insight that justified the surgical interventions.

This interpretation of Foerster's practice has to confront the aspect of his work in which he seems to come closest to the localization tradition: his mapping project.[48] From the mid-1920s onward, Foerster created maps of the brain, which at first glance appear to describe the ways in which the body was projected onto it. On the map different sections of the brain surface were inscribed with the body part to which they corresponded: foot, lower leg, upper leg, stomach, and so forth (e.g., fig. 4.3). The map recalls the localization project of a previous era, seeming to draw attention to functional areas of the brain that are dedicated to different parts of the body and different senses.

To understand these maps, however, we must place them in the context of Foerster's broader mapping project. Aside from his work on Wernicke's atlas, Foerster first constructed nervous-system maps at the level of the spinal cord and the periphery. Foerster's maps linked spinal roots with particular skin areas, which he called dermatomes. The posterior (dorsal) roots carried afferent sensory fibers, and the anterior (ventral) roots efferent fibers, which controlled vasodilation for delimited parts of the body surface. Figure 4.4 shows the dermatomes on the chest as beltlike and of roughly equal size, like their corresponding spinal-cord segments.

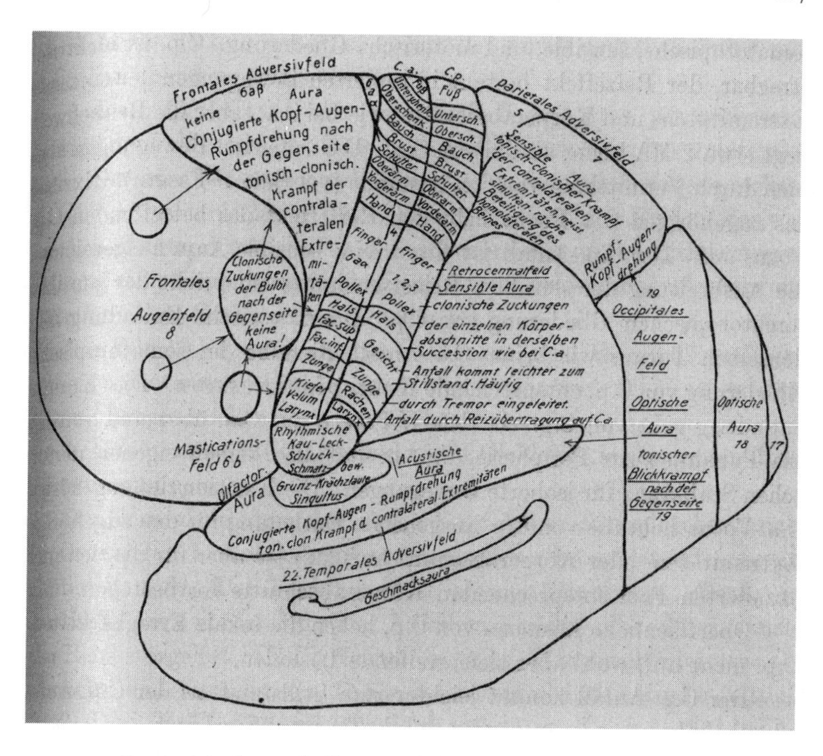

FIGURE 4.3. Foerster's brain map (epilepsy map). (Otfrid Foerster, "Die Pathogenese des epileptischen Krampfanfalles," *Deutsche Zeitschrift für Nervenheilkunde* [1926]: 15–53, on 43.)

Dermatomes were an obvious and relatively simple structure to map.[49] Even though the "metameric structure" of the human body could not be seen when looking at the surface of the body, there were "close connections [*Beziehungen*] between certain skin segments to the individual spinal cord segments and spinal roots."[50] Applying Charles Sherrington's method of "remaining sensibility" to humans, Foerster developed an outline of human dermatomes to complement the earlier research by British neurologist Henry Head, which was based on herpes zoster eruptions on the skin.[51]

The mapping was particularly helpful in Foerster's operations on patients with Little's disease or tabes. As we have seen there, he had to cut a particular dorsal root in order to end the attacks. Once the dura mater, the layer covering the spinal cord, was removed, however, "orientation about the kind of root, whether it was the fifth lumbar, or the first sacral root, [was] difficult."[52] As can be seen in figure 4.4 (right), all the dorsal roots looked roughly the same, and without a reliable method for distinguish-

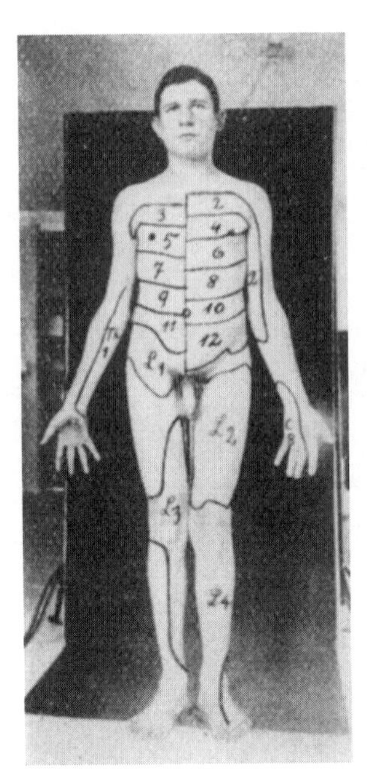

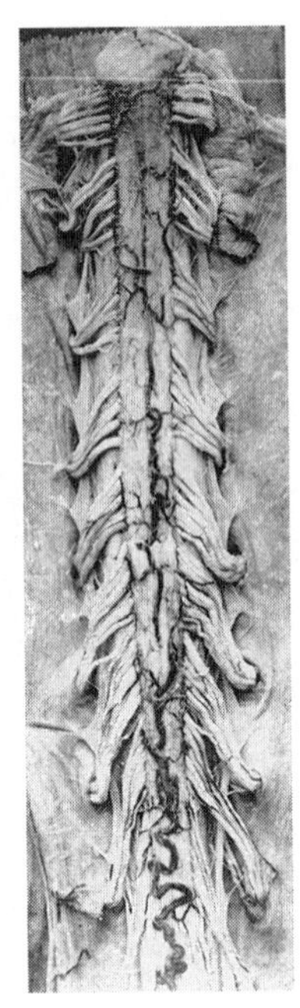

FIGURE 4.4. *Left*, dermatomes painted on skin by Foerster. *Right*, contemporary photograph of child's cadaver, showing spinal cord and dorsal roots. (Otfrid Foerster, "The dermatomes in man," *Brain* 56 [1933]: 1–39, on 20; van Gehuchten, "La radicotomie postérieure," 245–292, fig. 26.)

ing between them, the danger of cutting the wrong root was prohibitively high. In the early years of the operation, the general surgeon Alexander Tietze performed the procedure under Foerster's direction.[53] He drove a nail into the fifth lumbar vertebral arc to mark the site where the first sacral root leaves the dural sac.[54] It was a violent and potentially harmful technique. Foerster's "body mapping," a method of electrical stimulation, allowed for a much less invasive method for navigating around the spinal cord. He would apply a faradic (alternating) current to the dorsal roots, stimulating the vasodilator fibers running from the spinal cord to the periphery, thus causing a red patch to appear on the skin. This patch "in its form and extension exactly corresponded to the . . . dermatomes."[55] The location of the hyperemia when referenced to the dermatome map would indicate which dorsal root Foerster had stimulated and would provide a sure method for orienting him on the spinal cord. Foerster's faradization method dispensed with Tietze's nail.[56]

The same operational logic applied to the brain maps. They were used in epilepsy operations, which can be understood as an extension of Foerster's spinal cord work to the brain.[57] During World War I a large number of veterans received traumatic head wounds caused by bullets or shrapnel. Once the fighting had ceased, often after a delay of several years, these men started to suffer from late-onset traumatic epilepsy, and they began to present in large numbers at Foerster's clinic.[58] These patients experienced convulsions so frequent and severe as to render normal life impossible. The delay between the initial trauma and the onset of symptoms occurred because it was not the initial gunshot injury to their heads that caused the seizures, but rather the slowly forming scar tissue, which, as it contracted, pulled on neighboring, healthy tissue.[59] As the years passed, the effect became worse, finally demanding surgical intervention.[60] In surgery the doctor removed part of the cranial bone and then excised the scar tissue. With the scar tissue removed, the neighboring tissue was no longer adversely affected and the symptoms were reduced.[61]

But in order to excise the scar tissue and minimize collateral damage, Foerster had to know precisely where to cut. This was where the brain maps became useful. Again Foerster relied on electrical stimulation, carefully recording the responses when he applied his electrode to different parts of the brain. Indeed, though some of the sections of the brain map are marked with parts of the body, others have labels that would seem out of place on a functional map. They describe particular actions or phenomena that could be observed in the operating room. Alongside the "optical area" (*occipitales Augenfeld*), Foerster has written "optical aura" and, below this, "spasm of the gaze to the opposite side."

Not only did the maps indicate the phenomena that the surgeon could produce within the context of the operation; Foerster also specifically presented them as "epilepsy maps." The markings on the brain map corresponded to Foerster's descriptions of the patient's epileptic seizures, whose "character" he would outline in page-long descriptions in his publications: "They [the epileptic attacks] generally begin at the right mouth, then spread over to the right upper facial muscle, involving the left upper facial muscle as well, then move on to the right platysma [most superficial layer of frontal neck muscles] . . . , then to the thumb and finally to the index finger."[62]

If we look closely at these brain maps, we can discern several arrows, which indicate the direction in which the epileptic seizures would spread. When introducing his sections on various areas of the brain, such as the *vordere Centralwindung* or the *hintere Centralwindung* (the motor and the sensory strips lining the central sulcus), Foerster wrote: "The nature and course of the epileptic seizure is entirely characteristic for every area."[63] Each field was primarily the site of a characteristic epileptic seizure, rather than a static function. The maps were both pathological and concerned with movement.

Foerster's motivation for the production of such "behavioral maps" becomes evident in his exchange with his colleague Oskar Vogt.[64] In 1926 Foerster sent Vogt a brain map with arrows similar to that in figure 4.3.[65] Vogt, director of the famous Institute for Brain Research in Berlin Buch, had attempted to learn about brain function through the study of structural differences in his cyto-architectonic studies of the brains of higher animals.[66] As a result of his work, he developed various maps of the mammalian brain in which he inscribed the different cyto-architectonic areas (*Felderungen*). Since Vogt was ultimately interested in developing such a cyto-architectonic characterization of the human, not the animal, brain, he used the auxiliary of the homologous field: similarly structured cortical fields in closely related mammals were assumed to manifest the same reactions upon electrical stimulation. Based on this observation, Vogt developed homologous maps for the human cortex, using his research data from the long-tailed monkey. Foerster's electrical stimulation in humans was of great interest to Vogt, for the method could be used as a control for his own research. Vogt therefore asked Foerster in 1926 to indicate his stimulation results on a blank version of his homologous map. The map that Foerster returned to Vogt confirmed Vogt's research results, but at the same time, Foerster drew Vogt's attention to a fundamental difference between their maps. As Foerster pointed out, his map was an "epilepsy map" (*Epilepsiekarte*), indicating the spread of the epileptic impulse along the brain surface through a set of arrows.[67] If Vogt wanted to use it as a normal map, he would have to ignore the arrows: "[I ask you] not to pay attention to the arrows in my brain map[;] they only indicate the direction in which an irritation of extra-motor foci has an epileptogenic effect."[68]

Foerster's maps were not primarily an attempt to "localize" particular functions on the brain surface, pinpointing either those parts that controlled particular parts of the body or the sites of particular mental functions. His maps were at least as important in their function as tools for working out where the surgeon had to intervene in the context of the operation. The fact that Foerster's maps were auxiliary and subordinate to his surgical practice explains why he devoted so little space in his publications to them and the stimulation process from which they were produced. In contrast to Wilder Penfield, who, as we will see in chapter 6, devoted substantial space to the description of his surgical setups, Foerster would rather quickly refer to the "results of electrical stimulation" of a particular anatomical region on the "exposed cortex in the non-narcotized human."[69] Even though Foerster performed the electrical stimulations on a large scale—by the end of his career, he could draw on the results of almost three hundred "systematic explorations" of the brain—he did not seem to consider them central to his research.[70]

The Social Nervous System

In developing his model of the nervous system, Foerster, much like Freud, did not have to introduce alien elements into neuropsychiatry. Although his therapeutic turn marked his distance from the neuropsychiatric tradition—and, as I will show, provided him with the institutional capital to help make neurology independent from it—it is important to recognize that Foerster was exploiting resources inherent to the Meynert-Wernicke system. This is most clear in Foerster's writings from the 1930s, when he wrote a number of synthetic overviews of his work. Here he framed his discussion of nervous organization in social language, just as Meynert had done half a century earlier.[71]

Much like Meynert, Foerster suggested that the nervous system was composed of semiautonomous elements, which entered into a cooperative whole and thus could be described using social metaphors. Foerster wrote, "All the different [parts of the nervous system] represent a . . . cooperative society, each partner of which contributes a specific, more or less important, quota to the validity of the society."[72] He gave two main examples for the social organization of nerves, although he thought it to be a general principle of the nervous system. First, social organization could take place at the level of superimposed reflex arcs. Figure 4.5 shows what Foerster called a "reflex community" or "consortium" (*Reflexarbeitsgemeinschaft* or *Reflexarbeitsverband*). The various parts of the nervous system were "not independent of each other; rather, more or less all segments [*Abschnitte*] of the central nervous system form[ed] a great *Arbeitsgemeinschaft*." The different elements, the "spinal reflex arc, the different supraspinal subcortical elements, and the cortex [worked] closely together."[73]

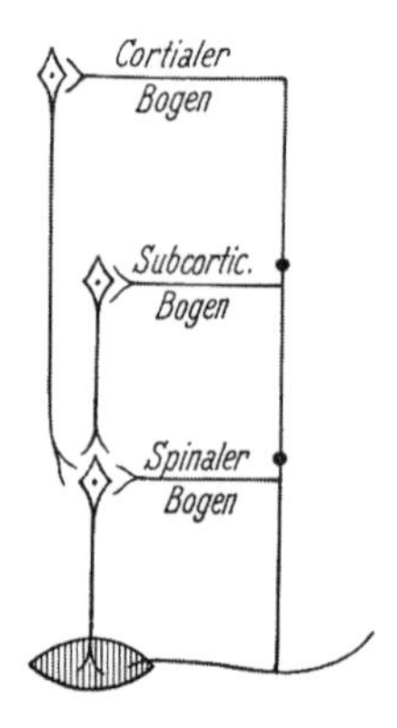

FIGURE 4.5. Reflex consortium (*Reflexarbeitsverband*) consisting of a spinal arc, a subcortical arc, and a cortical arc, illustrating the continuity between spinal cord and brain. (Foerster, "Symptomatologie der Erkrankungen," 5:88.)

In the case of Little's disease, it was the breakdown of such a *Reflexgemeinschaft* that caused the pathology.

Second, the *Arbeitsgemeinschaft* worked at the level of motility. Foerster emphasized that in addition to the *vordere Zentralwindung*, or motor strip, "many other cortical areas [were] the starting point from which motor impulses were sent to the muscles of the body." These cortical areas—including the *areae extrapyramidales*, the pyramidal tract as well as extrapyramidal pathways, among others—all worked together in the production of voluntary movement, building a large *Arbeitsgemeinschaft*, in which "each area contributes its particular share according to its own particular character."[74] By using social metaphors in his reflex model, ones that resembled those used by Meynert, Foerster was able to emphasize not simply the existence of different functional areas, but also the way in which function arose through the cooperation and combined effects of different parts of the nervous system.

The difference between Meynert and Foerster, then, can be located not so much in the constitutive elements of their model, but rather in the relative weighting and importance they attributed to each. As we saw in chapter 1, for Meynert the social metaphor was an important tool for describing the associative qualities of the nervous system, and thus for justifying the application of reflex principles to higher functions. It was a necessary condition of his pathological anatomical approach to psychiatry and of his localization project. In this way, however, it always remained subordinate to the methods of pathological anatomy. Foerster's innovation was to give this aspect priority, to make the systemic considerations dominant over the localization project that they had been called to serve.[75]

The Reliance on Plasticity: Reorganization and *Übungstherapie*

Foerster's use of social language in the latter part of his career draws our attention to another element of his model that was important throughout. For Foerster placed great emphasis on the ability of the nervous system to change and adapt to differing circumstances, and in particular to nervous damage. If a member of the nervous community was excluded, for example through a lesion in the nervous system, at first the entire organization broke down. Then the *Arbeitsgemeinschaft* would slowly begin to recover, as other elements took over the damaged nerves' roles. As Foerster argued, the nervous system was "not a machine made up of parts that stops working when a part fails, but it has an admirable plasticity and a remarkable adaptability not only to changed external conditions but also to attacks on its very substance."[76] Such plasticity and adaptation could help restore the lost function. True, this regained mobility was, as Foerster pointed out, often of a "certain kind"; owing to the anatomical limitations of the nervous system, it might be "stereotyped, fixed, rigid, immutable, unmodifiable."[77] Because Foerster remained wedded to the view that parts of the nervous system had certain tasks, no reorganization could be total, nor could lost function be completely recovered. But change was possible within certain limits, and loss of movement was never absolute.

The adaptability of the nervous system was crucial, because, as impressive as the recovery of Foerster's patients seemed, there was an obvious limitation to his technique. Though it was very successful at eliminating the major symptoms, it did so only by further damaging the nervous system. The misfiring reflex was not repaired but rather removed. A plastic nervous system, however, could make up for the loss. An example will illustrate Foerster's approach to treatment, an example that, in his own words, "present[ed] the extensive plasticity of the nervous system in a proper light." In the treatment of an irreparable lesion to the facial nerve, Foerster operatively connected a part of the healthy accessorius nerve to the facial nerve through a suture (fig. 4.6). The accessorius nerve normally supplied the trapezoid muscle, a muscle used to lift the shoulder. After the operation, the fibers of the accessorius nerve grew into the facial muscles, which thus became the effector organ of the nucleus of

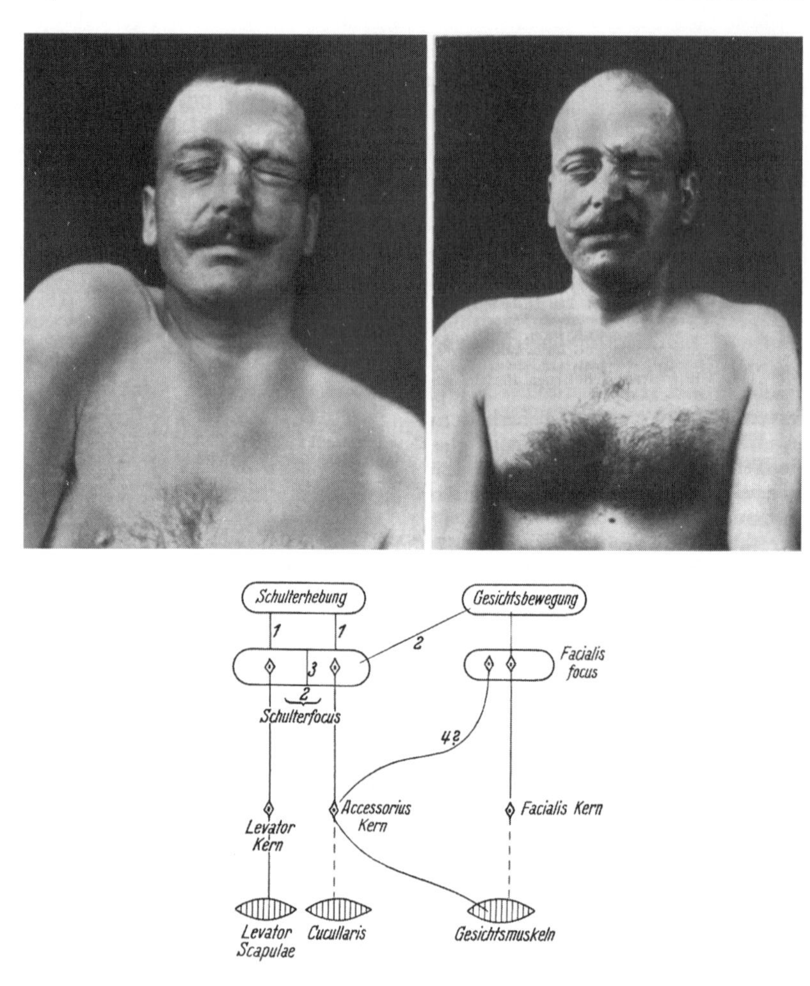

FIGURE 4.6. *Top*, voluntary movement of facial muscles, with (*left*) and without (*right*) movement of shoulder. *Bottom*, schema of reorganization after operation. (Foerster, "Bedeutung und Reichweite des Lokalisationsprinzips," 160–161.)

the accessorius nerve. For this reason, when the patient voluntarily lifted the shoulder, nervous excitation found its way to the facial muscles, causing them to contract.[78] The patient could use this new connection at will: "From the moment that the patient recognized this new possibility to set the facial muscles to work via the shoulder muscles, a connection in the cortex [was] put into action, which had existed before but had not been put to work, a connection between the cortical substrate of the movement images of the face and the motor cortical focus of the shoulder muscles of the precentral sulcus."[79] The patient was now able to move his face voluntarily.

This was of course a valuable development, but Foerster's therapy was premised on further changes. At the beginning, facial movement was accompanied by a lifting of the patient's shoulders, but after a while the patient learned how to suppress the shoulder movement and use his facial muscles independently. The restitution of function thus arose through a dissociation of the individual parts within the cortical motor focus of the shoulder. The elements belonging to the nucleus of the accessorius nerve broke away from the other elements; they "freed themselves from their previous forced cooperation."[80] The nervous system adjusted to the new situation, and both functions were restored.

In this case, the patient managed to regain function for himself, but Foerster considered that the doctor could facilitate the process. Through a series of exercises, the nervous system could be encouraged to reorganize. Surgical intervention was often just the first stage of the therapy; it could be followed by a regimen of *Übungstherapie* (literally "exercise therapy," which can be rendered in more modern terms as "physical therapy" or "neurorehabilitation"). In *Übungstherapie*, the patient would slowly learn to perform a particular movement that had been lost because of damage to the nervous system. It was such exercises that Foerster performed with Lenin during his stay in the Soviet Union.

For patients with disturbance of the gait (ataxia) in tabes, *Übungstherapie* involved daily exercise sessions that had to be closely supervised by the physician over extended periods of time, often months or even years. At first, when Foerster's finances were more limited, he practiced *Übungstherapie* with very little instrumental support. He emphasized that for *Übungstherapie* to be successful, "absolutely no special apparatus was necessary; with a few chalk marks one can always and everywhere improvise whatever necessary."[81] Later in his career, Foerster's working spaces were increasingly marked by the presence of equipment for his neurological gymnastics, such as parallel bars placed in the hospital wards so that patients could easily integrate their practice into their daily routines.[82]

During the sessions, the doctor would draw attention to "the smallest departure from the normal course of movement," such as lifting the leg too high or to the side when walking, and ask the patient to correct it.[83] Foerster emphasized that the appropriate exercise should be the

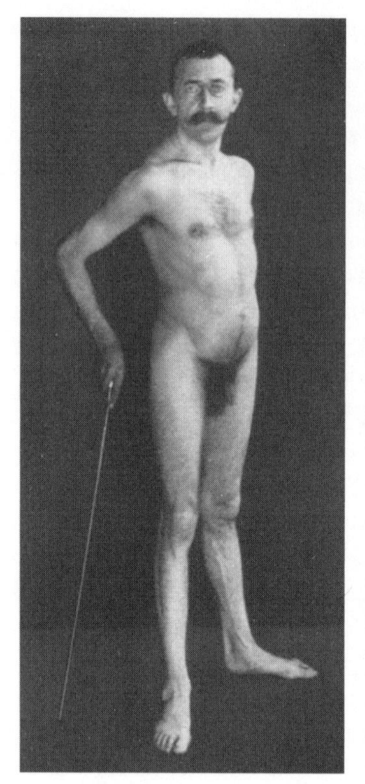

FIGURE 4.7. *Übungstherapie* in tabes. (Foerster, "Uebungstherapie," 8:397.)

disturbed movement itself; for example, a patient with gait disturbance should practice walking. For this, the patient needed insight into the components of walking; for example, he had to know that he needed to shift the body's center of gravity forward (fig. 4.7), and to lift up his trunk. The doctor told the patient how to achieve this, for example by plantar flexion of the posterior foot as shown in figure 4.7. Initially, the correct performance of movements required the patient's full attention, but the ultimate goal of *Übungstherapie* was to achieve a certain automatism. The patient was asked to "read the newspaper while standing" or "look at his watch." In many cases this goal was achieved.[84]

This complex process of compensation and restitution in *Übungstherapie* left its imprint on the relationship between doctor and patient. In Foerster's eyes, *Übungstherapie* should not be considered a doctor's intervention in the nervous system. Rather, it was a process of fostering and developing the nervous system's, and hence the patient's, own capacity for change. As Foerster argued, "*Übungstherapie* intervenes at the level of spontaneous restitution and supports and extends it."[85] The doctor did not repair the nervous community but instead tried to provide the conditions and the practice that would allow it to repair itself. To extend the social metaphor, *Übungstherapie* placed the doctor in the role of a social worker, helping a nervous community reconstitute itself after a traumatic event.[86]

Foerster did not himself use the social-worker comparison. His preferred metaphor was of the teacher (*Lehrmeister*). Just as a healthy person needed a *Lehrmeister* in order to learn a complicated skill such as playing the piano, the tabetic patient needed instruction from the doctor. The comparison worked both ways, and according to Foerster the first attempts at a new movement, for any person, had "strong resemblance to the ataxic movements of the patient with tabes." It was only through "careful repetition of the same movement" that learning took place: "walking, running, jumping, grasping, writing, and more complicated skills such as walking on a rope, or playing the piano, drawing, etc., have to be acquired through practice."[87] It is not surprising, then, that Foerster referred to his *Übungstherapie* patients as hard-working, deserving people who, if anything, had to be protected from overwork.[88] Foerster paid his patients considerable respect, because *Übungstherapie* required a working relationship between (almost) equal partners.[89] What was crucial, then, was that in both *Übungstherapie* and the learning of other skills, the *Lehrmeister* guided the movements and corrected errors, but the movements had to be performed by the student. In the practice of *Übungstherapie*, the distinction between activity and passivity was blurred, and it became unclear whether the patient was healing himself or the doctor was executing the cure.

Take the example of a girl with Little's disease. Immediately after the operation on the dorsal roots of her spinal cord, the girl was unable to walk. Nevertheless, through *Übungstherapie* she gradually regained the skill. In figure 4.8 she is being directed to shift her hip toward the center and stabilize it against the supporting leg.

Though the doctor or nurse directed the movements, it was central for the process that the girl perform the movements herself. Indeed, the goal of the treatment was for the patient to regain conscious and willed control. Moreover, Foerster believed that the very will of the patient was able to promote the regeneration of nerves. If the patient, often while "mo-

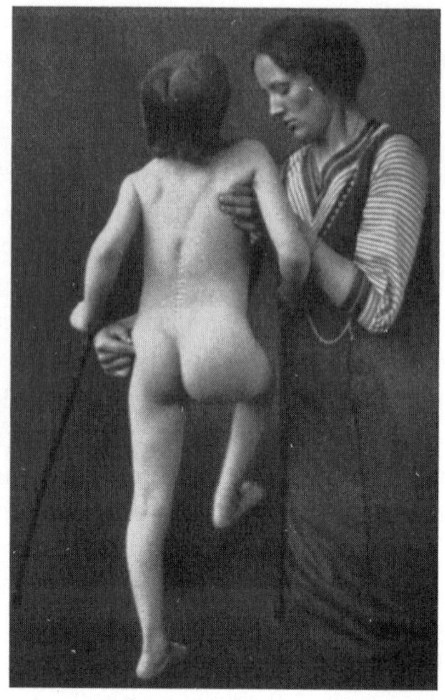

FIGURE 4.8. *Übungstherapie* in a young patient with Little's disease. Note that the surgical scar is still visible on the girl's back. (Foerster, "Uebungstherapie," 8:351.)

bilizing all of her will and under the continued encouragement from the doctor," performed those movements that the paralyzed muscles would normally execute, this would further the process of regeneration. According to Foerster, the impulse from the brain "forces the cells of the anterior horn to produce their neurites ever more quickly and more intensely than they would have been able to without this *vis a tergo* [accelerating force]."[90]

In certain moments during the treatment, the patient's insight into how best to aid the recovery even went beyond the doctor's. In cases of spastic torticollis, or wry neck, many patients were able to control their disease through what Foerster called a "special trick" (*Kunstgriff*): "Almost every torticollis patient has her own special trick, through which she is able to tame the spasm. One puts her finger on the cheek, another on the chin, a third on the nose, a fifth on the ear."[91] The *Kunstgriff* was "not a mechanical principle," as Foerster pointed out; the position of the head was "not violently" corrected, but gently improved, through the "relaxing effect of certain sensory stimuli" (fig. 4.9).[92]

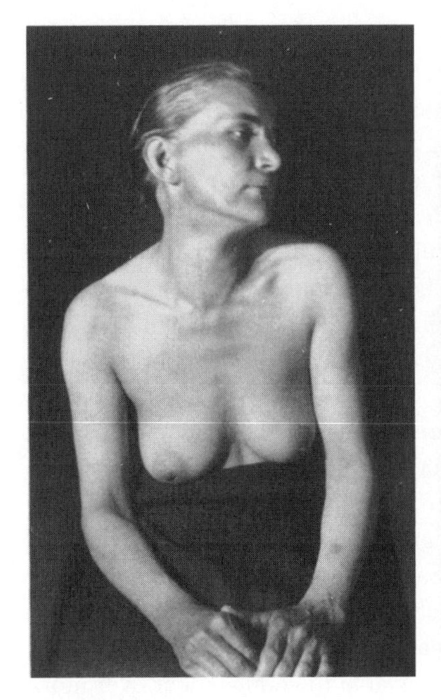
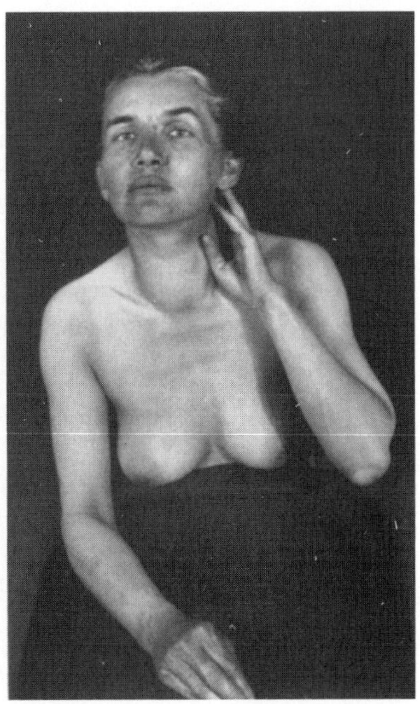

FIGURE 4.9. *Kunstgriff* against cramp of the neck in spastic torticollis. (Foerster, "Uebungstherapie," 8:372.)

In *Übungstherapie*, then, the doctor played only an auxiliary role. It was the patient who was required to will the actions and effect the therapy. In this way we could even draw a parallel to the practice of psychoanalysis. As we saw for both Foerster and Freud, pathology was the result of the organization of the nervous system, not the direct result of a lesion. For both it was this insight that opened up the possibility of a therapeutic intervention. The doctor could not repair the nervous damage, but he could help in the reorganization of nervous elements, to prevent the misfiring of a reflex arc or to reintroduce a repressed thought into the play of associations. But in both cases the treatment was not performed on but rather by the patient. The doctor in these situations could guide the treatment and encourage the patient through the most difficult aspects of the cure, but ultimately he could not make the patient work through the resistances or undertake the painstaking process of relearning a particular movement. In moving away from the simple lesion model that had marked the early neuropsychiatric tradition, both Freud and Foerster reintroduced the patient as an active participant in the cure.[93]

An Independent Neurology

For Foerster personally, his unorthodox turn to treatment became the key to his professional success. By being able to treat patients, Foerster bolstered his position within the hospital. Foerster's first institutional foothold was a small in-patient ward (*Bettenstation*) at the Agathstift, across the street from the Breslau Allerheiligenhospital and part of the neurological *Abteilung* (department) that he was granted in 1911. Even though these spaces were quite simple,[94] their very existence was a sign of his success: the department was one of only a few independent neurological departments in the country.[95]

When war broke out in 1914, Foerster was made consulting physician of the Sixth Army Corps and head physician of the Festungslazarett Breslau, a position that he kept until 1920.[96] During the four years of conflict, Foerster treated thousands of gunshot injuries to the nervous system. He reported almost four thousand cases of war-related peripheral lesions and almost four hundred cases of lesions to the spinal cord.[97] In large part owing to his increasing patient numbers, and despite budget constraints after the war, Foerster was able to commandeer ever more resources to support his work. In 1920 the city authorities recognized that Foerster's space at the Allerheiligen hospital was too limited, and he was offered new premises in the Wenzel-Hancke hospital, located in the southern new part of Breslau.[98] Even after the move to more spacious premises, Foerster continuously asked for more beds and more personnel, that is, assistant doctors, nurses, and technicians.[99] Moreover, he was not shy about seizing any and every opportunity to improve his working conditions. For example, in 1921 he asked for an empty room in the basement of the hospital, just below his department, "for the installation of electrical apparatus" and to use the power supply of that room for his experiments.[100]

A sign of Foerster's increasing reputation was the 1919 offer of the newly created *Ordinariat für Nervenpathologie* at Heidelberg. The offer initiated a bidding war between Breslau and Heidelberg. The medical faculty at Breslau "placed great value on keeping Professor Foerster for the city of Breslau," and made him both *Primärarzt* (senior physician) for neurology (*Neurologie*) at the Allerheiligen hospital, and *Ordinarius* for neurology at the University of Breslau.[101] As a result of these negotiations, he was granted a "disposition fund" (*Dispositionsfond*) that put him into a better financial position than most other department chairs.[102] The transfer of Foerster's *Abteilung* to the Wenzel Hancke hospital in 1921 was yet another consequence of his new sought-after status.[103]

The building and opening of the Neurological Institute at Breslau in 1934 under Foerster's direction can perhaps be seen as the culmination of Foerster's professional success. Funded by both the Rockefeller Foundation and the state and city government, the institute was a remarkable achievement during a time otherwise characterized by belt-tightening. Germany's economy had been devastated by the 1929 Wall Street crash because of its dependence on U.S. loans. The unemployment rate rose exponentially, until, by the beginning of 1933, it had reached 6 million, affecting one-third of the working population.[104] Further, as Foerster pointed out to Daniel O'Brien, his contact at the Rockefeller Foundation, in December 1931, Breslau had been hit particularly hard by the depression.[105] The economic troubles did place constraints on the new institute. Money had to be saved at various levels; the building was constructed out of a simple brick stone, and its furnishing was of the simplest though functional kind.[106] But that it existed at all was a testament to Foerster's growing reputation.

For Foerster the role of treatment in his professional success became a model for the development of neurology in Germany more widely. Just as his new treatments had provided the institutional support for establishing a neurological clinic and then an institute in Breslau, so too could the exploitation of neurology's therapeutic potential underwrite its independence from psychiatry. As Foerster's position became stronger locally, he became a spokesperson for the developing field: in 1913 he was made coeditor of the *Deutsche Zeitschrift für Nervenheilkunde*—the first German neurological journal, founded in 1891 by Wilhelm Erb; in 1919, Foerster also became editor of the *Zeitschrift für die gesamte Neurologie und Psychiatrie*, founded by Alois Alzheimer and Max Lewandowsky in 1910. Most importantly, from 1924 to 1932, Foerster was president of the Gesellschaft deutscher Nervenärzte, succeeding Max Nonne.

Foerster used his presidency of the Gesellschaft as a bully pulpit to promote the independence of the field. Each year he gave the opening addresses at the society's annual meeting. As he asserted in his 1928 address, it was his "primary mission in life . . . to help our specialty, neurology, to independence [*Selbständigkeit*]."[107] He was clear on the most appropriate strategy. For Foerster, a neurologist should not only be well versed in the anatomy, physiology, and pathology of the nervous system; he should also perform all diagnostic measures, such as encephalography and ventricular puncture, himself. In the realm of the "overall therapy," the neurologist should "gain as much independence and autonomy as possible." This included all surgical work on the nervous system, that is, operations on the spinal cords and peripheral nerves but also orthopedic and palliative interventions.[108]

Foerster promoted this vision at other conferences, most importantly at the First International Neurological Congress in 1931, in Berne. Here he authored a petition addressed to the state authorities of the participating nations to appeal for the independence of neurology, because the field was neglected in numerous countries and its future was, "at least in Germany, in serious danger." Foerster's proposal met with "enthusiastic applause [*lebhafter Beifall*]."[109] Foerster considered the congress an "important milestone on the path of neurology's development," especially because it bore witness to the breadth and importance of the specialty.[110] By that time, certain progress had been made toward independence, including the establishment of chairs of neurology at Hamburg, Breslau, Frankfurt, and Heidelberg. Nevertheless, to Foerster, these were only "partial successes"; neurologists in Germany were "still far away from the great final goal, the general recognition of neurology as a specialty [that is] equal to all other medical specialties."[111]

A great obstacle to institutional independence was the Great Depression. Apart from a few exceptions, such as Foerster's own institute, there was little appetite for the establishment of independent neurological clinics and institutes to add to the neuropsychiatric institutions that already existed. Political developments made the goal of an independent neurology even more distant. After Hitler's 1933 seizure of power, major neurologists, including Kurt Goldstein, had to leave Germany. In 1935 the Gesellschaft Deutscher Nervenärzte was shut down by the Nazis, who forcibly united it with the Deutsche Verein für Psychiatrie, to create the Gesellschaft Deutscher Neurologen und Psychiater. In this way they consolidated the alliance with psychiatry that neurologists like Foerster had for decades tried to break. As Ernst Rüdin, the new president of the society and the *Reichskommissar* for the Deutsche Gesellschaft für Rassenhygiene, explained, the merger was due to the "fundamentally new attitude of the German State to the art of healing." Because the Nazis assumed that serious psychiatric and neurological diseases were "hereditary [*erblich bedingt*]" it was clear that "prevention [was] better than cure." To meet the new goals of the Nazi government, psychiatric and neurological concerns had to be united.[112] This dual renunciation of specialization and therapy must have been a slap in the face for Foerster, who henceforth attended the meetings of the internists.[113]

In these new conditions, Foerster attempted, in the words of Percival Bailey, to make "his peace with the Nazi government."[114] In his speech at the Wiesbaden congress for internal medicine in 1939, Foerster laid his neurology in the service of the new German régime. He suggested that the human organism showed itself to be the "truly perfected national socialist state," a state in which members not only functioned by fulfilling their duties, responding to orders from "the leading office [*der führenden Stelle*]," but "remain[ed] subservient" even "without contact with the Führung" because "national socialist ideas [had] become so deeply ingrained."[115] He ended his paper: "Heil our Führer!"[116]

Despite this gesture of loyalty to the Nazis, after the seizure of power, Foerster was in a precarious situation. His treatment of Lenin in the early 1920s made him an easy target, and his wife's Jewish ancestry helped sideline him in Nazi Germany. When the North American neurosurgeon Wilder Penfield, who had worked with Foerster in 1928, revisited Breslau nine years later, he found a sick and old man, sitting alone in a "splendid Institute," his assistants drafted to military matters and his spirit broken.[117] The institute did not survive the war, and when Breslau became Polish in 1945, the Wenzel-Hancke hospital and institute were handed over to the Akademia Ekonomiczna for the renamed University of Wrocław.[118] Foerster's dream of an independent neurology in Germany would have to wait until the 1960s for its fulfillment.[119]

Forging Nerves, Forging Disciplines: Freud, Foerster, and the Medicine of Mind and Brain

In building up institutional capital through his new therapies, Foerster laid the foundations for an independent neurology in Germany, helping wrest it free from the neuropsychiatry to which it had previously been subordinate. In the (albeit interrupted) growth of neurology as a more autonomous specialty, we can see the characteristics solidify that make any comparison with psychoanalysis seem at first wrongheaded. Here, unlike Freud's case, the language of nerves remained on center stage, the concern with higher functions was mostly left to one side, and therapeutic salvation was entrusted to the surgeon's scalpel over the doctor's words.

But for all these differences, we can see a very similar engagement with the neuropsychiatric tradition, to which both Foerster's neurology and Freud's psychoanalysis could trace their heritage. Both recast the Meynert-Wernicke model by complicating simple localization of function, and both considered pathology, whether in hysteria or in tabes dorsalis, to be a breakdown in the normal organization of the nervous system rather than simply the result of nervous damage. For both men, "trauma" acted at one stage removed. Freud saw pathology in the displaced effect of a repressed memory on conscious life. Foerster saw motor disorder as the indirect result of a malfunctioning sensory nerve. Moreover, because both placed emphasis on the plasticity of the nervous system, they were able to formulate new ways to intervene in it and in this way help the patient correct nervous disorder: Freud's talk and Foerster's *Übungstherapie*. Further, and most importantly, both were able to achieve this reformulation by drawing on an aspect of the neuropsychiatric tradition that had previously been made subordinate to the localization project. For Foerster as for Freud, it was the systemic properties of a supercharged reflex—whether Foerster's *Reflexgemeinschaft* or Freud's webs of association—that could be used to challenge the localization paradigm, for which Meynert had deployed the reflex as an essential support.

Reading the development of neurology and psychoanalysis through the lens of the connective model of the reflex thus provides a new way of thinking through their relationship that does not begin with the oppositions stemming from the soma-psyche divide. Foerster's and Freud's reformulated reflex models were deployed in different contexts, to different institutional and professional ends, but they remained in many important ways profoundly similar. Likewise, emphasizing the systemic aspects of Foerster's system allows us to reconsider its relationship to the tradition of interwar neurological "holism," especially that of Kurt Goldstein. Though they are often presented as opposed, Goldstein and Foerster were united in their rejection of a simple localization of function, and both foregrounded the integrative workings of the nervous system.[120] Indeed, when the two men debated their views at the annual meeting at the Gesellschaft deutscher Nervenärzte in Dresden in 1930, a member of the audience observed that their papers did not "necessarily seem to contradict each other" but rather were "different ways of looking [*Betrachtungsweisen*] at the same problem."[121]

In this light, we can discern another, different, rupture in the mind and brain sciences. For in both psychoanalysis and neurology, other doctors continued Freud's and Foerster's work but moved beyond and denatured the connective model that had been so important for these earlier generations. In chapters 5 and 6 we will see how émigré psychoanalyst Paul Schilder and Canadian neurosurgeon Wilder Penfield developed forms of psychoanalysis and neurology based upon a fundamentally changed reflex model, one in which the sensory and motor elements were no longer considered as an integral whole but were considered separately, a move that made the word "reflex" no longer seem appropriate. And this, as we will see, had wide-ranging implications for the doctors' theory, practice, and relationship to their patients.

CHAPTER FIVE

Between Hospital and Psychoanalytic Setting

Paul Schilder and American Psychiatry, or How to Do Psychoanalysis without the Unconscious

On December 21, 1935, Paul Schilder (1886–1940), a Jewish émigré psychoanalyst from Vienna, walked down a festively decorated Manhattan street to post three letters. The first letter was addressed to Berggasse 19 in Vienna, written to the founder of psychoanalysis. Schilder had recently been the center of an "affair" at the New York Psychoanalytic Society, and an earlier appeal to Sigmund Freud had not yielded the response he had hoped for. Schilder thanked Freud for his "decisive answer." He regretted that Freud's "letter gives me a position in the psychoanalytic movement which I cannot accept" and suggested that he had no other choice but to withdraw from the New York society. Schilder was clear, however, that his respect for psychoanalysis was undiminished by the affair or Freud's role in it: Schilder concluded his letter by stating, "My personal admiration for you and my allegiance to psycho-analytic principles will not be changed."[1]

The second letter provided evidence of that admiration. Addressed to the Karolinska Institute in Stockholm, it contained his nomination of Freud for the Nobel Prize in Medicine. In the letter, Schilder emphasized Freud's double contribution: his psychoanalysis had allowed the "unique enrichment of medical science" and had "deeply influenced the cultural development in the 20th century." Primarily, however, Schilder emphasized the value of psychoanalysis as a "treatment of neurotic and psychotic disorders which produces cures in cases which cannot be cured by any other method." Thinking back perhaps to his time working with Julius

Wagner-Jauregg, who less than a decade before had won the Nobel Prize for his malaria fever therapy, Schilder asserted that this therapeutic aspect of the psychoanalytic method made it worthy of the prize.[2]

The third letter was local mail. It was written on behalf of Dr. Epstein, Schilder's former training candidate, urging the New York Psychoanalytic Society's educational committee to accept him for a control analysis (a form of continued psychoanalytic supervision) once his analysis with Schilder had finished. As the "affair" had cast doubt on Schilder's qualifications, Epstein's candidacy was in danger. Schilder hoped that an "innocent bystander" would not be harmed by the debate.[3]

The three letters demonstrate Schilder's ambivalent relationship to Freud and psychoanalysis. Schilder considered himself a psychoanalyst and had identified himself as such in his work. But in the hardening psychoanalytic orthodoxy in 1930s America, Schilder's colleagues were suspicious of the way he had developed psychoanalytical concepts, and this played a significant role in his resignation from the New York Psychoanalytic Society. In this chapter, I will examine this ambivalence, showing how it played out at institutional, clinical, and intellectual levels. Schilder's reformulation of psychoanalytic practice—and in particular, as I will suggest, his transformation of the reflex practice that underlay it—placed him at the center of an intercontinental and generational dispute over what constituted good practice in psychoanalysis.

Schilder, Psychoanalysis, and the Neuropsychiatric Tradition

Like Freud, Schilder was thoroughly trained in the medical sciences of his day.[4] As a medical student in Vienna from 1904 to 1909, Schilder worked on the perception of touch while in the physiological laboratory of Sigmund Exner, Freud's colleague at the neurophysiological laboratory of Ernst Brücke some thirty years before.[5] Schilder also studied at the pathological anatomical institute of Anton Weichselbaum for eighteen months; there he worked on general pathological topics, such as the amyloid degeneration of various organs and tetany in rabbits following extirpation of the parathyroid.[6] Further, in the period before the end of World War I, Schilder published widely on various neuropathological, physiological, and clinical topics. By that time he had gained a national and international reputation for his description of encephalitis periaxialis diffusa.[7]

FIGURE 5.1. Paul Schilder in his office at Bellevue, 1930s. (Courtesy of Brooklyn College Library Archives and Special Collections, Lauretta Bender Papers, box 27, Photographs [1912–48].)

After completing his medical degree, Schilder found points of contact with the Meynert-Wernicke tradition. In 1909 he became an assistant (*Volontärassistent*) to Gabriel Anton at the Psychiatrische und Nervenklinik in Halle, a position he held for three years.[8] Anton had succeeded Carl Wernicke as the director of the clinic after the latter's unexpected death in 1905, and a number of Wernicke's former assistants, such as Karl Kleist, were still active there. As Schilder noted in his "Vita," Wernicke's books and those of his student Hugo Liepmann "circulated freely" at Halle.[9] Anton had also studied with Meynert in Vienna.[10] In his work, Anton had taken up some key elements of the Meynert-Wernicke tradition, including Meynert's social language for describing the workings of the nervous system.[11]

In addition, from November 1918 to April 1928, Schilder worked as an assistant at the Universitätsklinik für Nerven- und Geisteskrankheiten in Vienna under Julius Wagner-Jauregg, Meynert's successor.[12] Wagner-Jauregg continued the biological orientation of Meynert's clinic. Under his leadership, the clinic became well known for its use of somatic therapies in the treatment of psychiatric disorders. From 1917 onward, Wagner-Jauregg and his co-workers were actively involved in the treatment of general paralysis of the insane through the use of malaria fever, a therapy for which, as we saw, Wagner-Jauregg later won the Nobel Prize.

Wagner-Jauregg's orientation encouraged a reading of the clinic's history that emphasized a particular vision of Meynert's work. Otto Poetzl, who was Schilder's co-worker and Wagner-Jauregg's successor in Vienna, cast the intellectual history of the clinic thus: "Psychiatry has to thank Meynert for the close relationship known between the physiology and pathology of the *brain* . . . however, Wagner von Jauregg did something equally important: psychiatry has to thank him for the close relationship between the physiology and pathology of the *body*."[13] This vision posited a link between Meynert's lesion model and Wagner-Jauregg's somatic treatments; both emphasized physical changes to the nervous system as a cause (and in Wagner-Jauregg's case, a treatment) for pathology. That is, Meynert was remembered in Wagner-Jauregg's clinic in a way that emphasized his pathological anatomy over the associative model that had been so important for Freud.

Given this self-understanding of the clinic's history, it is not surprising that Wagner-Jauregg should have been of two minds with respect to psychoanalysis. On the one hand, and in part a consequence of a common neuropsychiatric inheritance, Wagner-Jauregg's clinic was unusual in the number and prominence of psychoanalysts it employed: in addition to Poetzl and Heinz Hartmann, who made a name for himself as an ego psychologist, Anna Freud herself was a visitor at the clinic, going on medical rounds with Schilder.[14] Indeed, it was because of this openness to psychoanalysis that Schilder felt able to join the Vienna Psychoanalytic Society in 1919, shortly after taking up his position at the clinic; he delivered his first paper there in early 1920.[15] Moreover, Wagner-Jauregg allowed Schilder and Hartmann to lecture on psychoanalysis, which, according to Magda Whitrow, made his clinic "the only German-speaking clinic where a psychoanalyst was permitted to do so."[16] In a letter to Freud for his seventy-fifth birthday, Wagner-Jauregg took pride in his tolerance: "At my clinic other men, I only want to name Poetzl, Schilder, Hartmann, have written in the name of psychoanalysis, without me resenting them for it."[17] On the other hand, as his choice of words here indicates, Wagner-Jauregg's accep-

FIGURE 5.2. Julius Wagner-Jauregg (*center front*) and his clinic. Schilder is second left from Wagner-Jauregg. (Courtesy of Brooklyn College Library Archives and Special Collections, Lauretta Bender Papers, box 27, Photographs [1912–48].)

tance of his assistants' psychoanalytic interests can hardly be described as wholehearted. He tolerated that interest as long as it did not displace his assistants' other more directly somatic work. And as Schilder's interest in psychoanalysis grew, it is not surprising that he would look elsewhere, and particularly to America, for a more congenial environment to develop his ideas.[18]

The Body Schema—A Sensory Construction

While Schilder may have found Wagner-Jauregg's clinic increasingly unsuitable, given his burgeoning interest in psychoanalysis, his training there nonetheless left its mark on his work. In particular, though Schilder presented himself as a psychoanalyst, his appropriation of the neuropsychiatric tradition corresponded more closely to Wagner-Jauregg's interests than to Freud's. In short, Schilder cleaved more to Meynert and Wernicke's pathological anatomy than to Freud's criticism of localization. This is clearest in Schilder's most important contribution to psychoanalysis, his conception of body image.

The body image has a long history in the tradition I have explored in this book, and most of the figures I have discussed so far—Meynert, Wernicke, and Freud—describe a concept that, at first glance at least, looks very similar. As we saw in chapter 1, Meynert had developed a concept of the primary ego, constructed through a process of association. The sense of our bodies was the result of the repeated sensations that accompanied bodily movement. A similar process was at the heart of Wernicke's version, which drew explicitly on Meynert's example. As Wernicke detailed in his 1906 *Grundriss der Psychiatrie*, the "consciousness of the body" was constructed by a complicated process of tying together various sensory impressions.[19] He started with the sense of vision and moved through touch, hearing, and smell. From the various parts of the body, "messages" were sent to the projection areas of the brain, where they produced residuals, or "memory images" (*Erinnerungsbilder*).[20] These memory images were then, through a process of association, connected with one another. The "sum of the memory images of all organ sensations" formed the "content of the consciousness of the body."[21]

Freud's bodily ego (*Körper-Ich*) took a different tack. It appears, to my knowledge, for the first and only time in *The Ego and the Id*.[22] In this 1923 work, Sigmund Freud returned to the theoretical grounds he had laid out in *On Aphasia* and the *Project* by extending his findings to the second topology. For this reason, his formulation of the bodily ego should be understood in the context of his radicalization of association and his development of the concept of the unconscious. When explaining his claim that the "ego is first and foremost a bodily ego" in *The Ego and the Id*, Freud reiterated his argument against any necessary link between higher functions and consciousness. He claimed that "even subtle and difficult intellectual operations which ordinarily require strenuous reflection can . . . be carried out pre-consciously and without coming into consciousness. . . . Not only what is lowest but also what is highest in the ego can be unconscious."[23] For Freud, the ego was itself bodily, and so was in part unconscious: hence the iceberg model of the ego for which he is famous. The bodily ego was not the emergence of the body in the mind, as it had been for Meynert and Wernicke, but rather the embeddedness of the ego in the body, that is, its participation in the unconscious.

Given Schilder's psychoanalytic inclinations, it is tempting to root his body image in Freud's. But there are numerous problems with this connection, not least the fact that the two concepts were first elaborated in books appearing the same year. Schilder himself explicitly asserted the dependence of his work on Wernicke's. In a programmatic article, "Problems of Clinical Psychiatry" (1925), he referred directly to Wernicke's somatopsyche. Schilder argued that clinical psychiatry should "attach itself to Wernicke's great system." As he explained, Wernicke had "generalized . . . the experiences which aphasia and agnosia gave us, and understood the individual clinical pictures as the expression of disease of various brain systems." Of those three brain systems—the autopsyche, allopsyche, and somatopsyche—the latter gave a "neurophysiologically comprehensible system . . . that has as its basis the consciousness of one's own body."[24]

Schilder had developed these ideas in the 1923 volume *Das Körperschema*, published by the prominent medical press Julius Springer Verlag. Following the arguments of British neurologist Henry Head, Schilder described the body image as the "spatial image that everyone has of himself," which "includes the individual parts of the body and their spatial relationship to each other."[25] These elements were brought together by "past impressions" recorded in the cortex. In an account reminiscent of Meynert and Wernicke's work, Schilder explained how, on receiving a tactile sensation and simultaneously seeing one's left hand being touched, one learns to associate the sensation with that location. When the body schema was functioning correctly, on future occasions the same sensation would be experienced as sensation on the hand. The body schema was thus created from past impressions that helped organize and constitute new incoming ones. As Schilder wrote, the impressions "form organized models of ourselves, which can be called schemata; these schemata change the impressions which come from sensibility such that the final sensation of the position or the place enters consciousness already in relation to past impressions."[26]

Schilder drew conclusions about the workings of the body schema from a series of pathological cases. One of them was that of Barbara M., a forty-three-year-old patient who presented at the neurological clinic of the University of Vienna with a number of symptoms. Apart from a hemiplegia on the right side of her body after stroke on a luetic basis, she displayed severe disturbances in the localization of touch on her own body. As Schilder pointed out, "touch was localized wrongly on the right side in a bizarre way." If the patient was touched in the area of her right breast, "she first locates the impression on the shoulder, but after 4–10 seconds indicates a second impression, which she locates in the area of the elbow, a third on the thigh, a fourth on the back of the foot"[27]—a condition named polyesthesia. The patient also suffered from alloesthesia; that is, she perceived touch on the left (healthy) side of her body, with a similar delay, on the right side of her body, usually at symmetrical points. In this case, then, while sensitivity was preserved, the localization of touch was misplaced; the patient was unable to give the sensation any spatial meaning. In Schilder's interpretation, while the optical body schema was intact—as Schilder noted in his exam, the patient was "fully capable of indicating every point on the trunk when asked"[28]—the disturbance of the tactile body schema prevented the patient from aligning the raw feeling of touch with optically perceived parts of the body.

While Schilder concluded from this and similar cases that a tactile and an optical body schema existed,[29] another case suggested to him the existence of a third sensory dimension. Kurt Goldstein and Adhémar Gelb described a patient with a disturbed optical body schema. The patient accordingly had great difficulty locating a stimulated point on the body "despite intact surface or deep sensibility." The most distinctive feature of the case was that the patient could arrive at a nearly correct localization through "explorative twitches" (*Tastzuckungen*): "He moved a larger number of muscles in quick succession until he got close to the stimulated point." That is, he could relate sensory stimulation to location on the moving, but not the static, body. For Schilder this suggested the existence of a "kinesthetic body image." The "kinesthetic body image," Schilder proposed, explained how blind people were able to localize touch, although he noted that they did not need the actual movement each time; they could rely on "kinesthetic residuals."[30] The body schema, then, consisted of three parts, all of which were sensory—optical, tactile, and kinesthetic: "The impressions of tactile, kinesthetic and optic kinds are united in an overall image [*Gesamtbild*] and only by means of this overall image is the new individual event integrated [*eingeordnet*]."[31] If at least two of the three parts worked together, as we have seen, it was possible to localize touch on the body.

The priority of the sensory is especially clear in Schilder's attempts to explain motor disturbances, or "apraxia." That the chapter on the apraxias is by far the longest in *Das Körperschema* is testament to the difficulties inherent in making an otherwise sensory body schema relevant for the explanation of motor phenomena and their disturbances. As Schilder explained, in cases of body schema "apraxia," "otherwise accurate insight [is] not usable in action." Patients found it particularly difficult to perform reflexive actions (*reflexive Handlungen*). "If [the patient] is asked to show the right or left thumb, she fails almost regularly in a very characteristic way." And yet this was not a case of a disturbance of the optical body schema, because when asked to show the left side, she did so promptly: "She is fully capable of recognizing right and left."[32]

Schilder provided the following explanation: in a given movement, an image of an intended movement was "sent off" (*abgegeben*)[33] to the body, which resulted in innervation and action. However, the image was not the simple *Bewegungsvorstellung* (idea of movement) of the Meynert-Wernicke system. Freud, as we have seen, criticized the *Bewegungsvorstellung* for being a psychological contaminant of an otherwise physiological model; for Schilder the *Bewegungsvorstellung* was excessively motor. To get around this problem Schilder broke down the movement into its various elements. Here, he drew on Hugo Liepmann's *Bewegungsformel* (movement formula). A *Bewegungsformel*, according to Liepmann, "conveys the composition, the structure of an action," which clearly and sufficiently defines the action.[34] The *Bewegungsformel* was thus more complex than a *Bewegungsvorstellung*; it had a beginning and an end. This complexity was justified, however, because it could render the movement in sensory terms. According to Schilder, both the *Bewegungsbeginn* (beginning of movement) and the *Beendigung der Bewegung* (ending of movement) depended on the body schema. Both were static images of the body in a particular position, and as such were sensory. The *Bewegungsformel*, by constructing movement from such sensory impressions, allowed Schilder to remain faithful to the sensory nature of the body schema while deploying it to understand motor disturbances. Thus, even when he wrote about the "motor part of the body schema,"[35] the body schema remained, in his view, essentially sensory in nature.

Truncating the Reflex

As we have seen, a version of the reflex exam structured Freud's clinical and theoretical work. Freud considered the sensory and motor elements of the reflex to be parts of an irreducible whole. Indeed, insofar as he radicalized the associative parts of the Meynert-Wernicke model, Freud prioritized what lay between and bound together the sensory and the motor. For Freud, hysteria was caused by the pathological disjunction of sensory and motor pathways—because there were no ways for the affects associated with particular sensory memories to discharge themselves through appropriate motor pathways. The cure, then, resulted from the overcoming of this repression to allow that discharge. In both theory and practice, Freud insisted that the reflex had to be treated as an integral unit; its connectivity could never be broken.

Schilder, too, participated in the reflex tradition. His adoption of elements from the Meynert-Wernicke model allows us to draw a line of continuity between his and their work. But because Schilder understood the body schema as fundamentally sensory and thought that, to the extent that it comprised associations, these only associated sensations with other sensations, he effectively cut the reflex in half. The truncation of the reflex had considerable consequences for Schilder's later work: it entailed a dramatic revision of the psychoanalytic subject and exacerbated tensions between him and other elements of the psychoanalytic movement.

For the moment, it is the implications of this change for Schilder's practice that will occupy us. Traditional psychic reflex testing relied on objective events in the world: both stimulus and response could be perceived by a third party. Schilder's reliance on a truncated reflex, however, required him to access the patient's subjective experience directly. Take the example of a test for sensory disturbance. Schilder would apply a sensory stimulus, for instance, touching the skin at a specific location. Rather than waiting for a verifiable motor response, as in the old reflex exam, Schilder would ask the patient what she experienced. No further observation of external phenomena was needed.

In the testing of motor disturbances, a similar method was used. While sensory examination *ended* with the patient's subjective account, the examination of motor phenomena *began* at the level of the subject. The subject produced a *Bewegungsformel*, and this resulted in a specific movement. The patient could either be instructed verbally to move in specific ways, for example, to point at the examiner's eye or to handle a given object.[36] Schilder would then record the action produced and compare it to the doctor's initial command or to normal usage. As in the sensory case, in Schilder's practice one part of the reflex arc was tested in isolation from the other.

Schilder's truncation of the reflex was written into and reinforced by the structure of his examination. Generally, although the structure of the medical exam as presented in the case histories in *Das Körperschema* varied across patients, it moved from a general biographical and medical history to the patient's present complaints, and from there to a fairly brief general neurological exam (including, for example, the Wassermann test, examination of the cranial nerves, and basic reflexes). More often than not, the bulk of the case was a very detailed neurological examination specific to the patient's symptoms. For example, in a case of alloesthesia, the exam would include an extensive sensory exam (testing for perception of posture; perception of temperature, pain, and itching; perception of weight; and the localization of touch on the body). If motor disturbances were relevant, the exam might include testing for writing, drawing, praxias, reflexive actions, and the manipulation of objects.

Schilder would then record the results of this examination in a two-column notation system.[37] Depending on the patient's condition, the report would focus on either sensory or motor symptoms. In the example below, Schilder juxtaposed the objective stimulus (left) with the patient's subjective report of it (right):

Light touch of the inner side of the joint.	As if a thread were running down the left calf.[38]

Similarly, motor exams juxtaposed the subjective *Bewegungsformel* with the objective act that resulted:

Indicate right thumb:	Punches into the air with the outstretched right hand.
. . . .	
Thumb your nose:	Reaches for the nose of the examiner.[39]

In these examples, the examiner asked the patient to imitate his movements. Other means for initiating a particular movement included the verbal instruction to move in specific ways and the invitation to manipulate objects in a prescribed fashion, for example, driving a nail into the wall with a hammer or lighting a cigarette.[40] In his case notes, Schilder compared the subjective and objective accounts of the same event, essentially putting in relation the experience at the nervous center with the event at the nervous periphery.

Because it compared the subjective and objective aspects of the same event, judging them by their similarities, Schilder's system produced something that looks very much like a correspondence theory of truth. Unlike Freud's approach, and unlike reflex tests more generally, where there was no direct relationship of similarity between sensory stimulus and motor response (the patient's response instead had to be measured against a horizon of expectations), for Schilder the correct answer in the test was clearly marked by the identity between the subjective and objective sides. When the experience described by the patient exactly matched the sensory stimulus, or the patient performed precisely the action that the doctor had required, that is, when the statements in both columns were equivalent in meaning, Schilder concluded that there was no disturbance in the body schema. This possibility of correspondence was confirmed by a shorthand that Schilder used in his notation. If a motor and sensory test was completed correctly, Schilder marked it in the right column as "+":

[sensory] Pin pricks on the back of the right foot from top down:[41]	+.
[motor] With the left hand to the left ear:	+.[42]

Although Schilder attempted to break down the normal-pathological distinction in his work, his clinical practice was always structured by the idea of a correct answer and thus a clear definition of normality.[43]

Because Schilder compared the *meanings* of both columns, he paid little attention to the precise words used by the patients. It mattered *what* the patient said, not *how* he said it. The perhaps most arresting piece of evidence for this can be found in the way Schilder reported the patient's speech. Schilder often thought it legitimate to bypass the patient's particular formulation and distill the meaning in an objective third-person description:

Touch on bottom of the left toe:	On the left foot under the toes as if driven over with the needle very softly.
Touch on right ankle joint on the inner side:	Feels a soft movement in the shape of a half-arc from the toes to the ankle joint.[44]

Or, in another example, the distinctly medical "proximal-distal" appeared in both columns:

On the back of the foot of the left foot, stroke from inside to outside 3 cm	Now it goes from foot to toe left (that is proximal to distal).
On the back of the right foot, stroke 3 cm proximal to distal	Moved down on right foot from the ankle above in an arc.[45]

Schilder did not consider the translation of the patient's account into scientific language, losing its particular idiom and mode of expression, to entail any significant loss.[46]

As these examples show, Schilder remained content in his practice with the manifest meaning of the patient's speech. There was no attempt to read between the lines. The necessity to take the patients at their word had two seemingly contradictory consequences. First, even if there was subjective variation, or lack of cooperation, trust in the patient's report could not be challenged. One patient's "alloesthetic symptoms," for example, were "inconstant."[47] In another patient, for whom "stimuli of touch and the prick of a needle" generally "create[d] the impression of a stroke over the skin," "sometimes there develop[ed] at the touched or piqued part the impression of a complicated back and forth movement." Another patient was decidedly uncooperative: "The examination is rendered difficult through the adverse behavior of the patient[;] he emphasizes continuously that he does not want to stay here, that it does not help anyway."[48] These variations and difficulties were duly noted; they did not, however, lead Schilder to question his patient's account; it was essential to the very structure of his exam and could not be called into doubt.[49]

Second, Schilder sought out patients who would give him the most useful results. Schilder commented on the fact that two amputees with phantom limbs were "psychologically untrained,"[50] which rendered it impossible to ascertain whether their experience of their toes as directly attached to the stump was in the optical or tactile realm. The more sophisticated and scientifically trained the patient, then, the more reliable the account. For this reason, no subject was better than the doctor himself. British neurologist Henry Head, in collaboration with W. H. R. Rivers, had famously experimented on himself, cutting through the radial and external cutaneous nerve in his left arm and then suturing the ends back together. Schilder presented this experiment when discussing the role of the peripheral nervous system in the localization and the knowledge of sides of the body: "Head has, through the severing of peripheral nerves on himself, observed serious disturbances in localization, which, however, never lead to a confusion of sides. Here the determination of the side of the stimulus seems to be a function [*Leistung*], which was achievable even after the injury."[51] In his 1918 article "Viewpoints on General Psychiatry," Schilder confirmed his preference for such subjects by including two "self observations" and asking a colleague to observe his own "mood swings." Although he did not give us the professional background of his friend, from his activities—going to the hospital to work, examining "an interesting case"—it is fairly clear that he, like Schilder, was a doctor.[52]

As this analysis suggests, Schilder's concern with the manifest meaning of the patient's speech, which transmitted the contents of internal experience without distortion, was related to his dependence upon introspection. That is, he required the patient both to assess subjective experience accurately and to convey that experience to the doctor without loss. In the motor case, he had to trust the patient both to understand the doctor's command and to attempt to fulfill it as accurately as possible. Schilder expressed this confidence in the power of introspection in a 1923 paper devoted (ironically, as we shall see) to "Das Unbewusste." As he wrote there, "One's own experience can at any moment become an object. One can take it out of the living stream of experience, one can kill it[;] in one word, one can turn it into an object [*zum Gegenstand machen*]."[53]

In Schilder's exams, therefore, we see a decisive move away from the interpretive model that had structured Freud's work. Self-transparency was both a condition of Schilder's practice and, in consequence, an ideal of his psychoanalytic theory. Dividing the reflex in two, he no longer dealt with sensory stimulus and motor response, both objectively perceivable. Rather, in the sensory and the motor cases he needed to find the means to access an internal state: to produce a particular *Bewegungsformel* or to retrieve a particular sensory impression. Schilder posited the self-aware subject as this means. As such, the idea that the patient's utterances needed to be interpreted was foreclosed by the clinical practice. With a truncated reflex, the patient had to be taken at his word, both at the sensory and the motor level. In place of Freud's "dark continent," Schilder posited an autonomous, introspective subject.[54]

The Subject, Master in Its Own House, Yet Again: Schilder and Meyer

There are many reasons why Schilder might have left for America in 1928: the physician Erwin Stransky suggested that Wagner-Jauregg eventually withdrew his support; his biographer Dieter Langer conjectured that anti-Semitism in Austrian academia might have been a cause; Schilder's contemporaries even speculated that his "squeaky voice" stood in the way of academic success in Europe; Schilder himself implied that working conditions in Vienna were less than ideal. Moreover, it was relatively easy at the time for a European doctor to immigrate. In America Schilder's medical qualifications were recognized without problem.[55] The analysis of Schilder's divergence from psychoanalytic orthodoxy furnishes another possible motive. From this perspective, Schilder fell between two European stools. He was too Freudian for Wagner-Jauregg, and while Viennese psychoanalysis in the mid-1920s was a relatively open-minded environment, Schilder would have found more receptive conditions in the eclectic American scene.

Schilder's first contact with American medicine came through the prominent psychiatrist Adolf Meyer (1866–1950). A Swiss-German physician who immigrated to the United States in 1892, Meyer had received elite training in neuroanatomical research in various European universities. In America, after a short period of clinical work, he started work as a neuropathologist, working at hospitals in Illinois, Massachusetts, and New York. Aiming to connect his scientific knowledge with the realities of the clinic, from 1904 onward Meyer held key positions in clinical psychiatry, first at Cornell, then at Johns Hopkins, where he became the director of the newly built Henry Phipps Psychiatric Clinic in 1913. By the 1940s, he was considered the most influential figure in American psychiatry, and his students held positions in most of the leading medical schools.[56]

Meyer's training in neuropathology provided a point of connection with Freud's psychoanalysis.[57] When Freud gave his famous lecture on psychoanalysis at Clark University in 1909, during his only visit to the United States, Meyer was in the audience.[58] Meyer subsequently contributed to the reception of psychoanalysis in the United States. Among other things, he reviewed Freud's *Three Essays on the Theory of Sexuality* (published in German in 1905) in glowing terms, he proposed a dynamic explanation of schizophrenia that resonated with psychoanalysis, and, during his time as head of the New York Institute, he encouraged his staff members to familiarize themselves with psychoanalytic theory and to go to Europe to study it.[59]

Nevertheless, Meyer never joined a psychoanalytic organization and rejected key elements of psychoanalysis.[60] On the occasion of Freud's visit to Clark, Meyer gave a lecture that drew on Freudian vocabulary.[61] But even while remaining generally positive, Meyer criticized what he saw as Freud's excessive reliance on a sexual etiology.[62] Meyer's rejection of Freud's sexual explanations corresponded to his unease with the Freudian "unconsciousness," as Meyer referred to the key element of psychoanalytic theory in 1935.[63] In Meyer's early writings, the unconscious is conspicuous in its absence. In "Issues in Freud's Psychoanalysis," (1909–10) Meyer's most detailed discussion of psychoanalysis, it did not even merit a mention. Meyer came closest when he referred to "repressed undercurrent-like tendencies that are lived to completion in dreams only."[64] Similarly, his earlier review of Freud's work—a review of his *Fragment of the History of a Case of Hysteria* and *Three Essays on the Theory of Sexuality*, alongside works by Pierre Janet and James Jackson Putnam—avoids the term; again Meyer was content simply to refer to "undercurrents."[65] When he did refer to it, Meyer expressed his reservations about the unconscious by placing either the term or its definite article in quotation marks.[66] Later, Meyer became more explicit in his criticism. In 1938 he lamented the "overemphasis on an obligatory hypothesis of the unconscious" in Freud's work.[67]

Meyer's suspicion of the Freudian unconscious led him to criticize much psychoanalytic practice. He was particularly opposed to the length of psychoanalytic treatment and the expense that often accompanied it. He mocked the "now fashionable months of costly searching in an unconsciousness."[68] In opposition, Meyer emphasized the importance of introspection in his "psychobiology":

> The chief point in this conception of objective psychobiology is that it looks for an understanding of mentality which does not merely see intelligence tests or a reduction of man's life to sex and to the unconscious, but begins with, and turns back again to, a frank and reasonably balanced review of man's responsiveness and his positive and negative assets in the form of specific samples from the whole wide range of practical performance. It surveys, according to the extent of specific demand, the jobs and recreations, the interests and ambitions, personal, educational, civic and political, moral and religious, and the balance of actual performance and ambitions and opportunity.[69]

To Meyer, one did not have to dig into the depths of the mind that remained obscure to the patient, but could rather make use of the patient's introspective capacities. As historians have emphasized, Meyer's emphasis on "rational self-analysis" was linked to his embrace of a philosophy of common sense.[70] In what Ruth Leys referred to as his "positivism," Meyer rejected all that needed interpretation. Disease was directly accessible; one only needed to look.

The similarities in their training and outlook help us understand why Meyer extended an invitation in 1927 to Paul Schilder, as a "representative of European psychiatry," to join his clinic for a three-month lectureship.[71] Schilder stayed for a total of four and a half months, lecturing on somatic and psychological factors in psychoses and treating patients using psychoanalytic methods.[72] Meyer noted approvingly Schilder's independence from Vienna psychoanalytic orthodoxy: in the obituary he wrote for his colleague in 1941, Meyer remarked that "he had contact with the Freudian group [in Vienna], but was too widely active to limit himself to any one group."[73] Meyer's support for Schilder opened up new opportunities elsewhere in America. In 1929 Schilder accepted the prestigious position as research professor of psychiatry at New York University and clinical director of the psychiatric clinic at the Bellevue Hospital in New York. Aside from another short stint at the Phipps Clinic in 1930 (January to March), he remained at the Bellevue until his premature death in 1940.[74]

From *Körperschema* to Body Image

Schilder's early work on the body image already marked his departure from the Freudian version of psychoanalysis, but when he first came to America, it seemed that he had moved back into the fold. In 1935 Schilder published a translated and expanded version of *Das Körperschema* with the title *The Image and Appearance of the Human Body: Studies in the Constructive Energies of the Psyche* (henceforth *The Body Image*). The publication history of the books already gives a hint about their different orientations. Whereas *Das Körperschema* (1923) was published by Springer, the German academic medical press, *The Body Image* first appeared with the London publisher Kegan Paul, Trench, Trübner & Company (1935), as part of its "psyche" monograph series, that is, as a specialized psychological, though not specifically psychoanalytic, study. The second edition (1950) appeared in a series alongside psychoanalytic works, published by the New York–based International Universities Press.[75]

The Body Image comprises three parts.[76] The first part of the book, the "physiological basis of the body-image," directly follows Schilder's discussion in *Das Körperschema*, apart from a number of small differences such as the outsourcing of case studies to an appendix. At first glance, the two newly added sections seem to attest to Schilder's allegiance to Freudian psychoanalysis. They deal with the libidinous nature of the body image and its social structure. Here Schilder emphasized a sexual etiology of mental disorder, and the language he used is staunchly Freudian. Similarly, most of Schilder's examples in *The Body Image* look like typical psychoanalytic case histories. With their focus on the life histories of patients and their often generous length—most of them are five pages or more—they foreground the narrative element that marks Freud's own writings.

Moreover, in the revised book, Schilder related his body image directly to the Freudian unconscious. Schilder stated that some "psychic processes . . . remained in the background of consciousness," which he called the "sphere." The thought processes that took place in the sphere were "at first a general scheme, a mere diagram."[77] They were part of a general constructive process, in which thought processes developed from more "schematic" to better-developed stages. The impressions that constituted the sphere and especially those connected to the body image, "may rise into consciousness as images, but more often, as in the case of special impressions, remain outside of central consciousness. Here they form organized models of ourselves, which may be termed 'schemata.' Such schemata modify the impressions produced by incoming sensory impulses in such a way that the final sensation of position, or of locality, rises into consciousness charged with a relation to something that has happened before."[78] Because the schemata shaped those sensations that emerged in conscious life, they functioned like the unconscious. Indeed, Schilder was explicit about the connection. He argued that the schemata "certainly show the qualities which Freud ascribes to his system of the unconscious." Because of this, any "psychology which neglects the system of the unconscious, the sphere," would be "incomplete and superficial."[79]

And yet it is because Schilder seems to come so close to the Freudian model in his body image that the differences appear in such stark relief. Despite the apparent embrace of psychoanalytic themes and terminology, the gulf that separated Freud and Schilder, one that was reinforced by the differences between European and American psychoanalysis, remained large and imposing. As in the earlier book, and despite the attractive parallels between it and his own "sphere," Schilder explicitly rejected Freud's own formulation of the unconscious, and like Meyer, he used the term only in quotation marks.[80] Indeed, in direct contrast to Freud's account of the *Körper-Ich*, in *The Body Image* Schilder rejected the idea that elements of psychic life could be completely unconscious. They were instead, according to Schilder, characterized by awareness: "I do not think that there are any psychic processes which do not possess the quality of awareness. Nor do I think that we can speak of a psychic unconscious if unconscious is to mean that there is no awareness."[81] Similarly in his book *Mind: Perception and Thought in Their Constructive Aspects*, published posthumously in 1942, Schilder offered an account of "mind" that was structured like a psychology textbook of the time, with sections on perception and action and higher functions such as language and memory. Schilder mentioned the Freudian unconscious just once, and then only to criticize the concept.[82]

The sphere or body image, then, was less a new name for the Freudian unconscious than Schilder's replacement of it. As Schilder admitted with respect to the body image, "many of these psychic activities take place in the background of the consciousness and have, to put it briefly, a symbolic or a spheric character. They fall into the category of Freud's so-called 'unconscious.'" He was, however, quick to assert the difference: "But all these are psychic experiences."[83] Schilder's body image is best understood in terms of Meynert's *Verdunkeln*, a contingent and temporary darkening owing to a lack of attention: an unconscious without repression. Indeed, when describing the "sphere," Schilder preferred metaphors that presented it as an ever-present resource rather than the inaccessible depths of the mind. It was, he suggested once, a "reservoir" from which we would draw "under the urge of an actual situation to which we want to adapt ourselves."[84] It was in the same sense that Schilder described the sensory cortex as a "storeroom of past impressions."[85]

The shift in status had significant consequences. To see them play out, take a case from *The Body Image*: Schilder's version of Freud's *Rat Man*. E.M., a twenty-year-old male patient, suffered, among other things, from difficulties in urination and defecation. In particular, E.M. freely shared with Schilder the strong fear that animals would crawl into his anus while he was defecating.

Despite superficial similarities, the contrasts with Freud's rat-man case are striking. First, symbolization was predominantly restricted to the level of the body. Unlike Freud's case, where the rat punishment was the nexus of a number of associations in Lanzer's life, including his father's financial problems, concerns about a marriage, and a traumatic incident as a child, for Schilder such broader psychological considerations were not of primary importance for understanding the pathology. Pathology, for Schilder, rather stemmed from the libidinous development of his body image and the confusion of the genitals with other parts of the body. In this case, E.M. had libidinized his anus, and this was the root cause of his fantasy.

The distribution of libidinal energies was simply an extension of Schilder's old schema. As he wrote: "What goes on in one part of the body may be transposed to another part of the body. The hole of the female genital organs may appear as a cavity in another part of the body, the penis as a stiffness or as a piece of wood somewhere else. There is said to be a transposition of one part of the body to another part of the body. One part may be symbolic of the other."[86] Schilder formulated several "laws." For a body part to become symbolic of another, there had to be a relationship of resemblance: "Every protrusion can take the place of another. We have possibilities of transformation between phallus, nose, ear, hands, feet, fingers, toes, nipples and breasts; every round part can represent another—head, breasts, buttocks; every hole can be interchanged with another—mouth, ears (in some respects, eyes and pupils), openings of the nose and anus."[87]

Similarly, Schilder explained the Freudian concept of transference through the development of the body image. Schilder wrote, "It is true that it may be the patient's own genital which is transposed from one part of the body to others, but it may also be the sex-organs of another person brought symbolically into connection with different parts of the body."[88] Transference was thus the projection of the patient's body onto that of the therapist. In the analysis of E.M., "the analyst became the carrier of all the suffering about which the patient complained," a process Schilder called "narcissistic projection." Because E.M. considered Schilder to be in complete control of his body, this transference put him on the "way to recovery."[89] Though Schilder had embraced Freud's language of the libido, as the organization of his book indicates, that libido was structured by his old and un-Freudian conception of the body image.

The second contrast is a consequence of the first. For Schilder, the meaning of the symbol was exhausted by its libidinous referent. True, the patient's history was important for the case. A key event in E.M.'s life was the birth of his sister, when he heard his mother's cries of pain during labor. He imagined her being tortured and giving birth through the anus. This early fixation on the anal zone was reenforced by his parents' behavior; they showed great interest in his bowel movements and often gave him enemas. These, however, were not traumatic events that had been repressed, but rather accessible memories that helped explain the libidinous significance of the anal region.

For this reason, Schilder placed no particular importance on gathering the various conditions that had led to the disturbance in the libidinal body schema, nor on placing it within the complex constellations that made up the patient's mind. But this is precisely what Freud had done in the rat-man case; he had picked apart the rat punishment to unravel a web of stories. Nowhere in *The Body Image* did Schilder attribute to the symbolic the broader meaning it had for Freud. The closest Schilder got to it was his discussion of "tattooing, painting the lips, painting the face, bleaching or dyeing the hair," and, most importantly, cleanliness. He wrote: "There is no question that the meaning of all these changes in the appearance is not always in the consciousness[;] there is also a symbolic meaning. Psychoanalysis has shown that cleanliness is a tendency to overcome anal tendencies."[90] Schilder was referring here to the psychoanalytic interpretation of cleanliness according to which it pointed toward a delayed success in toilet training in early childhood.[91] After flirting with this psychoanalytic reading, however, Schilder returned to his model—in which one body part stood in for another at the level of libido—arguing that cleanliness "may also satisfy narcissistic tendencies" because it "may be a transformed masturbatory act."[92] In the act of washing, the washed body parts stood in for the genitals—they were cathected with libido—and thus the act of washing them corresponded to an act of masturbation. To misquote a common criticism of psychoanalysis, for Schilder sometimes a cigar was just a penis. Quite how it became so was no longer important.

As such, and third, the patient's confrontation with this history did not play any role in the treatment. Repression did not function the way it had in Freud's system, where particular memories had been expelled from consciousness. For Schilder, it instead worked at a predominantly affective level, because of the patient's "transformation of libidinous infantile tendencies" during adolescence. At this stage the "ego and ego-ideal" fought "against the perversion which is the expression of the infantile drive."[93] In E.M.'s case, it attempted to repress the patient's infantile obsession with the anus. But such repression was incomplete, and rather than fully overcoming the obsession, it transformed its affective charge. No longer the site of sexual pleasure, the anus became the location of itching and other neurasthenic symptoms. For this reason, Schilder did not remark on any resistances on the part of his patients. Memories seemed to come easily to the patients, and the analysis did not include the types of dramatic moments that had marked many of Freud's cases. Even if certain elements did not surface immediately in the analysis, they could be retrieved relatively easily, as "the patient himself realized that his dreams and associations led to these conclusions."[94]

Although Schilder's case histories show certain psychoanalytic features, they lacked, or fundamentally transformed, the central elements of Freud's system: resistance, repression, the complex of associations and overdetermination that made Freud's cases so rich. Schilder had replaced Freud's unconscious with his body image or sphere, and this switch at the heart of the system had significant effects in both the theory and the practice. The patient's symptoms were explained and his treatment initiated through the logic of the body image, not through the logic of the unconscious. But because this body image operated predominantly at the level of what Meynert and Wernicke had called the projection system, and not at the universalized association system that was so important for Freud, Schilder's cases seem to lack the complexity that characterize the older psychoanalyst's work.

The Schilder Affair

It is thus not surprising that some psychoanalysts were critical of Schilder's ideas, in particular of his attenuated unconscious. True, as we have seen in the approaches of such eclectics as Adolf Meyer, American psychiatry provided a receptive environment for someone like Schilder. And indeed similar tendencies came to dominate the American psychoanalytic scene later in the century. Ego psychology, popularized by Heinz Hartmann, an émigré psychoanalyst and former colleague of Schilder's under Wagner-Jauregg in Vienna (whom we encountered earlier in this chapter), became a constituent element of psychoanalysis in the United States from the 1940s onward. Hartmann emphasized autonomy and "ego strength" and de-emphasized, especially after World War II, the antirational elements within psychoanalysis.[95] But within the hardening factions of 1930s American psychoanalysis that informed postwar orthodoxy, Schilder encountered problems. He had joined the New York Psychoanalytic Society in the first years after his arrival in New York, simultaneously resigning from the Vienna Society. He came to regret this resignation, because his membership at the New York branch was to be short-lived.[96] Only a few years after his arrival, he became the center of an intergenerational controversy that has been labeled by historians the "Schilder affair." The affair began in 1933 and ended three years later, with Schilder's withdrawal of his membership.

The Schilder affair was the central act of a prolonged drama pitting an old guard of American psychoanalysts against a younger generation over the question of professionalization.[97] Both generations sought to increase the acceptance of psychoanalysis by tying it more closely to the medical profession.[98] The older generation of American psychoanalysts included such figures as William Anderson White, A. A. Brill, and Smith Ely Jelliffe, who, as Hale has shown, in the early part of the century "had represented psychoanalysis as a tiny avant-garde in American psychiatry." Brill in particular sought to cultivate a relationship with eclectic psychiatrists (such as Adolf Meyer), thus attempting to promote the connection of psychoanalysis with general medicine.[99] For his generation, psychoanalysis gained acceptance as an area of expertise within medicine, taught through supplementary training after a medical degree and specialization in neurology or psychiatry and adhering to the norms of the medical profession.

The younger generation agreed with Brill on the need for psychoanalytic practitioners to be trained physicians with a specialization in neurology or psychiatry. Often educated in Ivy League medical schools, they had come across psychoanalytic ideas through that training.[100] But their proposed path to medicalization differed from that of the pioneers. In order to improve the image of psychoanalysis in the United States, these analysts—a group of European-trained Americans such as Bertram Lewin, Ruth Mack Brunswick, and Lawrence Kubie—expressed the need to formalize psychoanalytic education to conform to the rising standards of professionalization within American medicine more generally.[101] Rather than asserting their scientific bona fides by downplaying the differences between psychoanalysis and other medical specialties, the new generation wanted to assert the autonomy and particularity of psychoanalysis within medicine.

The debate coalesced around the question of training analysis. Almost by necessity, the first generation of American psychoanalysts had not undergone analysis themselves; if they had, it had occurred on an individual and ad hoc basis. This was to change in 1925 when the International Psychoanalytic Congress stipulated that training was to become the responsibility of societies and institutes; all analysts were obliged to undergo both training and control analysis, overseen by educational committees.[102] For the new generation, training analysis would become the mark of true professionalism, a sign of both the legitimacy and the specificity of psychoanalysis within American medicine. Kubie, the most prominent proponent of the group, gave a summary of his views on the topic at the 1949 annual meeting of the American Psychoanalytic Association. He pointed out that, apart from learning about the experience of being analyzed, the procedure itself was of prime importance, because in the training analysis, the aspiring analyst would learn something "about himself," as well as about the workings of the unconscious: "It is important that he should have the experience of making discoveries of processes in himself of which he had been unaware. The subjective awareness of the emergence of material from unconscious to conscious levels gives the student a unique opportunity to know what the healing process is like. It is only in this way that the elusive concept of unconscious psychological forces can become a living reality to him."[103]

Brill's and Jelliffe's generation had, however, a markedly different attitude toward training analysis. As they understood it, a psychoanalytic practitioner was no different from any other medical doctor in his general ability to practice. As Brill had pointed out in his *Psychoanalysis: Its Theories and Practical Application*, the practitioner of psychoanalysis "should proceed in the same manner as in any other specialty." Training was of course important. Brill asserted that "to practice psychoanalysis without previous training in mental work is as dangerous as practicing surgery without a knowledge of anatomy." Warning about the dangers of lay analysts, he suggested those practicing psychoanalysis "should have a training in psychiatry and neurology."[104] But this knowledge was always the knowledge of the practitioner and never that of the patient. Brill pointed out that, apart from the medical training "in nervous and mental work," the aspiring psychoanalyst had to master Freud's key ideas, such as his theory of the neuroses, dream interpretation, and his writings on sexuality, as well as develop the appropriate judgment about which patients to take on and which to refuse.[105] But in all this training, the psychic life of the medical practitioner was never in question. Meyer went one step further, suggesting that, if an aspiring analyst needed to undergo analysis himself, this was a sign of poor mental health, which could endanger his medical practice. Framing the issue in his language of adjustment, in "Preparation for Psychiatry," a 1933 article about the training of psychiatrists, Meyer stated that he was "a bit suspicious of those who have been psychoanalytically trained chiefly because they needed to be psychoanalyzed on account of their own maladjustment or who had to be aroused by the lure of the revelations."[106]

Schilder quickly found himself at the center of the storm. Like the older pioneers, Schilder had not undergone analysis himself. Indeed, as he later wrote to Freud, denying its absolute importance in psychoanalytic training: "As you remember, I am not analyzed. I share this with rather important members of the Viennese group and especially with yourself.

Some might be inclined to consider this fact as detrimental, for one's psychoanalytic activities. I must confess I do not share this opinion."[107]

In line with the International Psychoanalytic Congress's 1925 guidelines, Schilder's psychoanalytic qualifications were called into question in 1933. It was decided that Schilder's case should be examined by a special committee. The augurs looked good for Schilder. The president of the New York Psychoanalytic Society was supposed to appoint the committee members, and at the time Brill was the president. Chaired by Smith Ely Jelliffe, the committee came to the conclusion that Schilder should retain his right to perform training analyses; it looked as though the older generation had won. The Psychoanalytic Society, however, overturned the decision, claiming that Brill had stacked the committee with his own supporters.[108] After a second special committee was installed and its report overturned as well, Schilder had had enough.[109] Three years after joining the New York Society and after several of his analysands had been admitted to the society,[110] Schilder felt scape-goated, treated like a "second rate member." After one last "appeal to the International Organization,"[111] Schilder resigned from the society, and even after being urged "to reconsider and withdraw" his resignation, he refused to return.[112]

It is clear that Schilder was the victim of larger forces at work in American psychoanalysis. If Schilder had not existed, the institutional and professional conflicts at the time would almost certainly have found another test case. But there were several reasons why the younger generation would have found Schilder particularly suitable. Indeed, in Brill's estimation, the question of training analysis was merely a pretext, and "Kubie and his gang [had] been hounding him [Schilder] for over two years." Despite his relative youth, Schilder aligned himself with the older generation of analysts; presumably owing to his contact with Freud, he considered himself something of a pioneer. As Brill wrote to Ernest Jones, the "research professor in the New York Medical School" was "very popular in the neurological and psychiatric organizations, and has contributed considerably to psychoanalysis. In my opinion he has done us a lot of good."[113] Schilder's positioning within psychoanalysis perhaps becomes clearest in his founding of the "Society for Psychotherapy and Psychopathology" in 1935, later called the "Paul Schilder Society." The fifteen founding members—all of them psychiatrists in New York—included a number of psychoanalysts of the old generation, most notably Brill.[114] Much in line with the ideas of the older generation introduced above, Schilder criticized the elitism of the Psychoanalytic Society in his "Outline for a Society for Psychotherapy and Psychopathology," with respect to both practitioners and patients. The technique and practice of psychoanalysis, he suggested, should be more broadly accessible to practitioners, and he urged them to move beyond "a group of socially selected cases" in their choice of patients.[115]

In light of Schilder's development of psychoanalytic concepts, we can suggest another reason. Beyond the seemingly practical question of training, the affair acted as a proxy for another debate about what constituted good psychoanalytic practice. For eclectic psychiatrists such as Meyer and Schilder—who, as we saw, appealed to psychoanalysts of the older generation—the unconscious remained a troubling and dispensable part of psychoanalysis. As for Schilder, even though he questioned the value of the normal-pathological distinction, in his practice there seemed to be the possibility of a correct answer: when the patient was able to give the appropriate introspective description of the doctor's stimulus. If the patient answered all questions correctly, treatment was not necessary, and surely if given the choice, it would be better to have doctors who passed the test. Schilder's contention that training analysis was not necessary was thus of a piece with his attempt to maintain the ideal of a self-transparent subject within the psychoanalytic paradigm.

For the second generation of psychoanalysts, this was itself a problem and spoke to deficiencies in Schilder's method. Freud, when contacted by Schilder for his opinion in the affair, affirmed the value of training analysis. If an analyst had not gone through analysis himself, it seemed, the depths of the unconscious could not be sufficiently explored.[116] Moreover, Freud criticized Schilder's psychoanalytic practice for being superficial: "An analysis of three or four months is generally not sufficient to clear up infantile amnesias, to reconstruct the psychic development, and to have the patient live through whatever of the repressed is capable of expression. Such a short analysis can hardly be more than a purely intellectual instruction."[117] The Schilder affair thus can be seen as a response to Schilder's distrust of an orthodox unconscious, which went against the convictions of the younger generation. It allowed them to frame the seemingly prosaic debate about training analysis in terms of psychoanalytic orthodoxy and heresy.

Conclusion

In chapter 3 I argued that Freud's radicalization of the Meynert-Wernicke model, in particular his attempt to universalize the process of association at the heart of Meynert's and Wernicke's reflex (thus emphasizing its connective qualities), provided him with the resources to posit the existence of an unconscious. This chapter lends support to that narrative by showing how Schilder's break with the reflex tradition, treating the two arcs separately, was of a piece with his de-emphasis and transformation of the Freudian unconscious. The patient's response was no longer a matter of interpretation as in the Freudian exam. Rather, access to the patient's experience was considered (and needed) to be simple and direct; Schilder had to posit a self-reflective and transparent subject. In so doing, he denatured Freudian psychoanalysis. Though he came to adopt many of its terms, they were fundamentally refigured in light of his adherence to the concept of the "sphere" as a replacement of Freud's unconscious.

Schilder's work remained guided by his earlier engagement with the neuropsychiatric tradition, in particular his development of Meynert's and Wernicke's body images. Here the focus remained predominantly at the level of the projection system. If association was invoked, it was purely between different sensory schemas; it constructed different ways in which the body could be projected into the mind. Of that mind and its secrets, Schilder had little to say. In Schilder, then, we have the curious case of a psychoanalyst who seems predominantly focused on the body, not as a complement to, but in opposition to a self-transparent mind. In chapter 6, we will turn to a neurosurgeon who, while developing a thoroughgoing somatic framework, made a similar self-awareness the precondition of his stimulation experiments on the sensory and motor cortex. But unlike Schilder, he thematized that mind directly and drew out more explicitly the consequences of his divided reflex model.

CHAPTER SIX

In the Operating Room

Wilder Penfield's Stimulation Reports and the Discovery of "Mind"

Wilder Penfield's homunculus, an illustration of the sensory and motor representation of the human body in the brain, is perhaps the most famous map in modern neuroscience (fig. 6.1). The figure's body parts are drawn in proportion to the cortical surface dedicated to them. For instance, the sensory nerves arriving from the hand terminate over a relatively large area of the brain surface, and therefore the hands of the homunculus are correspondingly large.[1] In comparison, the ends of the nerves stemming from the torso cover a much smaller area, which is why the trunk of the homunculus looks so slim.

The homunculus has often been presented as a single figure (fig. 6.1), but more properly it was two: there were sensory and motor versions (fig. 6.2).[2] This fact is crucial for understanding Penfield's work. Because the homunculi were located at the intersection of sensory and motor function, they can help us situate Penfield within the broader tradition of higher reflexes and the localization tradition with which they always had a fractious relationship. Penfield constructed two homunculi because unlike Foerster, but like Paul Schilder, he treated the two parts of the reflex arc as distinct, testing for sensory and motor responses separately. The experimental division between the sensory and motor investigations, a division that, as we shall see, was both the result and the structuring principle of Penfield's work, radically changed the parameters of neurological investigation. No longer engaged in the interpretive process demanded by sensory stimulus-motor response analysis, Penfield intervened with his electrode on the brain and relied on the patient to provide much of his

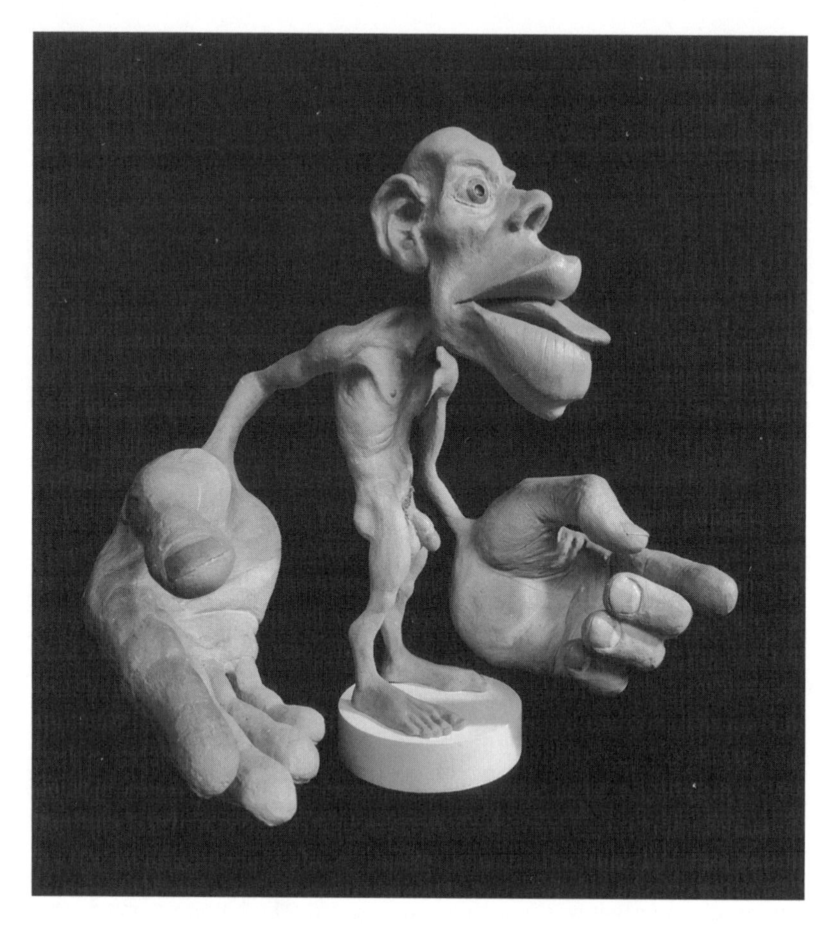

FIGURE 6.1. Museum version of Wilder Penfield's homunculus. (Courtesy of the Trustees of the Natural History Museum, London.)

experimental data. Only a self-transparent "mind" could access and describe the sensations produced by Penfield's cerebral probing. In his later work, when he became increasingly interested in the psychological implications of his brain explorations—which coincided with broader developments within the sciences of the mind and the brain at the time—he took this "mind" as the focus of his investigations. Penfield may not have elaborated a dualist metaphysics until the end of his career, imagining an exterior "mind" playing the brain like a "private computer,"[3] but that mind appeared much earlier as a working assumption in his experimental practice.

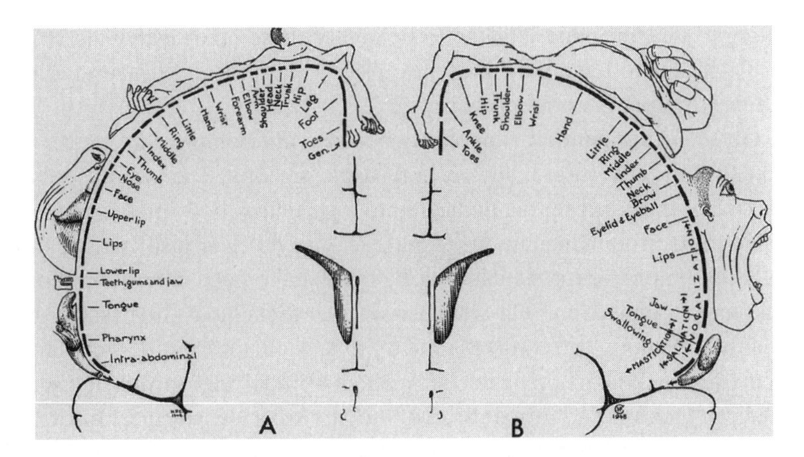

FIGURE 6.2. Sensory and motor homunculi. (Penfield and Rasmussen, *Cerebral Cortex of Man*, 44. Reproduced by permission of the Osler Library of the History of Medicine, McGill University.)

The Homunculus, Much Hated and Much Loved

Penfield's homunculus has an ambiguous status today. On the one hand, it is a mainstay of textbooks and popular science. In the neuroscience "bible" *Principles of Neural Science*, written by Eric Kandel and colleagues, it is presented to help students understand the function of the brain.[4] In addition, it has made frequent appearances in science writing aimed at a broader audience; Sandra and Matthew Blakeslee, in their 2007 introduction to brain and body mapping, called the homunculus the "$E = mc^2$ of neuroscience."[5] It is also a permanent museum exhibit at the London Natural History Museum and at the Nuremberg Tower of the Senses (see fig. 6.1).

The homunculus fascinates because it seems to provide a window onto our experience of the world. For example, Richard Dawkins and Yan Wong, in their book *The Ancestor's Tale* (2004), extended the homunculus principle to other animals and presented a number of species-specific proportional brain maps. Dawkins and Wong asked: "What is it like to be a star-nosed mole?" An examination of the "molunculus," they suggested, provides the answer: "You can see where the star mole's priorities lie. You can get a feel for the world of a star-nosed mole."[6] Similarly, Blakeslee and Blakeslee, also after discussing the homunculi of various animals—the cortical representation of the raccoon's forepaws, the mouse's whiskers, the star-nosed mole's nose, and the pig's snout—muse, "It is fascinating to imagine what it is like to be one of these other mammals. How would it feel to have your body awareness focused and distributed so differently from the primate norm?"[7]

On the other hand, it is precisely because the homunculus is read in experiential terms that many remain suspicious of it. The inference that the size of the cortical area dedicated to a particular body part is proportional to its prominence in our experience has no clear justification, and so many neuroscientists, including Penfield, have been careful to avoid asserting a connection.[8] For similar reasons, others have emphasized the slightly ridiculous, cartoonlike—and by extension, unscientific—qualities of the figure. When I first told Dr. William Feindel, curator of the Osler Medical Archives and former head of the Montreal Neurological Institute (MNI), of my interest in the homunculus—this was in 2007, when I was a graduate student doing research in Montreal—he exclaimed: "Ah, that silly thing!" Similarly, Queen Square neurologist G. D. Schott, in a 1993 editorial for the prominent *Journal of Neurology, Neurosurgery and Psychiatry*, criticized, among other things, the "purely artistic" nature of the representation.[9]

It would be wrong, however, to regard the two views on the homunculus simply as an expression of the divide between professional and popular science. For we can trace the ambivalence back to Penfield's own work. This becomes evident from Penfield's response to an attack on the homunculus from British neurologist Sir Francis Walshe, in a correspondence just after the war.[10] Walshe had called the homunculus a "rather deceptive monstrosity."[11] He disputed the relation between function and cortical size that he thought it implied. In particular, he criticized the fact that on the figure the index finger was exactly the same size as the little finger, which to him seemed unlikely, given the relative importance of those fingers. He wrote: "I surmise that you will be heartily tired of its horrific appearance—copied uncritically from text to text—before you see the last of it. You may even have to slay it yourself—an infanticide that might find extenuation."[12]

Penfield refrained from infanticide for another decade. From its first appearance in 1937 to Penfield and Jasper's major epilepsy book of 1954, the figure was a mainstay in Penfield's publications. In his correspondence with Walshe, Penfield blamed popular opinion for the stay of execution. The homunculus was, he said, "the only sort of thing that people in general understand. I would gladly kill the damn thing if I could, but that is never possible." But such an explanation does not account for the work and effort Penfield invested in the figure. The homunculus was not simply "copied uncritically from text to text." Over the years, Penfield presented the homunculus in ever new variations: the detached homunculus of his 1937 paper, a second edition with the figure spread out over the brain, and images featuring multiple cortical homunculi, the second sensory and supplementary motor cortex, and even a thalamic homunculus (see fig. 6.3 for some of these homunculi).[13]

Moreover, over the course of his career, Penfield provided a range of explanations for the continued presence of the figure. In his correspondence with Walshe, when the proportions of the figure were in question, Penfield de-emphasized their importance. At this time, he explained the value of the homunculus by pointing out how it called "attention to certain facts, such as the reversal of order of representation in the face and neck, as compared with the rest of the body."[14] At other times, it was precisely these proportions that Penfield foregrounded. In 1937 he emphasized that the homunculus illustrated the "size and sequence of the cortical areas," that is, both order *and* proportion.[15] It was because of the value he placed on the homunculus's proportions that Penfield, in response to Walshe's criticisms, revised the figure. In his 1950 book, he admitted to "minor inaccuracies now apparent in this figurine" and commissioned a professional artist, Hortense Cantlie, who had drawn the first version of the homunculus, to produce a second edition (fig. 6.2). Penfield was thus simultaneously deeply invested in the homunculus and seemingly at a loss for a workable justification of it.

However ad hoc his justifications for the figure might seem, the figure itself was not an ad hoc construction. Penfield's conflicted relationship to the homunculus stemmed from his surgical practice. As we will see, in his operations Penfield produced large quantities of data: sensory and motor responses to electrical stimulation of the cortex. This data needed to be organized and made meaningful. Penfield provided the order through a process of mapping that strongly favored a project like the homunculus. In addition, the experimental setup helps explain the constant temptation, both for Penfield and later for other writers, to read the homunculus as an expression of human experience. While the homunculus does not depict body parts in proportion to their prominence in our experience, because Penfield mapped the sensory cortex based upon the patient's subjective response to stimulation, experience played an irreducible part in its construction.[16]

Penfield's work can be subdivided into three major phases: his early training and research visit to Breslau in 1928; his work in Montreal from

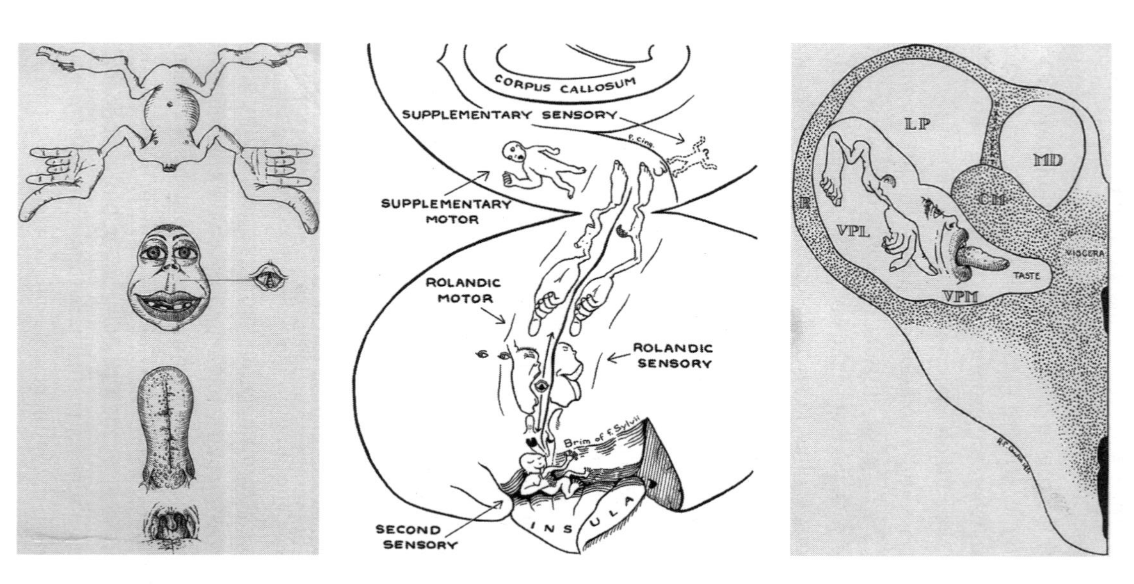

FIGURE 6.3. More homunculi (*left to right*: 1937, 1954, and 1954). (Penfield and Boldrey, "Somatic motor and sensory representation," 432; Penfield and Jasper, *Epilepsy*, 105 and 159. Reproduced by permission of the Osler Library of the History of Medicine, McGill University.)

1929 to 1945; and the period beginning at the end of World War II, when we see a rapprochement with the mind sciences. In this chapter I will focus mostly on the middle period. It was during this time that Penfield reoriented his work from its previous focus on epileptic therapy toward a broader brain-mapping project, laid down the institutional foundations for his research, and developed the experimental and notational techniques that provided the basis for the homunculus and his later turn to the study of "psychical" functions. But to understand this development, we first have to turn to Penfield's education and to the set of concerns that led him in 1928 to study under Foerster in Breslau.

Penfield in Breslau

Penfield's medical education was remarkably itinerant, crossing national and disciplinary lines. Most important for our purposes, it straddled and yet struggled to reconcile the clinic and the laboratory. Alongside Penfield's medical training—an MD from Johns Hopkins in 1918, a surgical internship with Harvey Cushing at the Peter Bent and Brigham Hospital in Boston, employment in New York at the Presbyterian Hospital with the surgeon Allan Whipple from 1921—he developed a research career. As a Rhodes scholar from 1914, Penfield worked at the neurophysiologist Charles Sherrington's laboratory in Oxford. He returned there after his surgical internship for a further year, and while in New York, he founded the Laboratory of Neurocytology in 1924. The same year, he took five months off from his New York position to learn Pío del Río Hortega's histological staining techniques in Madrid.[17]

Penfield's clinical and laboratory work at that time were not obviously related; his research in neuropathology had been focused on the nervous system of animals, not humans, and his surgical training at the Presbyterian Hospital, at least initially, had not emphasized surgery on the brain.[18] And yet, Penfield increasingly aimed to mobilize this scientific expertise for his clinical work. In his animal histopathological studies, Penfield had examined the process of brain scarring.[19] This research, he reasoned, had great therapeutic potential. A chance to test this hypothesis arose in January 1928, two months before his trip to Breslau. Penfield performed a radical operation on the eighteen-year-old William Hamilton. A falling chimney had struck Hamilton three years earlier, leaving a scar on his right forehead and adhesions and other changes to his brain. He had suffered from severe epilepsy ever since. In removing the patient's right frontal lobe, Penfield aimed to translate his "research conclusions into therapeutic action."[20]

The intervention was a success and freed the patient from his attacks, but the novelty of the technique left Penfield uneasy.[21] He did not feel prepared to repeat the procedure on a large scale. The problem was compounded by the situation of neurosurgery in America. At the time, it was still a young field. Although the profession found some institutional structure in organizations such as the Society of Neurological Surgeons and the Harvey Cushing Society, founded in 1920 and 1932, respectively, it was not until after World War II—with the creation of specialist boards, training programs, and neurosurgical journals—that neurosurgery became firmly established within the American medical landscape.[22] Surgeons from whom one could learn the techniques of brain surgery, beyond more standard procedures such as the treatment of tumors, were particularly hard to find. To gain such clinical experience, Penfield looked to Europe for inspiration, and the work of Otfrid Foerster seemed especially relevant. Funded by the Rockefeller Foundation, Penfield took his wife and four children to Breslau so he could work with Foerster for a six-month period in 1928.

As we have seen, Foerster had an excellent reputation as a clinician, and this was foremost in Penfield's mind while preparing for the trip. In his travel report for the Rockefeller Foundation, Penfield asserted, "Foerster's clinic is above all a clinic in which therapy takes first place."[23] The clinical emphasis set Foerster apart from most of his contemporaries, such as Max Nonne at the Eppendorf Hospital in Hamburg. As Penfield noted, Nonne's was "essentially a diagnostic clinic. There is here little or no therapy . . . the clinic, though admirable for the study of clinical signs and syndromes, accomplished little except to give rest to the weary."[24] Crucially, Foerster practiced neurosurgery, and it was at the heart of his work in Breslau. More specifically, Penfield knew that Foerster performed operations on the brains of epileptic patients that were close to the type he had performed on Hamilton.[25] While Penfield hoped to gain clinical expertise from his time in Breslau, Foerster saw the visit as a chance to gain greater expertise in neuropathology.[26] Foerster prepared a small microscopy laboratory for Penfield, in a corner of ward 8 of the Wenzel-Hancke hospital, to perform his histological studies.

The cooperation between Foerster and Penfield led to two joint papers, published in 1930 and 1931, which combined Penfield's histological understanding with Foerster's surgical know-how. In these papers, the two men developed their approach to the surgical treatment of epilepsy. All of the twelve patients that Foerster and Penfield discussed in their 1930 paper suffered from posttraumatic epilepsy: nine gunshot wounds, two birth defects, and one trauma caused by falling down stairs. In each case the injuries caused the brain tissue to form a scar, consisting of connective tissue, astrocytes, and blood vessels. Because such scars often contracted over time, they exerted a pull on the brain, which in turn could cause epilepsy. Clean excision of the scarred area would remove the epileptogenic focus without leaving another scar—Foerster and Penfield came to this conclusion independently before their encounter—thus freeing the patient from his epilepsy.

Penfield's operations notebook sheds light on his stay in Breslau. While his responsibility in the exchange with Foerster was for histopathology, his jottings demonstrate that he took a great interest in the process as a whole. In the notebook, each case merited an extensive discussion, usually including a brief history of the case, a short description of the seizure pattern, and a detailed description of the scar and neuropathology. It was, however, a final element that had the greatest impact on Penfield's future work: in the majority of cases, he also noted down the results of electrical stimulation of the brain.[27]

The Early Years in Montreal: A Combined Clinical and Research Hospital

When Penfield returned to North America, accepting a position as neurosurgeon at McGill, the skills he had learned in Breslau became central to his developing practice. The practice, however, was refigured in the institutional ecosphere that he constructed for himself in Montreal. Penfield outlined his plans in a letter to Edward Archibald, a thoracic surgeon at McGill who was instrumental in hiring Penfield. Penfield's "requirements for neuro-surgery at McGill" pertained to both the clinical and the research side of his appointment. On the clinical side, he demanded association with both the Royal Victoria and the Montreal General Hospital and, for example, asked for a room for perimetric and other clinical examinations. As for his research requirements, Penfield specified the creation of a laboratory of neuropathology, which, as he outlined, should be equipped with four to five small rooms, a secretary, and a technician.[28]

After his arrival in Montreal, Penfield did not lose sight of his plans. Five months after taking up his position, in January 1929, Penfield wrote another letter to Archibald, outlining his plan for a new "Institute for Neurological Investigation," which included a sketch of the planned building and a detailed outline of costs: he suggested the construction of a forty-bed hospital with seventeen laboratory rooms on seven floors in a new building attached to the Royal Victoria Hospital.[29] After an initial unsuccessful application to the Rockefeller Foundation in 1929, he gained funding two years later.[30] The foundation approved a grant of $1,282,652 to McGill University for the building of the new neurological institute and for the development of teaching and research. Of the sum, $1,000,000 was an endowment for the support of brain research; the rest of the grant covered building costs, along with $20,000 a year given by the province of Quebec and $15,000 a year given by the city of Montreal to cover the institute's running costs.[31]

In Penfield's institute, the link between clinic and research was instantiated in the close connection between operating theater and laboratory. A symbol of the exchange was the bell that Penfield had installed in the operating room to ring in the laboratory so that the fellows working there would be notified "when a lesion is completely exposed" to come to the operating room and observe.[32] We have traveled a long way in this story from Meynert's morgue in Vienna, but the bell is a reminder that we are dealing with the same desire to reconcile clinical and scientific approaches.

Penfield's research, and especially his mapping project, was structured by the investigation of epilepsy, in which, as William Feindel has pointed out, "the fields of cellular pathology, radiology, EEG, surgical physiology, neurochemistry and neuropsychology were all applied to advance the understanding of the pathophysiological process underlying focal seizures."[33] At Penfield's institute, epilepsy became a unifying, truly transdisciplinary disease. In the early years, however, only a minority of Penfield's patients were epileptics. For example, as we can see from Penfield's first "Report on Neurological Surgery" at the Royal Victoria Hospital from September 9 to December 31, 1928, he treated only three patients with traumatic or Jacksonian epilepsy in the period; the largest number of patients presented fractured skulls (13), tumors (12), and other conditions, such as hydrocephalus or even neurasthenia.[34] As Penfield's correspondence from the period suggests, he placed great significance on building up relationships with doctors, both English- and French-speaking, in order to create the referral network necessary for his project.[35]

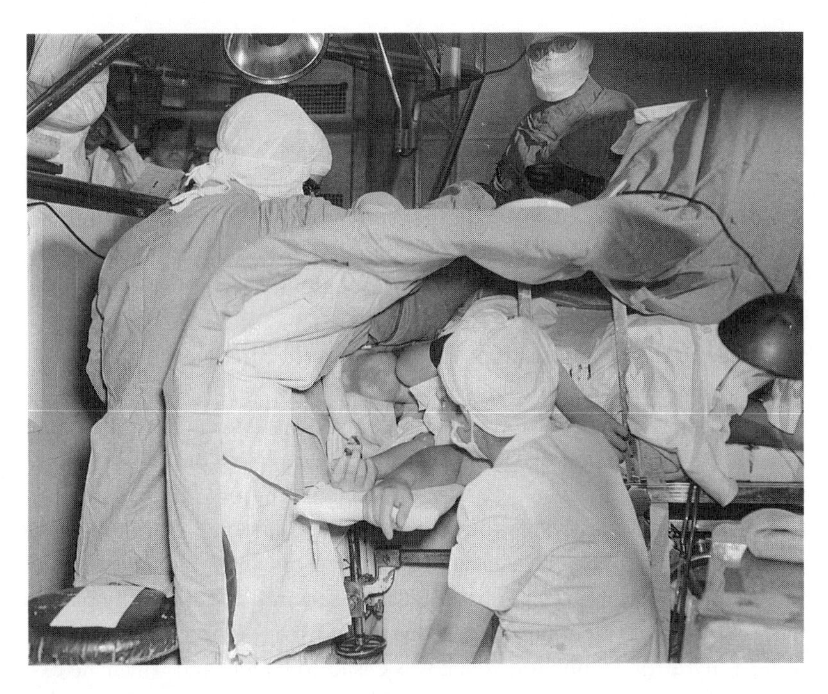

FIGURE 6.4. Operating room setup. (Courtesy of the Penfield Archive, Montreal Neurological Institute, McGill University.)

The Emergence of Penfield's Mapping Project

The shifting balance between Penfield's research and clinical duties had an impact on his surgical practice. In the Montreal context, the operation was a large and complex procedure, involving a number of people (fig. 6.4). An anesthetist (or sometimes a nurse) sat with the patient underneath a "tent" of surgical sheets held by a steel frame, to observe and comfort her if necessary.[36] In addition, the anesthetist had to be ready at all times to administer intravenous therapy or general anesthesia when needed.[37] In a neighboring room, divided from the operating theater by a glass wall, sat a stenographer, to whom the surgeon would dictate his insights during the operation through a microphone placed behind his back. An electroencephalographer with his equipment sat in a nearby chamber. He was able to see the operation site, and surgeons could communicate with him via microphone, as well as see the printed EEG record through the glass wall. EEGs were run in addition to stimulation before and during the operation to help locate epileptogenic areas.[38] A photographer was placed behind the glass wall, too. He controlled a mirror above the patient's head through which pictures were taken, a camera, and a synchronized photographic light. The surgeon and one or two assistants completed the roster of those in the room.

The goal of the operation was to surgically remove an "epileptogenic focus," the area of brain tissue that caused the patient's seizures, often the result of some previous damage. As Penfield pointed out throughout his writings, the operations—usually several hours long and tiring for both patient and surgeon—were not without danger, but often the benefits outweighed the risks. About half of his patients were cured fully of their epilepsy, while the great majority improved.

A considerable proportion of the operation was dedicated to the creation of a stimulation report. Initially (from the beginning of the operations in 1928 until 1934), Penfield regularly used two different kinds of current for the stimulation: galvanic (direct current) and faradic (alternating current), the first to map out the localization of sensory and motor function on the brain, the second to elicit light epileptic seizures that would help confirm the focus of offending brain tissue.[39] After 1935 Penfield switched to using a thyratron stimulator, which was easier to handle and which, he explained, was "now in use in physiological laboratories."[40] To make sure that stimulation in fact did take place, "the stimulating circuit is tested by means of one half watt neon lamp," which, if the circuit was intact, "glows red."[41]

The stimulation report comprised two elements: a brain map and a stimulation protocol. The surgeon would stimulate parts of the brain, which would be recorded with a number on the map, and the numbered response elicited would be noted in the patient's file:

14. Tingling from the knee down to the right foot, no numbness.
13. Numbness all down the right leg, did not include the foot.
12. Numbness over the wrist, lower border, right side.
11. Numbness in the right shoulder. . . .
(G) Flexion of knee.
18. Slight twitching of arm and hand like a shock, and felt as if he wanted to move them.
2. Shrugged shoulders upwards; did not feel like an attack.
(H) Clonic movement of right arm, shoulders, forearm, no movement of trunk.
(A) Extreme flexion of wrist, elbow and hand.[42]

As Penfield pointed out, "any questionable response should be checked by later restimulations," often without warning the patient, or by falsely announcing that the stimulation was to be carried out. All reliable responses had to be verified in this way. Then, said Penfield, "it [was] possible very quickly to be certain of the evidence."[43]

As initially conceived, and in line with Foerster's practice in Breslau, the stimulations oriented the doctor on the brain and guided his scalpel. The process focused on the individual case and recorded the particular responses gained after each stimulation. Over time, however, Penfield began to see in the stimulation results the possibility of a broader project, aimed at providing a generalized map of the cortex, detailing which areas were dedicated to each body part. Eventually this mapping project came to overshadow any utility the stimulations had in the particular epilepsy operation. This is not to say that Penfield lost sight of the clinical dimension of his operations. But, for him, the mapping possibilities of the procedure became increasingly central.

The changing function of the stimulation process can be seen in the prominence it was afforded in Penfield's protocols. In the early 1930s (1930–32), Penfield's practice followed Foerster's. Because the stimulations were simply intended to guide the doctor in the surgery, the stimulation protocol was provisional and dispensable, scribbled in pencil on the same sheet as hand drawings of the brain tissue. In the typed-up operation report, they were usually mentioned briefly in the "Procedure" section, though rarely discussed in detail. At times the simple statement "motor area was mapped out" sufficed.[44] Once they had played their role in locating the epileptic focus—a purely clinical task—they were no longer of interest. As the 1930s wore on, the stimulation reports appeared more regularly in the larger protocol, but their status remained ambiguous. Until 1935 they were sometimes listed under "Objective Findings" (a section in which adhesions of the meninges, the vascular situation, and any visible changes to the brain were discussed); sometimes under "Procedure" (the section that outlined the different steps of the operation); sometimes they had their own section, though that section did not have any fixed position within the protocol as a whole; and sometimes they did not make it into the main (typed) section of the protocol at all. Only from 1935 on were the stimulation reports firmly established in their own section (first called "Electrical Exploration," then, after 1936, "Stimulation Record"),[45] usually at the very end of the operation report.[46] And by this time, they had come to dominate Penfield's operation protocols. Eventually, "stimulation report" became a shorthand for the entire document.[47]

As the stimulation report became more prominent in Penfield's published work, its form changed (see fig. 6.5). As we saw, while in Breslau

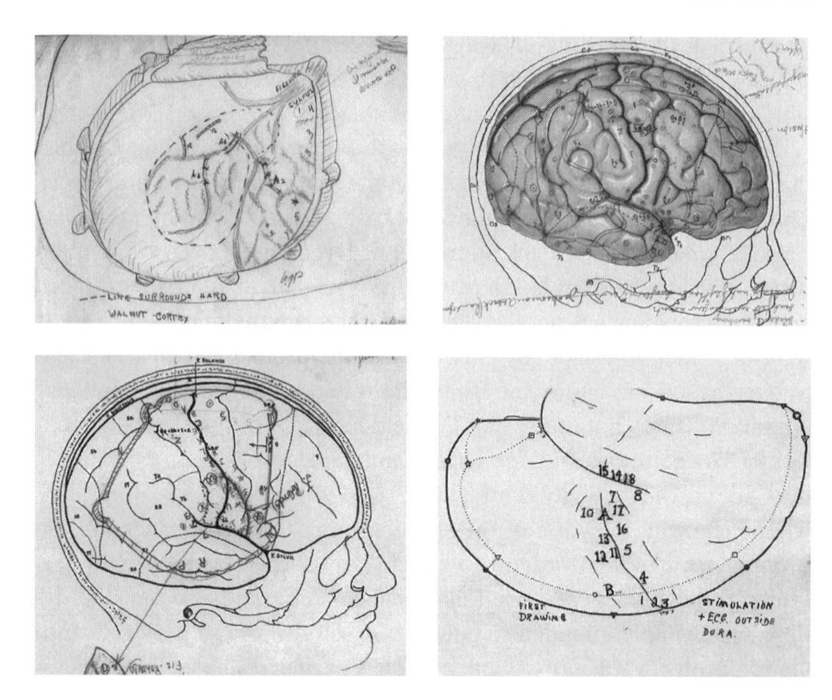

FIGURE 6.5. Brain charts, 1930–53. (Courtesy of the William Feindel Memorial Fund.)

Penfield largely worked as a neuropathologist, and in the early years back in America, his operation reports were dominated by characterizations of the tissue changes occurring in the course of epilepsy. In these reports, based on operations performed between 1930 and 1932, Penfield invariably included drawings of arteries and veins in the sketches accompanying his operation reports, carefully labeling them and sometimes pointing out their pathology, for example, "artery constriction."[48] Penfield sketched these simple pencil drawings on a white sheet of paper with a sterilized pen during the operation, based upon the open brain in front of him (fig. 6.5 top left). His charts containing the stimulation data were thus richly descriptive and highly particularized, recording the idiosyncratic shape and substance of each patient's brain.

Such a process of recording resisted any attempts to collate the data from multiple operations and construct generalized brain maps. Corresponding to Penfield's move toward brain mapping, from the mid 1930s we see a change in his operation records. Penfield's representations lost their rich anatomical detail—the veins and individual contours of the gyri disappeared—and became increasingly standardized. Later he preprinted a brain template on which to mark the stimulation points (the template underwent a number of iterations over time; see fig. 6.5). Though at this time he still occasionally described vascular changes in the protocols and included pencil drawings on his brain charts, Penfield usually left them out of his representations. Using the standardized template, it was simple to superimpose the results collected from numerous operations and build up a generalized map of the brain. In this way the homunculus was born.

Penfield and the Sensory-Motor Distinction

As the brain charts became more abstract, they looked less and less like the brain of the patient in front of him. Penfield had to find a way to relate the location of the stimulation points on the particular brain to the universal brain in his notes. And it is here that the central sulcus, a large fold on the brain surface, assumed a determinative function. Penfield commented on the process of producing a "standard human chart" of the brain in a 1937 article.[49] He would measure the distance of stimulation points from the three fissures (the Sylvian, median longitudinal, and, most importantly, the central sulcus). Because these fissures also appeared on the brain template, he could use this information to transfer the point to it. Data from both hemispheres were then summarized in a chart of the right hemisphere. As his notation showed, for Penfield the crucial piece of information was not the location of the stimulation point on the brain and its precise topographical relationship with other points, but rather its relationship with a limited number of fissures.[50] The importance of the central sulcus in Penfield's developing mapping project tracks the broader significance he gave to the sensory-motor distinction. Broadly speaking, sensory and motor responses were elicited from opposite sides of the central sulcus, and so early on, Penfield marked this anatomical structure as a dividing line between sensory and motor areas of the brain.[51]

In Penfield's practice, however, he confronted a number of challenges. First, especially because in the operation only a small part of the brain was exposed, it was not always obvious to Penfield which fissure was the central sulcus; so in his reports, he occasionally qualified the location through a question mark or drew the fissure and crossed it out again.[52] If the anatomical position was in doubt, the sensory and motor division of the stimulation responses took on a decisive role. In his early operations Penfield

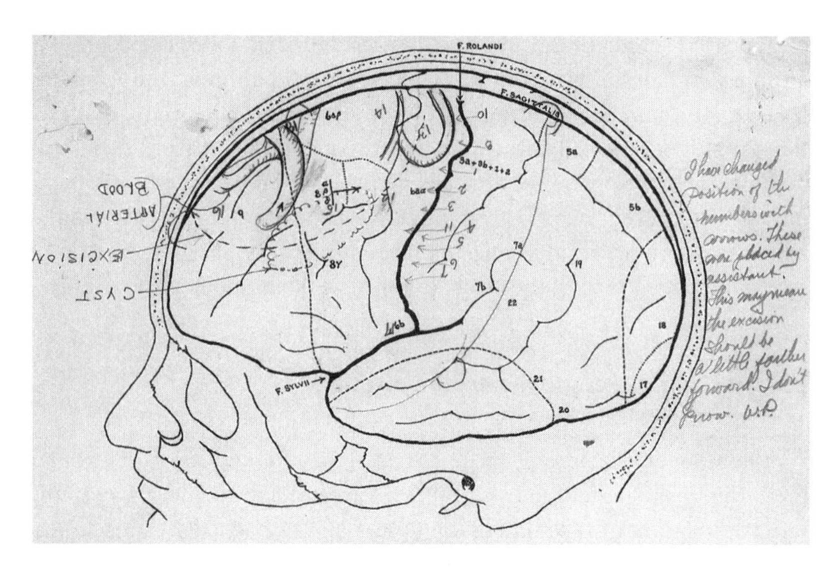

FIGURE 6.6. Brain chart of patient W.O., January 22, 1937. (Courtesy of the William Feindel Memorial Fund.)

would simply trust that a fissure was the central sulcus (or the fold he saw on the brain surface) if he had elicited motor and sensory responses from different sides.[53]

Stimulation results could thus cause Penfield to reassess his initial orientation. In 1937 he operated on W.O., who had a cerebral cicatrix and an arterio-venous aneurysm.[54] The typed protocol looked like any other produced around 1937. But looking at the brain chart (fig. 6.6), we see that Penfield made an interesting change to the recorded stimulation points on the brain surface. Stimulation points 1 to 11 (with the exception of 9, which had not been mapped) were all shifted toward the precentral side. The shift was indicated by arrows of about a half-inch in length. It is likely that the following happened: As Penfield mapped out the sensory strip of the brain, he obtained sensory results from stimulations 1 to 10, marking out the areas of arm and hand, face, and side of the body. When he moved to point 11, however, which was located just above point 3, that is, purportedly on the postcentral (sensory) side of the divide, he obtained a "marked twitching" at the "right side of face, mouth drawn downward, and slight phonation," that is, a motor response. He reasoned that point 11 was rather on the precentral side of the central sulcus,[55] and thus he moved point 11 over to the correct position on the brain template. Mak-

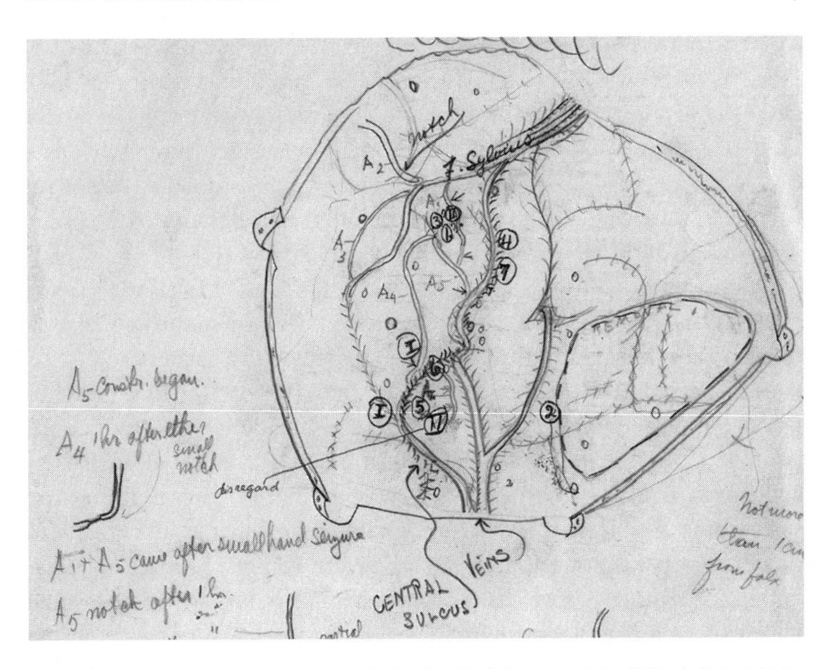

FIGURE 6.7. Brain chart of patient D.S., April 15, 1933. (Courtesy of the William Feindel Memorial Fund.)

ing up for the error systematically, he moved all stimulation points about the same distance.[56] Penfield did not comment on the change in the typed operation protocol. But he explained the arrows in a penciled note beside the operation chart: "I have changed position of the numbers with arrows. These were placed by assistant. This may mean the excision should be a little further forward. I don't know. W.P."

Second, it was not always clear that the division between sensory and motor responses, and thus the central sulcus, corresponded to a visible fissure in the brain. Take, for instance, the case of patient D.S. (1933) (fig. 6.7).[57] As usual, Penfield wanted to elicit a slight epileptic seizure so that he would be sure about the location of the epileptic focus that needed to be removed. Because the direct faradic stimulation of the motor area could lead to a powerful seizure, Penfield restricted himself to the stimulation of the sensory side of the central sulcus, which would normally produce only the first stage of the seizure.[58] For this reason the location of the central sulcus had great clinical significance. In D.S.'s case, Penfield performed a number of galvanic stimulations, ascertaining whether they elicited sensory or motor responses. Of the first seven (1 to 7) Penfield performed, six were motor: he received results from the mouth area (such as movements of swallowing or twitching of the mouth to a particular side), from the hand area (hand closure and extension), and eye movements. Only one stimulation (at point 3) resulted in a sensory response (sensation in the right side of the jaw and tongue), combined with a motor response (swallowing). Given these results and the topography of the brain, the location of the central sulcus seemed evident to Penfield.

When he moved from galvanic to faradic stimulation to elicit a seizure, however, the story became more complicated. He first stimulated an area of the postcentral (sensory) sulcus, which "gave rise to a small attack"—the patient's hand lifted, a movement that continued beyond the time of stimulation. For his second faradic stimulation, a stimulation of the mouth area, Penfield moved further down the central sulcus. He set his electrode exactly to the right of sensory point 3, that is just a little closer to the central sulcus, but still—as he believed, presumably following the topography of the brain—on the postcentral side of the divide. This first produced a sensation in the right jaw as well as chewing movements, just as Penfield would have expected from the location. Then, however, something unexpected happened. The stimulation "sent the patient off into a complete generalized seizure," after which the patient was comatose for a moment.[59]

On the basis of this result, it appears that Penfield went back to the operation drawing, (fig. 6.7)[60] and corrected the location of the point to move it over to the other side of the divide, a more appropriate location, given the fact that the stimulation elicited a full seizure. In his notes, Penfield remarked that while "working up" the central sulcus, he saw that it was "lying more posteriorly than we supposed."[61] In this case the distribution of sensory and motor responses was a better indicator of the location of the central sulcus than the physician's visual appraisal of the naked brain. It is not surprising, then, that later Penfield came to the conclusion that the central sulcus was "invisible."[62]

Third, increasingly, the stimulation data called into question the existence of a clear dividing line between the sensory and motor parts of the brain. In early cases there seemed to be a neat separation between the two, but as Penfield's work progressed, he remarked upon sensory responses from the motor side of the divide and vice versa.[63] At first he explained these anomalies by experimental malfunction, an error in the test rather than any real intermixing: early in his protocols, we see him referring to an "escape of current" to explain unexpected results.[64] In his 1937 paper with Edwin Boldrey, Penfield, in the historical section, took a more

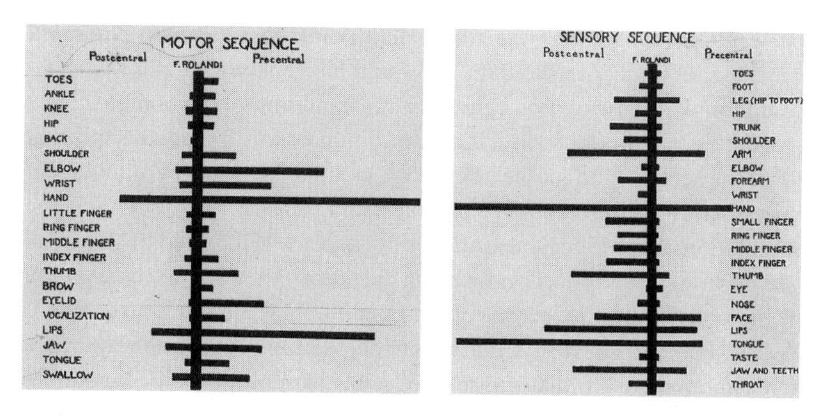

FIGURE 6.8. Proportional schema for motor and sensory responses. (Penfield and Boldrey, "Somatic motor and sensory representation," 431. Reproduced by permission of the Osler Library of the History of Medicine, McGill University.)

systematic approach to explaining outliers: he referred to the principles of "facilitation" (a response that was elicited by preliminary stimulation), "activation" (production of an "echo response" through stimulation in a different area), and "reversal" (a qualitative change of response, e.g., from flexion to extension) to explain unstable and changing responses to stimulation identified by Charles Sherrington and Graham Brown.[65] But by midcentury it had become clear to Penfield that the intermixing of sensory and motor stimulations was due to a systematic distribution. As he indicated proportionally in his schema (fig. 6.8), a certain percentage of sensory responses for each body part could be elicited from the motor side, and vice versa. Later, Penfield suggested the overall proportions to be 75:25 for somato-sensory responses and 80:20 for motor responses.[66]

For all three challenges, Penfield adopted strategies that allowed him to maintain the central sulcus's priority and to keep the distinction between sensory and motor responses as clean as possible. His efforts confirm the importance of the sensory-motor distinction in his work. In part that importance emerged from the experimental setup. The anatomical messiness of the sensory-motor distinction contrasted to its epistemological clarity. In the procedure, the difference between sensory and motor responses was marked socially. If the stimulation of the brain with the electrode elicited a sensation, it would be reported by the patient. If, on the other hand, stimulation provoked movement, this was noted by the observing anesthetist or nurse, who was a better witness in this matter. But as we have seen, that division was also necessary for Penfield's generalized mapping project. Only by relating the stimulation points to a common landmark was he able to collate the data and build his generalized maps. The assumption of a sensory-motor divide was a condition of the homunculus.

And as in Schilder's case, the assumption of this divide, the splitting up of the sensory and motor homunculi, created its own experimental demands. Only the introspective patient could provide sensory data. If the stimulation did produce a sensation, the patient was required to give a full and accurate description of it, paying particular attention to the location in the body where the sensation had been felt. Much later, after World War II, Penfield suggested that this process of introspection required a self-aware "mind," which was not reducible to sensory-motor function.[67] This "mind," Penfield asserted, was "aware of what is going on. The mind reasons and makes new decisions. It understands."[68]

Not only was the self-aware mind a condition of Penfield's experimental practice; it was also the product of the operational setup. Penfield went to great lengths to keep the patient conscious and lucid. Because the patient had to remain awake for the operation, Penfield did not use a general anesthetic.[69] Nevertheless, to prevent discomfort, Penfield used a local anesthetic, usually nupercaine with adrenaline.[70] Further, if intraoperative pain arose from meningeal arteries or tributary veins, or from traction on dural sinuses, Penfield managed it through injection of nupercaine into the relevant areas. Along with the anesthetist, Penfield monitored and managed pharmaceutically the patient's state of consciousness during the operation. Fine control of the patient's wakefulness was achieved with other substances: Penfield made frequent use of grains of codeine or caffeine, depending on the desired effect. In his operations, then, Penfield regulated the lucid and self-aware subject chemically.

For similar reasons, Penfield placed a priority on his "friendly relationship" with the patient.[71] The two could not see each other—the relevant area of the patient's brain was exposed to the physicians standing on the other side of the frame—but they talked to each other during the operation. The foundations for their rapport had been laid beforehand. On the day of the operation, Penfield took a careful personal history of the patient and examined him. This served a practical purpose, of course, but it also gave the patient and the doctor an opportunity to get to know each other. As Penfield noted, the surgeon "must take time for talk before and during the operation," in order to become "the patient's trusted friend."[72] The trust between doctor and patient was not always assured; Penfield asserted that children over ten "often make the best patients," because they were usually willing to cooperate and "make better witnesses than any but the most intelligent of adults—provided the operator is careful not to hurt them."[73]

At this early stage, Penfield was uninterested in higher functions. In 1936 he made "the prediction that at some time in the future much of the mental disease will be found to be due to true organic change primarily or secondarily within the brain."[74] For the time being, however, and certainly at the level of the clinic, he supported a clean division of psychiatry from the "neuro" disciplines.[75] Psychologist Donald Hebb's stay at the MNI in the late 1930s—he worked on brain function after surgical removals—was relatively short-lived.[76] But the fact that a self-aware patient was at the center of Penfield's operational practice meant that higher functions were never entirely alien to his practice, and as institutional conditions changed in Montreal, they moved to the center of his investigations.

Postwar Neurosurgery and the Turn to Psychology

Penfield's antipathies to psychological approaches were clear at the beginning of the war. For instance, in his 1942 article "Clinical Notes from a Trip to Great Britain," he challenged the reality of most "psychoneuroses," explaining them as the result of "malingering" or as a way for commanding officers to get rid of bad soldiers. The doctor should respond to these men's symptoms (diagnosed as "post-traumatic head syndrome" rather than the "shell shock" of the previous war) with "common sense methods and without resort to the complicated technics of psychoanalysis."[77]

Nevertheless, the experience of the war encouraged a shift in Penfield's position. The conflict afforded an expansion of Penfield's work both institutionally and disciplinarily.[78] Unlike the late 1920s and early 1930s, after Canada declared war on Germany on September 10, 1939, low patient numbers were no longer an issue for Penfield; rather, the institute was overflowing with the sick and wounded. The average daily patient population at the MNI increased from 48 patients in 1936 to 66 in 1942. Similarly, the number of operations had nearly doubled: 348 operations in 1935 and 700 in 1942. The MNI had "reached the breaking point" even before the "flow of wounded from overseas" began.[79] In this context, Penfield submitted an application to the Department of Pensions and National Health to build a military annex of thirty-two beds directly attached to the MNI. The application was accepted, and Penfield's new annex was completed in mid-1945; after the war it was integrated with little fuss into the MNI.[80]

As the institution expanded, so did its disciplinary ambitions. The war occasioned, according to Penfield, a "voluntary reorientation" of their research program, building on principles and skills already in place. The team of doctors and researchers at the institute worked on projects such as the study of motion sickness and the neurophysiology of acceleration to explain why pilots experienced "blackouts."[81] This did not mean that the work on epilepsy was discontinued. Operation numbers did not drop in the first year of war; they remained at a somewhat lower, but not drastically reduced, level for its duration.[82] Further, the research on epilepsy was reenergized by a higher number of patients with posttraumatic epilepsy, caused by the hostilities.[83] In the early postwar period, the MNI had more patients than ever and could claim competence in an ever-expanding range of pathologies.

Researchers at the institute also turned their attention toward psychology, a development encouraged by the Rockefeller Foundation. Though the director of the Rockefeller Foundation's Division of Medical Sciences, Alan Gregg, had originally supported Penfield's Montreal Neurological Institute because it was a project "of the strictest possible 'real' scientific focus," he also had a long-standing interest in psychiatry and related fields. Indeed, Gregg, during his tenure, allocated about two-thirds of all expenditures to psychological projects;[84] he was involved, along with others, in the funding of Stanley Cobb's Department of Psychiatry at the Massachusetts General Hospital, a department favorable to psychoanalysis with several psychoanalysts among its staff, as well as the Chicago Institute for Psychoanalysis, headed by the émigré psychoanalyst Franz Alexander.[85]

Gregg's orientation was eminently clear to Penfield, who discussed it at length in the 1967 book *The Difficult Art of Giving: The Epic of Alan Gregg*.[86] As Penfield pointed out, Gregg "aimed his research support at the circle that surrounded psychiatry," giving money to psychiatry's "sister specialties—neurophysiology, neurology and psychology" in what he called his "strategic plan." The MNI was part of this plan, and Gregg's pressure meant that the separation of the "psy" from the "neuro" disciplines there would not be permanent. In 1950 Brenda Milner, a young psychologist from England, arrived at the institute. Focusing on cognition and memory research, Milner joined the MNI as a doctoral student under the neuropsychologist Donald Hebb. Having completed graduate work in Cambridge with Frederic Bartlett before moving to Montreal with her husband, she was interested in cognition and memory, which she began to study in Penfield's patients.

With Milner's work, neuropsychology became firmly established within the larger research program at the MNI. The new prominence of the field had more than a little to do with Milner's work on H.M., probably the most famous patient in the history of neuroscience, whose memory and cognitive function she systematically studied and who became the textbook case demonstrating hippocampal function.[87] H.M. suffered from severe loss of long-term memory after the removal of large portions of both temporal lobes in an epilepsy operation conducted by Connecticut doctor William Beecher Scoville.[88]

Moreover, larger shifts within medicine made the rapprochement seem more likely. The newly confident and expansive MNI participated in a broader development in the somatic mind sciences after the war. The 1950s saw research into learning, stress, and sleep (with the first description of REM in 1953), alongside discoveries in cell physiology such as synaptic excitation or the mechanism of the action potential and research into signaling molecules such as nervous growth factor and various neurotransmitters including GABA and serotonin.[89]

These developments had a number of effects. Not least, they encouraged a development of psychiatry that aligned it more with Penfield's project. With the 1952 introduction of chlorpromazine, the first antipsychotic drug, a new somaticism was ascendant in psychiatry.[90] Not only did the development of the "major tranquilizers" lead to great restructurings in the care of psychiatric patients—de-institutionalization and so forth—it also set off a whole series of discoveries of additional psychoactive drugs, most importantly, antidepressants. This opened up the possibility for a reconceptualization of psychiatry as a biological science, with treatment acting upon the brain.

Montreal was, in fact, a site of much research activity along these lines. Donald Hebb, who had worked at the MNI in the 1930s, had published in 1949 his groundbreaking *The Organization of Behavior*, in which he developed his theory of learning. Moreover, Heinz Lehmann at the Allan Memorial Institute, the psychiatry department at McGill, contributed the first research publications on chlorpromazine in 1953 and, three years later, on the antidepressant imipramine.[91]

Recasting the Voice: Penfield, Speech, and Aphasia

The 1950s thus saw the emergence of a self-confident psychiatry, recast for neurological appropriation. In this environment, Penfield too gradually turned his attention to higher functions, moving beyond the cortical mapping that had, until then, made his name. As we have seen, Penfield's surgical practice presented him with sensory and motor responses. We saw how hard Penfield worked to maintain a clear distinction between them, finding means to preserve the integrity of the central sulcus as a dividing line between sensory and motor areas, in order to match the clarity of the distinction in his clinical practice. The sharp division between motor and sensory responses conditioned the surgeon's relationship to his patients. Motor responses were unproblematic: the anesthetist under the tent with the patient would observe any movement and describe it to the operating surgeon. Sensory responses were more difficult to deal with, since they could not be objectively observed; only the patient had access to them. For this reason, the patient's means of communicating became crucial to the endeavor. Penfield had to rely on the patient's voice as an unproblematic and transparent medium, capable of communicating internal experiences without distortion.

This understanding of the voice was strongly shaken in 1935, when Penfield first produced a vocalization response: "When area 5 (marked 5 on the drawing) was stimulated the patient called out "Oh" in a somewhat groaning tone. This was definitely involuntary vocalization. . . . In all the area was stimulated thirty-one times, each time producing vocalization. . . . So far as I know this is the first example of true vocalization as the result of electrical stimulation of the human brain."[92] Penfield's excited response (repeated in the published version) was an anomaly; rarely did he make claims about the priority of obtained responses in a stimulation report. But we can trace it to Penfield's relief in being able to categorize the vocalization as a motor response, even though the sound, "oh," might have been considered willed. In this case, the vocalization showed itself to be a purely motor phenomenon because it could be reproduced reliably on stimulation (thirty-one times). As Penfield's practice developed, he formulated more criteria: vocalization and word arrest (a second kind of motor speech) could be elicited from both hemispheres. They did not follow the rule of "ideational speech," which was to be located on only one hemisphere—usually the left for right-handed people.[93] Finally, Penfield referred to studies according to which vocalization could be produced in animals as well, with dogs barking and cats purring upon electrical cortical stimulation.[94]

As we have seen, in his surgical practice, Penfield read the patient's words as an expression of an internal sensory state. They could not be seen as motor responses to electrical stimulation without corrupting the sensory-motor distinction that was essential for his research program and mapping project. This was why the study of vocalization was so important to Penfield. It allowed him to relegate speech distortions to the realm of motor phenomena and thus cordon them off from the meaning-communicating function of speech. It was through such "ideational," or later "psychical" (read: nonmotor) speech that patients relayed to Penfield the results of his sensory cortex stimulation.[95] Psychical, unlike motor speech, transmitted the self-reflective experience of the patient; it was a revelation of the mind.

The twinning of motor and ideational speech was mirrored in another phenomenon elicited by Penfield's practice: what he called "double consciousness." In the electrical stimulation of the motor cortex, the patient experienced her responses as in some way foreign: "When the motor convolution is stimulated, the patient may be astonished to discover that he is moving his arm or leg. He may be surprised to hear himself vocalizing, but he never has the impression that he has willed himself to do those things."[96] The same was true for sensory responses. Whether they were somato-sensory, auditory, or visual, the patient always insisted on the crude and elementary character of the sensation: "He is never under the impression that he has touched an external object. He considers it an artifact, not an ordinary sensation."[97] Just as motor vocalization created a division between the acting body and the passive mind, in double consciousness the patient felt alien from the stimulated part of the brain. At such times, Penfield later wrote, "the patient's mind, which is considering the situation in such an aloof and critical manner, can only be something quite apart from neuronal reflex action."[98] The experience of "double consciousness" suggested that the "mind" remained out of reach for Penfield's electrode.

It is then no coincidence that at about the same time that Penfield identified vocalization, he also started to turn his attention away from the sensory-motor cortex; the separation of motor speech (controlled in the cortex) from psychical speech, suggested that the latter would have its own location. Penfield's earlier investigations had brought him only to the threshold of mind. To locate that mind, he would have to look elsewhere.

From the beginning of the postwar period, Penfield expanded the scope of his stimulations. Figure 6.9, from Penfield and Rasmussen's 1950 *The Cerebral Cortex of Man*, outlines the different cortical areas in which stimulation interferes with speech or produces vocalization.[99] In the beginning, Penfield focused on stimulating the motor area on the precentral gyrus (or the area just anterior to it). For example, in the early 1930s, Pen-

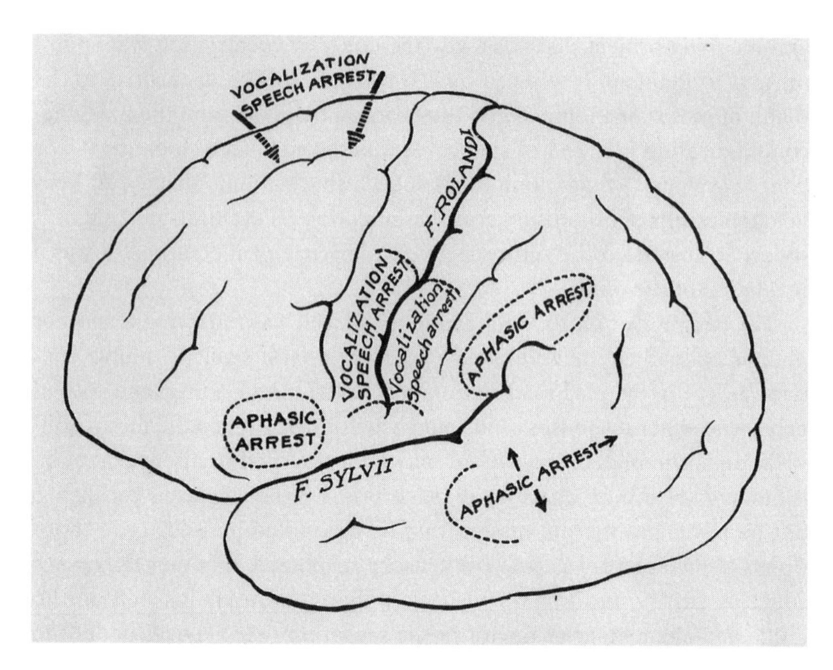

FIGURE 6.9. Cortical areas whose stimulation affects speech. (Penfield and Rasmussen, *Cerebral Cortex of Man*, 107. Reproduced by permission of the Osler Library of the History of Medicine, McGill University.)

field stimulated points on the precentral gyrus almost exclusively.[100] Starting in 1935, but on a regular basis only after World War II, Penfield turned his attention to other speech areas.[101] For instance, he began to outline the superior frontal area, another motor speech area.[102] But from the late 1940s, he often completely neglected to explore the precentral gyrus.[103] Instead he focused directly on mapping "psychical" speech areas, such as Broca's area,[104] the parietal aphasic arrest area, and the temporal aphasic arrest area.[105]

In Penfield's explorations of these areas, he elicited a set of responses that was qualitatively different from his stimulations of the sensory-motor cortex. He recalled later, "I was more astonished, each time my electrode brought forth such a response. How could it be? This had to do with the mind!"[106] These could no longer be understood in sensory-motor terms. As he wrote in 1959, it made sense to call the new areas the "interpretive cortex to distinguish them from sensory and motor areas."[107]

The complex and intriguing nature of the responses can be seen in the following case. In early February 1951, Penfield operated upon a young

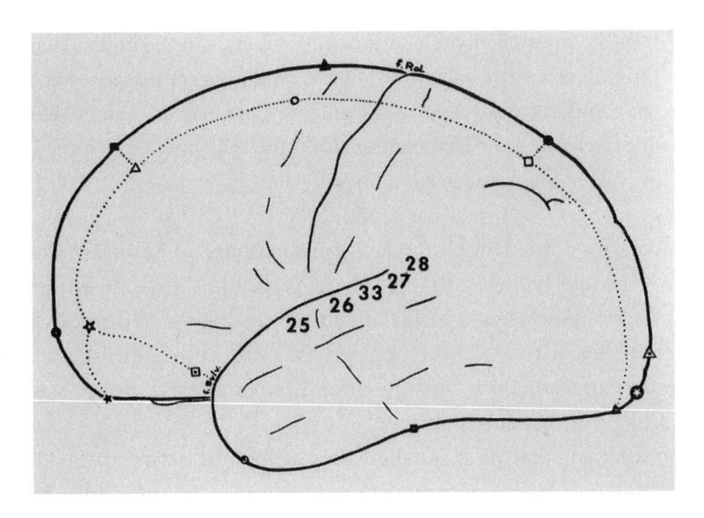

FIGURE 6.10. Evidence of "psychical hallucinations." (Penfield and Perot, "Brain's record," 626. Reproduced by permission of the Osler Library of the History of Medicine, McGill University.)

woman. Y.N. had begun to have seizures four years prior to the operation, at age twenty.[108] In her seizures, she would habitually experience a sensation in her back, combined auditory and visual hallucinations, and a feeling of fear, as well as a generalized motor seizure. More specifically, she would hear the voice of a man talking to her, which frightened her considerably. Electrical stimulation would make the patient undergo a similar experience; in fact, this was the earliest case I could find in which Penfield reported "psychical hallucinations" after stimulation. It is not surprising, then, that Penfield included the case of Y.N. in his 1963 summary paper discussing psychical phenomena.[109] In the paper he also included a brain chart, which clearly outlined the location of five points of stimulation from which psychical responses were elicited (fig. 6.10):

26. "I could hear someone talking, murmuring or something."
25. "There was talking or murmuring, but I cannot understand it." Apparently this resembled the voice she heard before her attacks.
27. Silence, then a loud cry. "I heard the voices and then I screamed. I had a feeling all over."
28. "I hear something." She did not see anything. The electrode was kept in place. When asked what she heard, she said, "A buzzing-like." Then she added "A man buzzing or murmuring." When asked where he was she said, "In the back."

27. Repeated thirteen minutes after last stimulation at this point. After a time the patient said that she was trying to hear what someone was saying, but she could not make it out. She said it was one man doing the talking.
33. She began to sob, "That man's voice again! The only thing I know is that my father frightens me a lot."[110]

Drawing on John Hughlings Jackson's account of "dreamy states" in epilepsy, Penfield believed that psychical responses could be either experiential hallucinations, in which the subject "re-lives a period of the past although he is still aware of the present," or interpretive illusions, in which the person experienced a "misinterpretation or altered interpretation of the meaning of present experience."[111]

In parallel to what he has called the distinction between the "motor" and "ideational" aspects of speech, Penfield figured the difference between sensory and psychical responses as that between sensation and perception: the latter involved "interpretation." He wrote: "Interpretation, as I am using the term, is part of the process that converts sensation into perception. It represents a further stage in the total integrative action of the brain."[112] In his "psychical" responses, it appeared that Penfield had discovered a deeper and more foundational mechanism. We should not be surprised to see in Penfield's more philosophical musings, just as with Schilder, a reassertion of "mind." As Penfield's psychical explorations—and his lack of interest in the work of most psychiatrists—demonstrate, this mind was to be approached somatically, located in different parts of the brain, where the simple data of sensation were brought together and combined in the complex process of perception. But it is also clear, as Penfield himself elaborated, that he maintained a dualistic philosophy, sharply distinguishing mind and body.[113] Unlike the sensations of the cortex, which were always experienced as in a way exterior to the mind, other brain structures might present the mind as it was.

Rehabilitating the Homunculus

The two main developments I have outlined in this chapter provide a frame for understanding the continuing attraction of the homunculus in Penfield's work, and they demonstrate that it was closely integrated into his central practice of brain stimulations.

First we see a progressive dematerialization of his brain maps, which led in the direction of the cartoonlike homunculus and allowed the collating of data that helped construct it. This dematerialization tracks a growing emphasis on the stimulation reports in Penfield's work and a revalorization of Penfield's mapping project over the clinic. Second, we see an increasing interest in the psychical, an interest emerging out of Penfield's investigation of the voice, which was integral to his sensory-motor exploration but which encouraged the positing and then thorough investigation of different "psychical" areas. Penfield's interest in the psychical was consonant with the assumption that it was possible to assess the patient's experience, an experience that the patient would be able to communicate freely and without distortion.

Strictly speaking, the homunculus only measured the average cortex area devoted to a body part, and cortical proportions did not necessarily give insight into human experience. But we can see why this was a tempting and plausible reading. As the point at which the reflex arc had been cut, the area mapped by the homunculus could easily come to be seen as some kind of window into the psychical realm. At the sensory cortex, one seemed to be investigating the boundary between body and mind; bypassing the nerves linking the peripheral sense organs to the brain, here one could affect it directly and hear the thinking, self-reflexive patient respond. Indeed, as I have tried to suggest in this chapter, the means by which Penfield gathered the data necessary for the sensory homunculus required him to posit the existence of a "mind" that could respond to the stimulation. Experimentally at least, the sensory cortex was an experiential construct. For this reason, the experiential and psychical reading of the homunculus was not a later anomaly, which we can disregard as simply the misinterpretation by later museum curators and popular science writers. Rather, it was encouraged by Penfield's own experimental practice.

A look at Penfield's practice also provides a perspective on his unease with the homunculus and helps explain why in his later work he continued to reproduce and rework it. The (sensory) homunculus provided meaningful information to the extent that it could be seen as lying at the border between mind and body. But as Penfield expanded his search beyond the sensory-motor cortex in the search of the mind, the mind remained elusive and retreated ever deeper into the brain. Though he had initially treated the temporal lobe and the "psychical responses" elicited there with great excitement, toward the end of his career he no longer considered it the site of the mind.

Penfield had been using a simple brain map since 1940, but in the postwar period he began to explore beyond its preset limits, and we see the introduction of brain maps that charted additional areas in 1949 and 1951, respectively—see figure 6.11 for additionally drawn "folds" of the inner part of the hemisphere. Figure 6.11 (top) shows real, paper folds that included the insular region of the temporal lobe. For our purposes, figure 6.11 (bottom) is especially interesting, because it represents an enlargement of the area of the temporal lobe. Penfield began to include results from deep stimulation—in which he would insert the electrode several centimeters into the brain, along the folds of the island of Reil—in his stimulation reports and would often obtain psychical responses from these areas.[114] In 1952, just as he was publishing his findings about these "psychical" areas, Penfield suggested another location for the mind, moving deeper into the brain, what he termed the "centrencephalic system," a neuronal structure in the diencephalon, or higher brain stem.[115]

But no matter how deep into the brain he probed, Penfield remained unsatisfied. He was unable to escape the phenomenon of double consciousness, which, as before, indicated to him that he remained outside of the mind. In the stimulation of speech and "psychical" areas, the patient reported a "doubling of awareness."[116] The patient recognized the stimulated memories for what they were and remained aware of his surroundings in the operating room.[117] From this, Penfield concluded that although the responses pointed him to the mind, it was not to be found within the neuronal circuits that he had stimulated with his electrode. He admitted in 1975, "What the mind does cannot be accounted for by any neuronal mechanism that I can discover." Even the "highest brain-mechanism" represented only the meeting point of mind and brain, the "psychicophysical frontier."[118]

We can thus think of Penfield's explorations of the brain—from the early treatment of the sensory-motor cortex to the later deep stimulations of the temporal lobe—as a constant if unsuccessful search for "mind." His investigations of the sensory cortex during the 1930s had suggested to him the existence of a higher integrating system. He relied on this system to assess the effects of his electrical stimulation of the sensory cortex and relay that information to him. Penfield's later work was the attempt to investigate this system, the "mind," directly. But as he began to roam more widely with his electrode, the mind continued to elude him. First he discovered the "psychical," then he probed deeper until he posited the brain stem as the seat of consciousness. But these too disappointed him in the end. Whenever Penfield stimulated a cortical area with his electrode, it appeared that he had only reached a way station within a larger neuronal circuit, never an actual "center" of integration itself. Penfield never

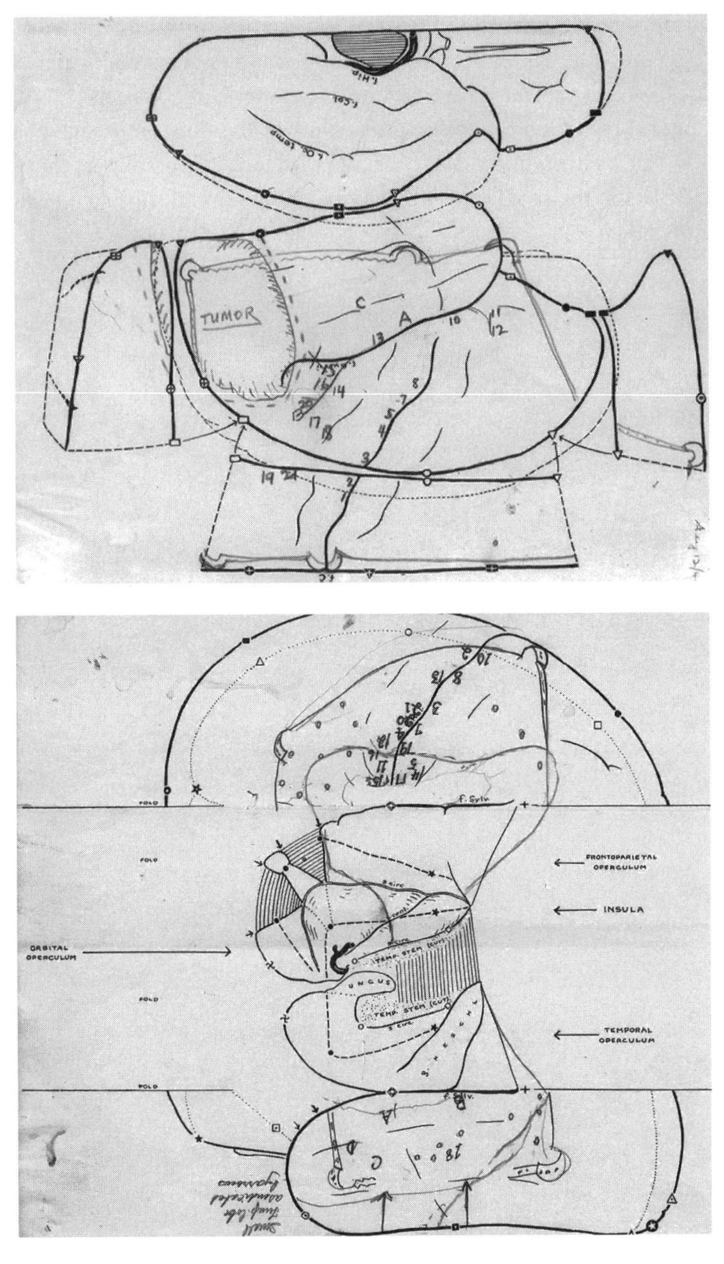

FIGURE 6.11. Brain maps, 1949 (*top*) and 1951 (*bottom*). (Courtesy of the William Feindel Memorial Fund.)

distanced himself from the project of localization, as Freud did most decisively (indeed, Penfield's name came to stand in for a confirmation of the localization project after a period of wrong-headed "holism").[119] But Penfield, in his failed attempts to trace down the mind, must have come close to the frustrations and discontents with localization expressed by his counterpart in the psychoanalytic camp. The real mind slipped constantly from his grasp.

Epilogue

At the seventy-sixth Annual Meeting of the American Neurological Association (ANA) in Atlantic City, New Jersey, June 18–20, 1951, neurologists, psychiatrists, and psychoanalysts came together to discuss the possibility of a productive exchange between their different fields. They called their meeting a "Symposium on Brain and Mind," a title that indicates both how close their interests were and how far apart their approaches remained. Wilder Penfield, who had just completed his yearlong term as president of the ANA, presented a paper titled "Memory Mechanisms." The paper treated Penfield's investigations of the cerebral cortex, where electrical stimulation had helped epileptic patients recall and relive past memories, what Penfield at that time still called "psychic responses." Penfield's paper was one of the highlights of the event, and it elicited enthusiastic commentaries from a number of participants. The first and longest response came from the psychoanalyst Lawrence Kubie, who celebrated the success of Penfield's paper in glowing terms: "I can sense the shades of Harvey Cushing and Sigmund Freud shaking hands over this long-deferred meeting between psychoanalysis and modern neurology and neurosurgery."[1]

Kubie not only interpreted Penfield's experiments as providing neurological evidence for the psychoanalytic claim that neurosis was related to hidden memories. A practicing psychoanalyst, Kubie was stunned by how quickly the neurosurgeon could evoke "precisely that type of re-experiencing of the past which the analyst has to struggle for days and weeks and months and years to achieve." Citing these impressive results, Kubie urged his audience to "bring the two disciplines [neurosurgery and psychoanalysis] closer together" and suggested a series of concrete interventions to encourage such a rapprochement: brain researchers might conduct a "psychoanalytic survey" or a "battery of psychological tests" before operating on the brain.[2] Kubie followed up on these suggestions. At around this time, he asked Penfield whether he could sit in on brain operations; Kubie huddled underneath the tent with the patients, talking to them and recording their speech with a Dictaphone.[3]

One can sense in Kubie's words a certain optimism that two giants of the mind and brain sciences might enter a productive dialogue. At the time of the symposium, both neurology and psychoanalysis had significant cultural capital, especially in the United States.[4] World War II had taken a heavy toll on American mental health, and psychoanalysis had risen in prestige not only because it could explain the neuroses and psychosomatic disorders that had emerged in the war, but also because it offered effective treatment. The psychoanalytic approach to neuroses provided a powerful alternative to the somatic treatments then in vogue in psychiatry.[5] Even outside medical circles, psychoanalysis gained increasing renown. It figured in best-selling books like Benjamin Spock's 1946 *Baby and Child Care*, in novels, in women's magazines, and in Hollywood movies. Freud himself achieved a level of fame that is rare for doctors, and Ernest Jones's heroic portrayal in the widely read biography of him, published later in the decade (between 1953 and 1957), made him a fixed point in midcentury American culture.

Neurosurgery was also a field on the rise. The 1951 symposium capped off a thirty-year growth in the profile of neurosurgery. As Delia Gavrus has argued, neurosurgeons cultivated a sense of elitism first in the creation of specialized organizations such as the Society for Neurological Surgeons, which then spilled over into popular culture. In the interwar period especially, the neurosurgeon became a heroic figure idolized in the popular media, and neurosurgery was picked out as the epitome of scientific medicine. Moreover, the use of advanced technological techniques, including Penfield's electrical stimulation, which he showcased at the symposium, and the electrocorticographic and electroencephalographic methods presented by his MNI co-worker Herbert Jasper, contributed to the sense of prestige and exclusivity.[6] The fact that the neurosurgeon seemed to be probing the mysterious boundary between body and mind only added to his mystique.[7]

But while Kubie was enthusiastic about productive collaboration, the sentiment was not reciprocated; Penfield was less interested in collaborating with psychoanalysts than Kubie was with him. At the 1951 meeting, Penfield did not respond to Kubie's commentary and suggestions. He restricted his response to the questions asked by other participants, all of whom matched him in their somatic orientation.[8] On Freud and Cushing shaking hands, he remained silent. And while Penfield tolerated Kubie's presence in his operating room, it was clear that he considered this more professional courtesy than an invitation to interdisciplinary dialogue.

Beyond the Two Cultures?

How can we account for this asymmetrical relationship between neurosurgery and psychoanalysis? Not by using the type of simple opposition that divides psychoanalysis and neurosurgery according to a mind-brain duality. After all, Kubie, a man trained in both neurology and psychoanalysis, thought the two fields existed in at least productive proximity. Moreover, Penfield's dualism did not cast his neurosurgery as a purely somatic enterprise, but rather integrated the mind into it: the mind was crucial to Penfield's operations and experimental practice and later became an (albeit elusive) component of his research. Penfield was suspicious of psychoanalysis not because he excluded reference to the mind from his work, but rather because he adhered to a particular version of it, an adherence that explains his engagement only with a certain type of mind science.[9] As we saw, for Penfield the transparent mind was opposed to the opaque body, which left little room for an unconscious. Freud's *Körper-Ich* would have seemed to him a contradiction in terms.

In this way Penfield's indifference to psychoanalysis can be explained in the same terms as Schilder's fractious relationship to psychoanalytic orthodoxy; Schilder's body image and Penfield's homunculus are analogous constructions. Both mapping projects seemed to study how we experience our bodies, and both resulted, I argued, in large part from Schilder's and Penfield's common attempt to treat the two reflex arcs separately. Psychoanalyst and neurosurgeon alike tested sensory and motor functions, not together, as had been the case for the earlier reflex tradition, but in isolation, and this allowed them to imagine and study the projection of the body onto the mind. The importance of introspection for the testing of (especially) sensory function provides context for both doctors' assertion that the subject was self-transparent, whether in Schilder's reformulation of the Freudian unconscious or Penfield's later meditations on "mind." Because both doctors cut the reflex in half, providing similar articulations of the relationship between localization and the reflex, they both produced concepts of mind that were out of sync with those of traditional Freudians.

Neurology-neurosurgery and psychoanalysis surely differ in radical ways: one takes place in the hospital, the other generally in private practice; one treats patients with movement disorders, the other neurotics; one acts through manipulation of the body or with a surgical knife, the other through the spoken word; in one case the doctor wears a white coat, in the other a three-piece suit. But it would be a mistake to see these differences as the results of an ontological opposition between the two disciplines' objects of research: soma versus psyche.

That presentation, I have argued, is unconvincing even in the case of Penfield and Kubie. But if we go back to an earlier generation of neurologists and psychoanalysts, it is even less plausible. In this book, I have shown that Freud, at least initially, did not consider himself to be working on the mind as such; instead, he figured his psychological practice as a way of intervening on the nervous system. Understood in reflex terms, his psychoanalysis does not seem qualitatively different from the range of therapies being developed in neurology. Indeed, between Freud and Foerster we can discern a number of striking parallels. Both denied that there was a direct correlation between lesion and symptom—the principle upon which pathological anatomy had been built. Instead they examined how nervous damage created broader systemic disorder. Freud figured "trauma" as the cause of hysterical symptoms, symptoms that not only did not directly express, but might also hide, the underlying etiology. Similarly, Foerster understood neurological diseases like tabes to be a disruption of the *Reflexgemeinschaft*, where damage to a sensory reflex arc produced motor malfunction. This reformulation allowed both doctors, in contradistinction to the "therapeutic nihilism" of the neuropsychiatry in which they had been trained, to recast their interventions on the nervous system as therapeutic. Foerster thought he could alleviate symptoms by isolating the refractory member of the "nervous community," surgically interrupting its connections to the rest of the nervous system and then promoting the reformation of the nervous community through *Übungstherapie*. Freud thought it was possible to rewire pathological associations through talk therapy, using the patient's own transferences and resistances to effect the cure. The fact that both treatments were based on assumptions about the "plasticity" of the nervous system, its ability to remake nervous connections in appropriate circumstances, explains the reshifting of the doctor-patient relationship in their work. Foerster, just like

Freud, placed the doctor in an ancillary role, helping the patient through the difficult task of healing himself.

The different institutional and therapeutic forms that psychoanalysis and neurology adopted can be traced in large part back to the part of the nervous system on which the two disciplines focused. Psychoanalysis remained concerned primarily with higher functions and examined the complex interaction of nervous associations in the brain, while neurology, at least at first, was constructed to help explain and treat lower nervous function. It is important to remember that neither doctor considered this distinction absolute: Freud as much as Foerster refused anything more than a provisional and pragmatic value to the opposition between higher and lower function. Nevertheless, a provisional and practical opposition was sufficient to produce significant institutional and practical implications. The differences between psychoanalysis and the neuro disciplines can be understood better as the historically contingent ramifications of this relatively minor difference in focus than as the consequence of conflicting metaphysics.

Given the proximity between Freud and Foerster, on the one hand, and Schilder and Penfield, on the other, the clearest ruptures we can discern seem to be not *across* disciplines, but *within* them. Rather than remaining fixated on the terms of mind and body that are common both to certain forms of neuroscience today and the criticisms of "reductionism" that are often leveled against them, it is perhaps more helpful to consider other frames and models for thinking about the disciplinary interaction.

And as we enter the "connectomic era" in neuroscience, the localization-connection dyad might prove to be particularly valuable. For while, according to the categories of mind and body, neuroscience seems to be directly opposed to psychoanalysis, the two disciplines look less alien to each other when one considers how both tend to mobilize a connective principle to challenge the claims of localization. This does not mean that modern neuroscience will necessarily be receptive to psychoanalysis per se. That option has been explored, and the results are not encouraging. In the late 1990s, a group of neuroscientists including the Nobel laureate Eric Kandel and the neuropsychologist and historian of psychoanalysis Mark Solms created what has come to be called "neuropsychoanalysis" (with its own eponymous journal founded in 1999). But despite efforts to cultivate a symmetrical relationship between neuroscience and psychoanalysis, it is the former that gives legitimacy to the latter. As Nima Bassiri has observed, "while psychoanalysis might *inform* neuroscience conceptually, it is neuroscience that can ultimately *ground* psychoanalysis scientifically."[10]

Nevertheless, the surprising proximity between the two can help reframe anxieties that neuroscience is overrunning other academic disciplines, especially in the humanities. Fears about disciplinary imperialism often couple anxieties about the integrity of the "colonized" discipline with a sense of the disciplinary robustness of the invader.[11] But as we have seen in this book, the formulation of a new relationship between the connective and the localizing models has often been accompanied by disciplinary instability closer to home; when Foerster and Freud recalibrated the neuropsychiatric model, they contributed to its fragmentation. In the 1880s and 1890s, the prioritizing of the connective model of the nervous system over the project of functional localization did not (in the short term, at least) herald a period of neuro-hegemony. If this history is any indication, the working out of a fully connective model in the twin American and European projects over the next decade could have as profound implications for the disciplinary status of neuroscience as it does for other fields that currently fear its influence. For it is surely not inconceivable that, just as it did 120 years ago, the sustained analysis of the nervous connections in the brain might generate among neuroscientists a greater attentiveness to the importance of society, context, and indeed history in the understanding of the self. Neuroscience might open itself to the humanities rather than simply engulfing them.[12]

This does not mean that there are no problems in the way in which neurological and other scientific ideas today have been used both in the humanities and in popular culture.[13] In a world where education is increasingly seen as a means for economic growth, and the prestige of the sciences (including neuroscience) has never been greater, attention, resources, and students have been directed toward them. This sidelining of the arts and humanities, whose value can be measured in economic terms only with great difficulty, can have a detrimental impact on education and society more broadly: currently the humanities provide the most vigorous approach for understanding the world in its historical complexity and engendering empathy for other people within and outside our own culture, both of which are key to a healthy democratic citizenship.[14]

But the story I have laid out here should encourage us to remain true to humanistic principles when examining these issues, being as attentive to the complexities and tensions within scientific texts as we are to those in literary ones. The interconnected histories of the neuro disciplines and psychoanalysis should make us wary of figuring the impact of neuroscience on modern culture using the oppositional model of neuroscience versus psychoanalysis, *Natur-* versus *Geisteswissenschaften*, the nomothetic versus the idiographic, brain versus mind, a hypostatization of the "two cultures" model that can be as debilitating as it is ahistorical. For the relationship between the neuro disciplines and the humanities has a history, and when we recognize the complexity of its past, we liberate ourselves from the fixed categories that seem to determine its future.

Acknowledgments

I came to the history of science along an indirect path. After an extended period of training in medicine and in neuroscience, I embarked on a PhD in the history of science at Harvard, where I relied on the patience and goodwill of a wonderful group of mentors who did not tire of teaching the subject to a neophyte. My first debt of gratitude is to Anne Harrington, whose wisdom and unwavering support helped me greatly in the process of becoming a historian. Charles Rosenberg scrawled notes on almost every single document I produced as a graduate student. He was an invaluable guide in my development as a scholar. I was lucky to work with Shigehisa Kuriyama during the final stages of my dissertation writing. His creativity and perceptiveness, and his insistence on the importance of scholarly presentation, left its mark on this work and on my broader academic ideals. Alison Simmons provided a rigorous and imaginative philosophical supplement to the historical perspectives of my other advisers, and I am grateful to have benefited from her sharpness and insight. No less important, the stimulating graduate community at Harvard helped induct me into the field; I enjoyed and gained immensely from the camaraderie of the individuals in my cohort—Jennifer Dariani (née Clark), Ben Hurlbut, Dániel Margócsy, John Mathew, Amber Musser, Chitra Ramalingam, Bill Rankin, and Kara Swanson. I also want to thank the circle of friends fascinated with the history of the mind sciences at Harvard, in particular Scott Phelps, Phil Loring, and Kevin Stoller, for their thoughtful perspective and many conversations. Scott deserves particular thanks. We work on very similar themes and topics, and I relied and continue to rely on his insight, encyclopedic knowledge, and generosity. The following people supported my graduate education in ways too numerous to list: Ellen Bales, Jimena Canales, Paul Cruickshank, Alex Csiszar, Peter

Galison, Jeremy Greene, Daniela Helbig, Sarah Jansen, David Jones, Natalie Köhle, Deborah Levine, Aaron Mauck, Grischa Metlay, John Murdoch, Sharrona Pearl, Chris Phillips, Alisha Rankin, Lukas Rieppel, Henning Schmidgen, Matt Stanley, Elly Truitt, Marga Vicedo, Marco Viniegra, Adelheid Voskuhl, Juliet Wagner, Oriana Walker, Debbie Weinstein, Alex Wellerstein, Beth Yale, Nasser Zakariya, and Huicong Zhang.

I was lucky to move to an equally supportive and stimulating environment at Princeton. I thank Bill Jordan, the History Department chair, for granting me a full-year sabbatical and generous maternity leave. I benefited from talking to my departmental colleagues, who have provided feedback on my work and broadened my historical horizons. The list is long, but I should mention Alison Isenberg, Phil Nord, Andy Rabinbach, and Wendy Warren. Princeton is also home to a very dynamic history of science community: Graham Burnett, Benjamin Elman, Dan Garber, Charles Gillispie, Tony Grafton, Federico Marcon, Sue Naquin, Jennifer Rampling, Eileen Reeves, Emily Thompson, Janet Vertesi, and the extraordinarily bright set of graduate students who come together at the weekly Program Seminar. It is a privilege to work with such thoughtful and inspiring scholars. Four people deserve particular recognition. Erika Milam read my work, and our conversations over coffee, lunch, and ice cream sustained me during the final stages of this book. Angela Creager, Michael Gordin, and Keith Wailoo managed to make time in their busy lives to read my manuscript, in full, several times. They are exemplars of collegiality and scholarship, and I am profoundly grateful for their support. In addition to these, I want to recognize the friendship and support of Christiane Frey, Shel Garon, Eleanor Hubbard, Rob Karl, Judy Laffan, Mike Laffan, Perry Leavell, Jon Levy, Leo Nguyen, Barbara Oberg, Beth Rabbit, and Rebecca Rix.

In writing this book, I have drawn on the expertise of an international community in the field of the mind and brain sciences. I benefited from the generosity and insights of Geneviève Aubert, Cornelius Borck, Gerhard Fichtner, John Forrester, Michael Hagner, Rainer Herrn, Albrecht Hirschmüller, George Makari, Andreas Mayer, Frank Pillmann, Michael Schröter, and Fernando Vidal. I was particularly lucky to begin this project when a group of young scholars were embarking on research in the mind sciences: Nima Bassiri, Stephen Casper, Delia Gavrus, Susan Lamb, Sophie Ledebur, and Frank Stahnisch. Their example and support have sustained me through the process of writing. Stephen Casper deserves special acknowledgment for reading almost all of the manuscript, at various stages, and providing invaluable feedback.

I have presented material from this book at a number of venues: at the annual meetings of the History of Science Society and the American Association for the History of Medicine, and at colloquia and talks at Johns Hopkins, the ETH Zurich, University College London, the IAS Konstanz, Clarkson University, Princeton University, the University of Pennsylvania, the Center for the History of Psychiatry at Weill-Cornell, the Max Planck Institute for the History of Science, HPS Cambridge, the University of Toronto, and the University of Vienna. Many of my ideas were tested and refined on these and other occasions. In addition to the people I have mentioned, I want to recognize the input by Hans Adler, Warwick Anderson, Mitchell Ash, Babak Ashrafi, Jesse Ballenger, Dietz Bering, Carin Berkowitz, Jeremy Blatter, Nathaniel Comfort, Heimo Ehmke, Mary Fissell, Diana Fuss, Fred Girod, Jeremy Greene, Johannes Haushofer, Vanessa Heggie, Nick Hopwood, Stephen Jacyna, Eva Johach, Bernhard Kleeberg, Albrecht Koschorke, Kenton Kroker, Howard Kushner, Nicolas Langlitz, Susan Lanzoni, Ruth Leys, Randall Packard, Tom Quick, Tobias Rees, Gabriele Schwab, Skúli Sigurdsson, Max Stadler, Daniel Todes, and Monika Wulz.

From inception to completion, this project received funding from many institutions. I am grateful to the Center for European Studies at Harvard University, the Krupp Foundation, the Schmittmann-Wahlen Foundation, the Max Planck Institute for the History of Science, the ACLS and Mellon Foundation, the Johanna and Alfred Hurley *61 P76 P82 P86 University Preceptorship in History, the Princeton University Committee on Research in the Humanities and Social Sciences, and the Institute for Advanced Study in Konstanz, Germany. I completed the manuscript while in Konstanz, where I found both an interdisciplinary conversation that enriched my thinking and the space and time necessary to get those thoughts on the page. The intellectual community at the institute is a testament to the work of its director, Rudolf Schlögel, and secretary, Fred Girod.

In collecting materials for the book, I relied on the help of many archivists and librarians. In Europe, I would like to thank Ruth Koblizek (Vienna), Kornelia Drost-Siemon (Göttingen), Sylvia Loesch (Düsseldorf), and the staff at the State Archive in Breslau and at the Institute of Psychoanalysis in London. In North America, I am indebted to a wonderful group at McGill University and the Montreal Neurological Institute: Duncan Cowie, Christopher Lyons, Sandra McPherson, Lily Szczygiel, and Ann Watson. I also want to thank the staff at the Alan Mason Chesney Medical Archives at Johns Hopkins University, Marianne LaBatto at the

Brooklyn College Archives and Special Collections, Leonard Bruno at the Library of Congress Manuscript Division, and the staff at the New York Academy of Medicine Rare Book Room. Many of the papers on which I have drawn were not readily accessible to researchers, and this work owes much to those individuals who were generous enough to allow me to see unsorted files or private collections: Geneviève Aubert (Louvain), Dr. Frank Pillmann (Halle), Dr. Richard Leblanc (McGill), and especially the late Dr. William Feindel, Michael Feindel, and the William Feindel Memorial Fund.

I would also like to thank the editors of *Modern Intellectual History* and the *Bulletin of the History of Medicine* for permission to reproduce "The Disappearing Lesion: Sigmund Freud, Sensory-Motor Physiology, and the Beginnings of Psychoanalysis," and "Exercises in Therapy—Neurological Gymnastics between *Kurort* and Hospital Medicine, 1880–1945," which appear here in revised form.

Margarete Ritzkowsky worked with me on deciphering several almost illegible manuscripts, and Alice Christensen, Christian Jany, and Felix Rietmann provided invaluable research assistance. In particular, Alice's bibliographic and research skills and her *Sprachgefühl* were of enormous help in the process. I also want to thank the two anonymous reviewers for their thoughtful comments, and the team at Chicago: my editor, Karen Darling, who gave helpful feedback at the right moment and shepherded me efficiently through the publishing process; and my copyeditor, Lois Crum, for her meticulous work at the final stages.

Finally, my family. I am grateful to Sofie and Lulu for giving up daytime napping only when this book was almost finished; they make life delightful. My parents, Reinhard and Helga, were patient with a daughter who gave up medicine for a more precarious career path. They have been understanding throughout. In the past three years, their dedication to grandparenting has helped me keep my head above water. My mother-in-law, Anstice, despite her growing flock of grandchildren, has always preserved a share of her tremendous energy to give the Guenther-Barings hands-on help and moral support. I thank them all, as well as those whose friendship has long since made them part of our family: my dear friend Nina Osterfeld in Germany, and in Princeton, Martha King, Paul Wilderson, and Louis and Lydia Hamilton. I feel privileged to have them all in my life. My most profound debt is to my husband, Edward Baring, to whom I dedicate this book. He has been my toughest critic and my greatest support. His generosity, kindness, and wit have been an inspiration, and I am deeply grateful for his love and the life we share.

Notes

Archives

ABPS	Archives of the British Psychoanalytical Society, Institute of Psychoanalysis, 112a Shirland Road, London W9 2BT.
AMC	Alan Mason Chesney Medical Archives of the Johns Hopkins Medical Institutions, 5801 Smith Avenue, Suite 235, Baltimore, MD 21209.
BCA	Brooklyn College Archives and Special Collections, 2900 Bedford Avenue, Brooklyn, NY 11210.
COVA	Cécile and Oskar Vogt Archive, Universitätsstr. 1, building 22.01/22.03, 40225 Düsseldorf, Germany.
CWPF	Carl Wernicke Patient Files, Universitätsklinik und Poliklinik für Psychiatrie, Psychotherapie und Psychosomatik, Julius-Kühn-Str. 7, 06112 Halle/Saale, Germany.
DMEHM	Department of Medical Ethics and History of Medicine, Göttingen University, Humboldtallee 36, 37073 Göttingen, Germany.
HMC	History of Medicine Collections, Institute for the History of Medicine, Vienna University, Josephinum, Währingerstrasse 25, 1090 Vienna, Austria.
LOC	Library of Congress Manuscript Division, 101 Independence Avenue SE, Room LM 101, James Madison Memorial Building, Washington, DC 20540.
NYAM	New York Academy of Medicine, Rare Book Room, 1216 Fifth Avenue at 103rd Street, New York, NY 10029.
OLHM	Osler Library of the History of Medicine, 3rd floor, 3655 Promenade Sir William Osler, Montreal, Quebec H3G 1Y6.
RFA	Rockefeller Foundation Archives, 15 Dayton Ave., Sleepy Hollow, NY 10591.
RFAB	Royal Film Archives Brussels, Ch. de Forest, 62/5, 1060 Brussels, Belgium.
SAB	State Archive Breslau, Ulica Pomorska, 250-215 Wrocław, Poland.
SRMNI	Stimulation Reports held at the Montreal Neurological Institute, Montreal, Quebec.
VUA	Vienna University Archive, Postgasse 9, 1010 Vienna, Austria.
WUA	Wrocław University Archive, Karola Szajnochy St. 10, 50-076 Wrocław, Poland.

Introduction

1. See the growing recent scholarship on neuroscience, e.g., Suparna Choudhury and Jan Slaby, eds., *Critical Neuroscience: A Handbook of the Social and Cul-*

tural Contexts of Neuroscience (Chichester, West Sussex, UK: Wiley-Blackwell, 2012); Fernando Vidal and Francisco Ortega, eds., *Neurocultures: Glimpses into an Expanding Universe* (Frankfurt: Peter Lang, 2011); Nikolas Rose and Joelle Abi-Rached, *Neuro: The New Brain Sciences and the Management of the Mind* (Princeton, NJ: Princeton University Press, 2013); Melissa Littlefield and Jenell Johnson, eds., *The Neuroscientific Turn: Transdisciplinarity in the Age of the Brain* (Ann Arbor: University of Michigan Press, 2012); Steve Fuller et al., "Neurohistory and history of science," Isis Focus Section, *Isis* 105.1 (2014): 110–154. See also the writings of Raymond Tallis, e.g., "Think brain scans can reveal our innermost thoughts? Think again," *Observer*, June 1, 2013, www.theguardian.com/commentisfree/2013/jun/02/brain-scans-innermost-thoughts, accessed July 7, 2014.

2. William R. Uttal, *The New Phrenology: The Limits of Localizing Cognitive Processes in the Brain* (Cambridge, MA: MIT Press, 2001), 147, 103. See also William R. Uttal, *Distributed Neural System: Beyond the New Phrenology* (Cornwall-on-Hudson, NY: Sloan, 2009).

3. Uttal, *New Phrenology*, 107.

4. For scholarly work on neuroimaging, see Joseph Dumit, *Picturing Personhood: Brain Scans and Biomedical Identity* (Princeton, NJ: Princeton University Press, 2004); Anne Beaulieu, *The Space Inside the Skull: Digital Representations, Brain Mapping and Cognitive Neuroscience in the Decade of the Brain* (Amsterdam: University of Amsterdam, 2000); Kelly A. Joyce, *Magnetic Appeal: MRI and the Myth of Transparency* (Ithaca, NY: Cornell University Press, 2008); Michael Hagner, ed., *Der Geist bei der Arbeit: Historische Untersuchungen zur Hirnforschung* (Göttingen: Wallstein, 2006); Cornelius Borck, "Recording the brain at work: The visible, the readable, and the invisible in electroencephalography," *Journal of the History of the Neurosciences* 17 (2008): 367–379. See also M. R. Bennett and P. M. S. Hacker, *Philosophical Foundations of Neuroscience* (Malden, MA: Blackwell, 2003).

5. Alison Abbott, "Solving the brain," *Nature* 499 (July 18, 2013): 273.

6. Olaf Sporns, "The human connectome: A complex network," *Annals of the New York Academy of Sciences* 1224 (2011): 110. See also Olaf Sporns, *Discovering the Human Connectome* (Cambridge, MA: MIT Press, 2012); and Sebastian Seung, *Connectome: How the Brain's Wiring Makes Us Who We Are* (Boston: Houghton Mifflin Harcourt, 2012).

7. Sporns, "Human connectome," 110. Similarly, in another project, the five-year NIH-sponsored Human Connectome Project, scientists have worked toward the goal of building a "network map" of the human brain, mapping systems across the brain in order to "elucidate the neural pathways that underlie brain function and behavior." The project was launched in 2009. Human Connectome Project, www.humanconnectome.org.

8. Egidio D'Angelo, "Toward the connectomic era," *Functional Neurology* 27.2 (2012): 77. These developments have been under way for a while. Since 2002, theo-

retical and experimental neuroscientists have met annually for the Brain Connectivity Workshop, in an attempt to move beyond "isolated structure-function correlations" and to develop an approach for explaining "how the behaviour of neuronal systems results from the interactions of their elements." Klaas Enno Stephan et al., "The Brain Connectivity Workshops: Moving the frontiers of computational systems neuroscience," *Neuroimage* 42.1 (2008): 2. It is not clear whether any meetings took place after 2007.

9. Roth prefers the language of "center-networks." Hans-Peter Krüger, ed., *Hirn als Subjekt? Philosophische Grenzfragen der Neurobiologie* (Berlin: Akademie, 2007), 29. For a similar perspective, and a discussion of the literature, see J. C. Marshall and G. R. Fink, "Cerebral localization, then and now," *Neuroimage* 20, Suppl. 1 (2003): S6. All translations from the German are mine, unless otherwise indicated.

10. My understanding of the internal tension within the localization tradition is informed by Sigmund Freud's sharp critique, as formulated in his 1891 *On Aphasia* (see chapter 3). Another critique from within neurology, with a similarly sharp eye for the internal contradictions within Wernicke's work, can be found in the work of Norman Geschwind. See Howard Kushner, "Norman Geschwind and the use of history in the (re)birth of behavioral neurology," *Journal of the History of the Neurosciences* 24 (2015): 173–192.

11. The classic account of the epistemic conditions for this rise of "scientific medicine" is Michel Foucault, *The Birth of the Clinic: An Archeology of Medical Perception*, trans. A. M. Sheridan Smith (New York: Pantheon Books, 1973).

12. This helped psychiatry gain acceptance in a time when memories of Gall's phrenological project were still fresh.

13. In the nineteenth century, there were two main currents of the localization project, differentiated by the practice in which they grounded their work: first, Fritsch and Hitzig's project, and second, neuropsychiatry, the focus of this study. Michael Hagner, "Lokalisation, Funktion, Cytoarchitektonik: Wege zur Modellierung des Gehirns," in *Objekte, Differenzen, Konjunkturen: Experimentalsysteme im historischen Kontext*, ed. Michael Hagner, Hans-Jörg Rheinberger, and Bettina Wahrig-Schmidt (Berlin: Akademie, 1994): 121–150. For histories of localization, see esp. Michael Hagner, *Homo cerebralis: Der Wandel vom Seelenorgan zum Gehirn* (Berlin: Berlin, 1997); Anne Harrington, *Medicine, Mind and the Double Brain: A Study in Nineteenth-Century Thought* (Princeton, NJ: Princeton University Press, 1987); Robert Young, *Mind, Brain, and Adaption in the Nineteenth Century: Cerebral Localization and Its Biological Context from Gall to Ferrier* (New York: Oxford University Press, 1990); Susan L. Starr, *Regions of the Mind: Brain Research and the Quest for Scientific Certainty* (Stanford, CA: Stanford University Press, 1989); Olaf Breidbach, *Die Materialisierung des Ichs: Zur Geschichte der Hirnforschung im 19. und 20. Jahrhundert* (Frankfurt: Suhrkamp, 1997); Samuel Greenblatt, "Cerebral localization: From theory to practice: Paul Broca and Hugh-

lings Jackson to David Ferrier and William Macewen," in *A History of Neurosurgery in Its Scientific and Professional Contexts*, ed. Samuel Greenblatt (Park Ridge, IL: American Association of Neurological Surgeons, 1997): 137–152. For a history of psychiatry in Germany at the time, see Eric Engstrom, *Clinical Psychiatry in Imperial Germany: A History of Psychiatric Practice* (Ithaca, NY: Cornell University Press, 2003).

14. Wilhelm Erb and Carl Westphal first (independently of each other) described the canonical test for the knee-jerk reflex in the mid 1870s. Wilhelm H. Erb, "Ueber Sehenreflexe bei Gesunden und bei Rückenmarkskranken," *Archiv für Psychiatrie und Nervenkrankheiten* 5 (1875): 792–802; Carl Westphal, "Ueber einige durch mechanische Einwirkung auf Sehnen und Muskeln hervorgebrachte Bewegungs-Erscheinungen," *Archiv für Psychiatrie und Nervenkrankheiten* 5 (1875): 803–834.

15. Since my research focuses on a particular use of the reflex tradition, I do not give a full history of it as a theoretical and practical principle. I leave aside, for example, its formulation and development in the history of experimental psychology and neurophysiology, although there are overlaps and that too is an important history, involving such figures as Descartes, Willis, Hall, Müller, Whytt, Legallois, Bell, Magendie, Prochaska, Sherrington, Watson, and Pavlov. Rather, I concentrate on the way in which the reflex was appropriated by neuropsychiatrists, and later neurologists and psychoanalysts, in order to understand the more complex functioning of the nervous system, not just at the level of the spinal cord, to which initially the reflex had been restricted, but also as a conceptual tool for understanding brain function, or the interactions of multiple nerves. For the literature on the reflex, see Ruth Leys, *From Sympathy to Reflex: Marshall Hall and His Opponents* (New York: Garland, 1990); Franklin Fearing, *Reflex Action: A Study in the History of Physiological Psychology* (New York: Hafner, 1930); Georges Canguilhem, *La formation du concept de réflexe au XVIIe et XVIIIe siècles* (Paris: Presses universitaires de France, 1955); Georges Canguilhem, *Die Herausbildung des Reflexbegriffs im 17. und 18. Jahrhundert*, trans. Henning Schmidgen (Paderborn: Fink, 2008); Edwin Clarke and Stephen Jacyna, *Nineteenth-Century Origins of Neuroscientific Concepts* (Berkeley: University of California Press, 1987); Edwin Clarke and C. D. O'Malley, *The Human Brain and Spinal Cord: A Historical Study Illustrated by Writings from Antiquity to the Twentieth Century* (San Francisco: Norman, 1996); Marc Jeannerod, *The Brain Machine: The Development of Neurophysiological Thought*, trans. David Urion (Cambridge, MA: Harvard University Press, 1985), chaps. 2 and 3; Daniel Todes, *Pavlov's Physiology Factory: Experiment, Interpretation, Laboratory Enterprise* (Baltimore: Johns Hopkins University Press, 2001); Daniel Todes, *Ivan Pavlov: A Russian Life in Science* (New York: Oxford University Press, 2014). On higher reflexes, see esp. Clarke and Jacyna, *Origins*, 124–147; and Tom Quick, "Techniques of life: Zoology, psychology and technical subjectivity (c. 1820–1890)" (PhD diss. University College London, 2011).

16. The few books on German neurology include Bernd Holdorff and Rolf Winau, eds., *Geschichte der Neurologie in Berlin* (Berlin: De Gruyter, 2001); Klaus Zülch, *Die geschichtliche Entwicklung der deutschen Neurologie* (Berlin: Springer, 1987). The literature on the history of neurology in other national contexts is richer; see chapter 4.

17. E.g., Albrecht Hirschmüller, *Freuds Begegnung mit der Psychiatrie: Von der Hirnmythologie zur Neurosenlehre* (Tübingen: Edition Diskord, 1991).

18. In Meynert's lab, Freud had, among other things, developed a gold staining method to visualize fiber connections in the nervous system, which was in line with the aims of his larger project; see chapter 3.

19. In arguing for a continuity between Freud's early and later work, I see my work to be in line with recent historical scholarship on the history of psychoanalysis. See John C. Burnham, "The 'New Freud Studies': A historiographical shift," *Journal of the Historical Society* 6.2 (2006): 213–233. For a full discussion of the relevant literature on Freud, see chapter 3.

20. It is in this straightforward sense that I am using "genealogy": as the study of family resemblances across disciplinary divides.

21. By bringing together models from the neurological and psychoanalytic traditions, I attempt in this book to think through the distinction that Jan Goldstein has made in her impressive book between a "vertical" and a "horizontal" self. Jan Goldstein, *The Post-Revolutionary Self: Politics and Psyche in France* (Cambridge, MA: Harvard University Press, 2005): 1–17.

22. On the literature on medical specialization, see esp. George Weisz, *Divide and Conquer: A Comparative History of Medical Specialization* (New York: Oxford University Press, 2006); George Rosen, *The Specialization of Medicine with Particular Reference to Ophthalmology* (New York: Froben Press, 1944); Rosemary Stevens, *Medical Practice in Modern England: The Impact of Specialization and State Medicine* (New Haven, CT: Yale University Press, 1966); Rosemary Stevens, *American Medicine and the Public Interest* (Berkeley: University of California Press, 1998); Hans Eulner, *Die Entwicklung der medizinischen Spezialfächer an den Universitäten des deutschen Sprachgebietes* (Stuttgart: Ferdinand Enke, 1970); Charles Rosenberg, *The Origins of Specialization in American Medicine, An Anthology of Sources* (New York: Garland, 1989). For a detailed discussion of the literature on the history and sociology of specialization, see Weisz, *Divide and Conquer*, xii–xv. See also Fernando Vidal's discussion of "discipline" in *The Sciences of the Soul* (Chicago: University of Chicago Press, 2011), 3–8.

23. On the transnational dimensions of neurology and neuroscience, see Stephen Casper, "Anglo-American neurology: Lewis H. Weed and Johns Hopkins neurology, 1917–1942," *Bulletin of the History of Medicine* 82 (2008): 646–671; Frank Stahnisch, "German-speaking émigré neuroscientists in North America after 1933: Critical reflections on emigration-induced scientific change," *Österreichische Zeitschrift für Geschichtswissenschaften* 21 (2010): 36–68.

24. This, of course, is not to say that they present the boundaries as stable and unchanging, nor that they fail to take into account the often fluid nature of the "neuro" and "psy" fields. See Stephen Jacyna, *Lost Words: Narratives of Language and the Brain, 1825–1926* (Princeton, NJ: Princeton University Press, 2000); Mitchell Ash, *Gestalt Psychology in German Culture 1890–1967* (Cambridge: Cambridge University Press, 1995); Elizabeth Lunbeck, *The Psychiatric Persuasion: Knowledge, Gender, and Power in Modern America* (Princeton, NJ: Princeton University Press, 1994); Jan Goldstein, *Console and Classify: The French Psychiatric Profession in the Nineteenth Century* (Chicago: University of Chicago Press, 1987); Gerald Grob, *The Mad among Us: A History of the Care of America's Mentally Ill* (New York: Free Press, 1994); George Makari, *Revolution in Mind: The Creation of Psychoanalysis* (New York: HarperCollins, 2008); Roger Smith, *Between Mind and Nature: A History of Psychology* (London: Reaktion Books, 2013); Kurt Danziger, *Constructing the Subject: Historical Origins of Psychological Research* (Cambridge: Cambridge University Press, 1990); Stephen Casper, *The Neurologists: A History of a Medical Specialty in Modern Britain, c. 1789–2000* (Manchester, UK: Manchester University Press, 2014); Nancy Tomes, *A Generous Confidence: Thomas Story Kirkbride and the Art of Asylum-Keeping, 1840–1882* (Cambridge: Cambridge University Press, 1983); Delia Gavrus, "'Men of strong opinions': Identity, self-representation, and the performance of neurosurgery, 1919–1950" (PhD diss., University of Toronto, 2011).

25. Alison Winter, *Memory: Fragments of a Modern History* (Chicago: University of Chicago Press, 2012); Kurt Danziger, *Marking the Mind: A History of Memory* (Cambridge: Cambridge University Press, 2008). For other books that construct their stories around guiding threads, but which critically examine other disciplinary divisions, see Andreas Mayer, *Sites of the Unconscious: Hypnosis and the Emergence of the Psychoanalytic Setting* (Chicago: University of Chicago Press, 2013); Allan Young, *The Harmony of Illusions: Inventing Post-Traumatic Stress Disorder* (Princeton, NJ: Princeton University Press, 1995); Cornelius Borck, *Hirnströme: Eine Kulturgeschichte der Elektroenzaphalographie* (Göttingen: Wallstein, 2005); Mical Raz, *The Lobotomy Letters: The Making of American Psychosurgery* (Rochester, NY: University of Rochester Press, 2013); Hagner, *Homo cerebralis*; Kenton Kroker, *The Sleep of Others and the Transformation of Sleep Research* (Toronto: University of Toronto Press, 2007); Jesse Ballenger, *Self, Senility and Alzheimer's Disease in Modern America: A History* (Baltimore: Johns Hopkins University Press, 2006); Nadine Weidman, *Constructing Scientific Psychology: Karl Lashley's Mind-Brain Debates* (Cambridge: Cambridge University Press, 1999); Anne Harrington, *The Cure Within: A History of Mind-Body Medicine* (New York: Norton, 2008); Jonathan Metzl, *Prozac on the Couch: Prescribing Gender in the Era of Wonder Drugs* (Durham, NC: Duke University Press, 2003); Albrecht Hirschmüller, *The Life and Work of Josef Breuer: Physiology and Psychoanalysis* (New York: New York University Press, 1989); Andrew Lakoff, *Pharmaceutical Reason: Knowledge and Value in Global Psychiatry* (Cambridge: Cambridge Uni-

versity Press, 2005); Scott Phelps, "Blind to their blindness: Indifference, anosognosia, and the denial of illness, 1880–1960" (PhD diss., Harvard University, 2013).

26. Roger Smith, *Inhibition: History and Meaning in the Sciences of Mind and Brain* (Berkeley: University of California Press, 1992).

27. The literature is massive, but important books include Charles Taylor, *Sources of the Self: The Making of the Modern Identity* (Cambridge, MA: Harvard University Press, 1989); Jerrold Seigel, *The Idea of the Self: Thought and Experience in Western Europe since the Seventeenth Century* (Cambridge: Cambridge University Press, 2005); Goldstein, *Post-Revolutionary Self.*

Chapter One

1. Karl Jaspers, *General Psychopathology*, trans. J. Hoenig and Marian W. Hamilton (Chicago: University of Chicago Press, 1963), 17. Jaspers was not the first to use the term *brain mythology*. I was able to find one earlier use, in A. Mayer, *Psychological Bulletin* 4 (1907): 171, quoted in Arnold Pick, *Die agrammatischen Sprachstörungen: Studien zur psychologischen Grundlegung der Aphasielehre* (Berlin: Springer, 1913), 1:40. As Albrecht Hirschmüller has pointed out, the term cannot be found in contemporary critiques of Wernicke or Meynert. Hirschmüller, *Freuds Begegnung*, 18n22. As we shall see, the criticism captured by the phrase was relatively common currency by the end of the nineteenth century, and so we can surmise that the term *brain mythology* was used before Jaspers, Pick, and Mayer put it into print.

2. Jaspers, *General Psychopathology*, 17–19.

3. Ibid., 18.

4. Ibid., 4.

5. Ibid., 17.

6. However, as we shall see, at least for Goldstein the situation was more complicated.

7. Foucault, *Birth of the Clinic*, 10, 90–91.

8. For Foucault, pathological anatomy and the clinical anatomical method emerged because they fit the new medical gaze, rather than vice versa as in the traditional story.

9. For other accounts of Paris medicine, see Erwin Ackerknecht, *Medicine at the Paris Hospital, 1794–1848* (Baltimore: Johns Hopkins Press, 1967); Caroline Hannaway and Ann La Berge, *Constructing Paris Medicine* (Amsterdam: Rodopi, 1998).

10. Axel Bauer, "Die Formierung der Pathologischen Anatomie als naturwissenschaftliche Disziplin und ihre Institutionalisierung an den deutschsprachigen Universitäten," *Würzburger medizinhistorische Mitteilungen* 10 (1992): 322; Erna Lesky, *The Vienna Medical School of the 19th Century*, trans. L. Williams and I. S. Levij (Baltimore: Johns Hopkins University Press, 1976).

11. Rokitansky was criticized in 1846, however, by a young Rudolf Virchow for embracing humoral pathological elements in his explanation of pathology as changes in the blood. Rudolf Virchow, "Rokitansky: Handbuch der allgemeinen pathologischen Anatomie (Wien 1846)," *Medicinische Zeitung* 15, 49–50 (1846): 237–238, quoted in Felicitas Seebacher, *"Freiheit der Naturforschung": Carl Freiherr von Rokitansky und die Wiener Medizinische Schule: Wissenschaft und Politik im Konflikt* (Vienna: Österreichische Akademie der Wissenschaften, 2006), 48n159. Later, Virchow acknowledged Rokitansky's work as the Linnaeus of pathological anatomy, an ironic turn toward the old classificatory medicine that in Foucault's narrative was displaced by the pathological anatomical method. R. Virchow, *Wiener medizinische Wochenschrift* 5 (1855): 417, quoted in Lesky, *Wiener medizinische Schule*, 132. See also Seebacher, *"Freiheit,"* 47–48.

12. Carl Rokitansky, *Handbuch der allgemeinen pathologischen Anatomie* (Vienna: Braumüller and Seidel, 1846), 4, 2.

13. Erna Lesky, introduction to *Carl von Rokitansky: Selbstbiographie und Antrittsrede*, by Carl Rokitansky, ed. Erna Lesky (Vienna: Böhlaus, 1960).

14. Max Neuburger, *Die Wiener medizinische Schule im Vormärz* (Vienna, 1921), 307, quoted in Lesky, *Selbstbiographie*, 90n42.

15. Lesky, *Selbstbiographie*, 90n42, 51.

16. Lesky, *Wiener medizinische Schule*, 130–131.

17. Lesky, *Selbstbiographie*, 53. On the early history of clinical *Prosekturen* and the foundations of pathological institutes, see Johannes Pantel and Axel Bauer, "Die Institutionalisierung der Pathologischen Anatomie im 19. Jahrhundert an den Universitäten Deutschlands, der deutschen Schweiz und Österreichs," *Gesnerus* 47 (1990): 303–328. For Lesky, the inception of the *Ordinariat* marked the peak of pathological anatomy's ascendance, but this judgment may be a little premature given Rokitansky's post-1848 institutional success.

18. Lesky, *Wiener medizinische Schule*, 137–138.

19. A commentator in *Die Gartenlaube* speaks of his "utter harmlessness from a political and any other point of view." Prof. Richter, "Eine Stätte, von wo Licht ausging," *Die Gartenlaube* 47–48 (1863): 747–750, 757–759, here 750. Whereas Rokitansky was not considered a political radical before 1848 (or thereafter), his assistant Georg Maria Lautner was actively involved in revolutionary politics, which meant that he was carefully observed by the authorities. For a fuller account of Rokitansky as a politician, see Peter Urbanitsch, "Zwischen Revolution und Konstitutionalismus: Rokitanskys Weg in die Politik," 147–161; and Helmut Rumpler, "Carl von Rokitansky als Exponent des österreichischen Liberalismus," 163–185, both in *Carl Freiherr von Rokitansky (1804–1878): Pathologe, Politiker, Philosoph, Gründer der Wiener Medizinischen Schule des 19. Jahrhunderts*, ed. Helmut Rumpler and Helmut Denk (Vienna: Böhlau, 2005).

20. Urbanitsch, "Zwischen Revolution und Konstitutionalismus," esp. 153.

21. Elisabeth Springer, *Geschichte und Kulturleben der Wiener Ringstrasse* (Wiesbaden: Franz Steiner, 1979), esp. 40–76.

22. Carl Schorske, *Fin-de-siècle Vienna: Politics and Culture* (Cambridge: Cambridge University Press, 1979), 24. See esp. his chapter on the Ringstrasse, 24–115. On the defortification of the German city, see Yair Mintzker, *The Defortification of the German City, 1689–1866* (Cambridge: Cambridge University Press, 2012).

23. Elisabeth Lichtenberger, *Wirtschaftsfunktion und Sozialstruktur der Wiener Ringstrasse* (Vienna: Böhlaus, 1970), esp. 17–52. For Vienna at the turn of the nineteenth century, see Steven Beller, *Rethinking Vienna 1900* (New York: Berghahn Books, 2001); and Schorske's classic account, Schorske, *Fin-de-siècle Vienna.*

24. Richter, "Eine Stätte," 758.

25. Springer, *Geschichte und Kulturleben*, esp. 120–121.

26. Erna Lesky, *Meilensteine der Wiener Medizin: Grosse Ärzte Österreichs in drei Jahrhunderten* (Vienna: Wilhelm Maudrich, 1981), 80; Erna Lesky, "Carl von Rokitansky (1804–1878)," *Neue österreichische Biographie ab 1815* 12 (1969): 38.

27. Richter, "Eine Stätte," 747.

28. Engstrom, *Clinical Psychiatry*.

29. Wilhelm Griesinger, *Die Pathologie und Therapie der psychischen Krankheiten für Aerzte und Studierende* (Stuttgart: Krabbe, 1845). The slogan "mental disease is brain disease" in this formulation cannot be found in Griesinger, although he gives the gist of it in his 1845 book (1). Cf. Heinz Schott and Rainer Tölle, *Geschichte der Psychiatrie* (Stuttgart: Beck, 2006), 70.

30. Cf. Engstrom, *Clinical Psychiatry*, 90. As scholars have emphasized, they adapted Griesinger's program to their own ends; see, e.g., Otto Marx, "Psychiatry on a neuropathological basis: Th. Meynert's application for the extension of his venia legendi," *Clio Medica* 6.1 (1971): 141. Hirschmüller, *Freuds Begegnung*, 28.

31. Helmut Gröger, "Zur Entwicklung der Psychiatrie in der Wiener Medizinischen Schule," in *Gründe der Seele: Die Wiener Psychiatrie im 20. Jahrhundert*, ed. Brigitta Keintzel and Eberhard Gabriel (Vienna: Picus, 1999): 33–34.

32. Theodor Meynert to the "medicinische Professoren-Collegium der Universität Wien," Antrag auf Habilitation, June 10, 1864, Beilage E, Curriculum vitae, Personalakt Theodor Meynert (MED PA 689), Vienna University Archive (hereafter VUA).

33. Hirschmüller, *Freuds Begegnung*, 97; Theodor Meynert to the "medicinische Professoren-Collegium der Universität Wien," Antrag auf Habilitation, June 10, 1864, Beilage E, Curriculum vitae, Personalakt Theodor Meynert (MED PA 689), VUA.

34. Hirschmüller, *Freuds Begegnung*, 97–98; Der niederösterreichische Landes-Ausschuß an das löbliche k. k. Professoren-Collegium der medizinischen Fakultät zu Wien, April 12, 1866, Personalakt Theodor Meynert (MED PA 689), VUA. Meynert competed for the post with the assistant of pathological anatomy Ferdinand Schott. He won the post with a vote of 13 to 4.

35. Rendered "brain-trunk" in the English translation of Meynert's textbook. Theodor Meynert, *Psychiatrie: Klinik der Erkrankungen des Vorderhirns begründet auf dessen Bau, Leistungen und Ernährung* (Vienna: Braumüller, 1884); Theodor Meynert, *Psychiatry: A Clinical Treatise on Diseases of the Fore-Brain Based upon a Study of Its Structures, Function, and Nutrition*, trans. Bernard Sachs (New York: Putnam's, 1885). See, e.g., Meynert, *Psychiatry*, 174, which corresponds to Meynert, *Psychiatrie*, 161.

36. Theodor Meynert, "Die Blosslegung des Bündelverlaufes im Grosshirnstamme," *Oesterreichische Zeitschrift für praktische Heilkunde* 11 (1865): 5–8, 25–28, 85–89, 153–156, 184–186, 437–440.

37. Theodor Meynert, "Die Medianebene des Hirnstammes, als ein Stück der Leistungsbahn zwischen dem Vorstellungsgebiete und den motorischen Hirnnerven," *Allgemeine Wiener medizinische Zeitung* 51 (1865): 411, 419.

38. These diseases were "primärer Irrsinn" (primary insanity), "Übergangsformen" (transitional forms), "einfacher Blödsinn" (simple idiocy), "paralytischer Blödsinn" (paralytic idiocy), alcoholism, and epilepsy. Theodor Meynert, "Das Gesammtgewicht und Theilgewichte des Gehirnes in ihren Beziehungen zum Geschlechte, dem Lebensalter und dem Irrsinn, untersucht nach einer neuen Wägungsmethode an den Gehirnen der in der Wiener Irrenanstalt im Jahre 1866 Verstorbenen," *Vierteljahresschrift für Psychiatrie in ihren Beziehungen zur Morphologie und Pathologie des Centralnervensystems, der physiologischen Psychologie, Statistik und gerichtlichen Medizin* 2 (1867): e.g., 159.

39. N.B.: in Vienna, the Psychiatric Clinic until 1875 was located outside the university context. But with Meynert working there, the growing hegemony of academic psychiatry was clearly felt. On the history of psychiatry in Austria, see, e.g., Martina Gamper, ed., *Psychiatrische Institutionen in Österreich um 1900* (Vienna: Ärzte, 2009); Leslie Topp, "The modern mental hospital in late nineteenth-century Germany and Austria: Psychiatric space and images of freedom and control," in *Madness, Architecture and the Built Environment: Psychiatric Spaces in Historical Context*, ed. Leslie Topp, James E. Moran, Jonathan Andrews (New York: Routledge, 2007), 241–261; Leslie Topp, "Psychiatric institutions, their architecture, and the politics of regional autonomy in the Austrian-Hungarian monarchy," *Studies in History and Philosophy of Biological and Biomedical Sciences* 38 (2007) 733–755; Eric Engstrom and Volker Roelcke, eds., *Psychiatrie im 19. Jahrhundert: Forschungen zur Geschichte von psychiatrischen Institutionen, Debatten und Praktiken im deutschen Sprachraum* (Mainz: Schwabe, 2003). A little later, the principles of pathological anatomy were increasingly followed in Prussia as well, e.g., by Virchow. See Engstrom, *Clinical Psychiatry*.

40. Hirschmüller, *Freuds Begegnung*, 74. As Hirschmüller points out, the critique was also motivated because Meynert was preferred over the more established candidate Max Leidesdorf. The formal precondition for the appointment was the extension of Meynert's *Habilitation* in the anatomy and physiology of the nervous system to psychiatry, which was granted in 1872. Wien am August 1, 1872, für den Minister für Kultus und Unterricht, Personalakt Theodor Meynert (MED PA 689), VUA.

41. Leopold Wittelshöfer, "Notiz," *Wiener medizinische Wochenschrift* 27 (1870): 578.

42. Carl Rokitansky to "löbliches med. Professoren-Collegium," January 17, 1873, Personalakt Theodor Meynert (MED PA 689), VUA.

43. Ibid., January 18, 1873.

44. Separat-Votum, Prof. Dr. Schlager, 20.1.1873, Personalakt Theodor Meynert (MED PA 689), VUA. On the conflict between Meynert and Schlager, see also Sophie Ledebur, "Das Wissen der Anstaltspsychiatrie in der Moderne: Zur Geschichte der Heil- und Pflegeanstalten *Am Steinhof* in Wien" (PhD diss., University of Vienna, 2011), 38–39. On difficulties in leadership at other psychiatric institutions, see also chapter 2 of this book and Eric Engstrom, "Assembling professional selves: On clinical instruction in German academic psychiatry," in Engstrom and Roelcke, *Psychiatrie im 19. Jahrhundert*, 120–121.

45. Separat-Votum, Prof. Dr. Schlager, January 20, 1873, Personalakt Theodor Meynert (MED PA 689), VUA.

46. Cf. Hirschmüller, *Freuds Begegnung*, esp. 71–76.

47. Lesky, *Vienna Medical School of the 19th Century*, 107. See also Robert J. Miciotto, "Carl Rokitansky: Nineteenth-century pathologist and leader of the New Vienna School" (PhD diss., Johns Hopkins University, 1979), 23.

48. Lesky, introduction to *Selbstbiographie*, 22.

49. Rokitansky, *Handbuch*, 6.

50. This is not to say, however, that Rokitansky and Skoda assumed a simplistic relationship between lesion and symptom. They also accounted for diseases for which it was impossible to find a solid lesion. For those, they located the disease process in the blood, thus returning to a quasi-humoral explanation of disease. Cf. Lesky, *Vienna Medical School*, 110–111.

51. Richter, "Eine Stätte," 759n2.

52. In his book, Skoda placed the findings of percussion and auscultation on a pathological anatomical basis. Josef Skoda, *Abhandlung über Perkussion und Auskultation* (Vienna: J.G. Ritter von Mösle's Witwe and Braumüller, 1839).

53. Lesky, "Carl von Rokitansky," 42.

54. Lesky, *Vienna Medical School*, 334.

55. Theodor Meynert, "Ein Fall von Sprachstörung, anatomisch begründet," *Medizinische Jahrbücher* 12 (1866): 152–187.

56. When the actual reference is ambiguous or general, I have alternated between the male and female pronouns for patients.

57. Ibid., 154.

58. The lesions were encephalitis (inflammation of the brain) in an anatomical region named the insula, and an edema near the anatomical structure of the cuneate fascicle, both caused by an embolus that had traveled into the insular artery. Meynert, "Sprachstörung," 156.

59. Ibid., 166, 167.

60. Theodor Meynert, "Studien über das pathologisch-anatomische Material der Wiener Irren-Anstalt," *Vierteljahresschrift für Psychiatrie* 3 (1868): 381–402.

61. The majority of Meynert's cases, however, came from the Vienna *Irrenanstalt*. Meynert, "Gesammtgewicht und Theilgewichte des Gehirnes."

62. Theodor Meynert, "Skizzen über Umfang und wissenschaftliche Anordnung des psychiatrischen Lehrstoffes," *Psychiatrisches Centralblatt* 1 (1876): 3.

63. Indeed, in an 1868 paper, Meynert criticized the "low barometer reading of exactness" of (alienist) psychiatry. Theodor Meynert, "Ueber die Nothwendigkeit und Tragweite einer anatomischen Richtung in der Psychiatrie," *Wiener medizinische Wochenschrift* 18 (1868): 573–576, 589–591, quotation on 575.

64. Meynert, "Gesammtgewicht und Theilgewichte des Gehirnes"; and Meynert, "Studien über das pathologisch-anatomische Material."

65. Theodor Meynert, "Ein Fortschritt in der psychiatrischen Diagnostik," *Mittheilungen des Wiener medicinischen Doctoren-Collegiums* 3 (1877): 249. Meynert had begun to "denominate" according to Griesinger's system. But according to Meynert's own observations, "from year to year . . . the numbers of melancholics and manics decreased" (267); these were important elements in Griesinger's system, which made Meynert question Griesinger's approach.

66. Meynert to "löbliches Dekanat," July 25, 1873, Medizinische Dekanatsakten, MED S 17.4, VUA.

67. Vorschlag Leidesdorf von Jakob Weiß und Wilhelm Svetlin als Assistenten von Meynert, June 16, 1875, Medizinische Dekanatsakten, MED S 17.4, VUA. If we look at Meynert's late nosological system, however, we are surprised to see that despite his earlier programmatic claims, the system is largely conservative, drawing for the most part on traditional diagnoses. In the end, although he did go through a process of redefining the categories based on his physiological insights, his changes were not very great. See, e.g., Theodor Meynert, "Die acuten hallucinatorischen Formen des Wahnsinns," *Jahrbücher für Psychiatrie* 2 (1881): 185.

68. Charles Rosenberg, "Contested boundaries, psychiatry, disease, and diagnosis," *Perspectives in Biology and Medicine* 49.3 (2006): 411.

69. Cf. Eric Engstrom, "Neurowissenschaften und Gehirnforschung," in *Geschichte der Universität Unter den Linden, 1810–2010*, ed. Heinz-Elmar Tenorth (Berlin: Akademie Verlag, 2010): 777–797, on 778.

70. Bell, at least initially, aimed to use his understanding of the spinal cord to study the brain. Cf. Clarke and Jacyna, *Origins*, 111.

71. Fearing, *Reflex Action*, 121.

72. Theodor Meynert, "Das Zusammenwirken der Gehirntheile," talk given at the Tenth International Medical Congress in Berlin, 1890, in *Sammlung von populär-wissenschaftlichen Vorträgen über den Bau und die Leistungen des Gehirns* (Vienna: Braumüller, 1892), 207.

73. For a discussion of Gall, phrenology, and localization, see Roger Cooter, *The Cultural Meaning of Popular Science: Phrenology and the Organization of Consent*

in Nineteenth-Century Britain (Cambridge: Cambridge University Press, 1984), esp. 1–11; Harrington, *Medicine, Mind, and the Double Brain*, esp. 6–34; Young, *Mind, Brain, and Adaptation*, 9–100; Clarke and Jacyna, *Origins*, 33–46, 212–307.

74. More precisely, in his 1861 publication, Broca emphasized that he dealt not with a motor but with an intellectual function. Paul Broca, "Remarques sur le siège de la faculté du langage articulé, suivies d'une observation d'aphémie (perte de la parole)," *Bulletin de la Société de l'Anatomie de Paris* (1861) 36: 335. On the 1861 debate at the Anthropological Society, see Jacyna, *Lost Words*, 53–80.

75. It is in his aphasia book that Wernicke's reflex model was first—and perhaps most clearly—presented. Another clear exposition of the mental reflex is in Carl Wernicke, "Aphasie und Geisteskrankheit," talk given at the Verhandlungen des Congresses für innere Medizin, Wiesbaden 1890, in *Gesammelte Aufsätze*, 153–60, first published in 1890.

76. Broca, "Remarques sur le siège."

77. However, of course, he did that too. There as well he drew on Meynert, more specifically in his identification of the first temporal gyrus as sensory equivalent to Broca's area of motor speech. Meynert in 1866, referring to a case of aphasia that had come to dissection, had suggested the existence of a *Klangfeld* (sound territory) in the area of the temporal lobe. Wernicke, drawing on his own anatomical evidence, picked up on Meynert's argumentation and closed the circle of the reflex arc. Meynert, "Sprachstörung."

78. There was a third type of aphasia, transcortical aphasia, in this simple model. I will consider this in chapter 2.

79. Gustav Fritsch and Eduard Hitzig, "Ueber die elektrische Erregbarkeit des Grosshirns," *Archiv für Anatomie, Physiologie und wissenschaftliche Medizin* 37 (1870): 300–332. On the episode, see also Hagner, *Homo cerebralis*, 273–279.

80. Fritsch and Hitzig, "Erregbarkeit," 312. However, there are indications that Fritsch and Hitzig did not fully embrace the term (311 and 332).

81. E.g., David Ferrier, "Experimental research in cerebral physiology and pathology," *West Riding Lunatic Asylum Medical Reports* 3 (1873): 30–96; Hermann Munk, *Ueber die Functionen der Grosshirnrinde* (Berlin: Hirschwald, 1881). Fritsch and Hitzig themselves did not talk about "sensory centers" but rather divided the brain into "motor" and "nonmotor" centers. Fritsch and Hitzig, "Erregbarkeit," 310.

82. Hall believed that the excito-reflector system was connected to the spinal cord and the medulla oblongata but was physiologically (and perhaps anatomically) separate. Marshall Hall, *Memoirs on the Nervous System* (London: Sherwood, 1837), 49.

83. See especially Marshall Hall, "On the reflex function of the medulla oblongata and medulla spinalis," *Philosophical Transactions of the Royal Society of London* 123 (1833): 635–665; and Hall, *Memoirs*. For an excellent discussion of Hall's work and the early history of the reflex concept, see Leys, *From Sympathy to Reflex*.

84. Apart from those discussed here, Volkmann and Pflüger are notable examples in the German tradition. In the British context, William Carpenter, Richard Grainger, Thomas Laycock, and John Hughlings Jackson conceived brain functions in sensory-motor terms by extending the scope of the reflex. Cf. Clarke and Jacyna, *Origins*, 124–147. See also Quick, "Techniques of life," esp. chap. 2.

85. For details on the experiment, see Johannes Müller, *Handbuch der Physiologie des Menschen*, vol. 1 (Coblenz: Hölscher, 1833).

86. As Hall and others at the time pointed out, the medulla oblongata, located between the brain and the spinal cord, was responsible for respiratory function. Like the medulla spinalis, it was involved in reflex function. Hall, *Memoirs*, 35.

87. Müller, *Handbuch*, 1, 698.

88. This emphasis on "real sensations" corresponded to a principle of Müller's for which he was much better known: the law of specific nerve energies, according to which different sensations were caused by the different nervous structures that carried them, not by the stimuli provoking the sensations in the first place.

89. Müller, *Handbuch*, 1, 699. Hall did not address the exact process by which excitation moved from the excitatory to the reflector part of the reflex arc. His notion of tonus, however, suggests that more than one muscle was involved. Hall, *Memoirs*, 93–94.

90. Müller, *Handbuch*, 1, 699–700.

91. Many theorists of the reflex at the time drew on the concept of tonus (e.g., Marshall Hall, Johannes Müller). They usually referred to the tone of the muscles of the body and sphincter muscles, which disappeared when the spinal cord was removed. E.g., Hall, *Memoirs*, 31, 94; or Müller, *Handbuch*, 1, 783–789.

92. Wilhelm Griesinger, "Ueber psychische Reflexactionen: Mit einem Blick auf das Wesen der psychischen Krankheiten," *Archiv für physiologische Heilkunde* 2 (1843): 84.

93. Theodor Meynert, "Beiträge zur Theorie der maniakalischen Bewegungserscheinungen nach dem Gange und Sitze ihres Zustandekommens," *Archiv für Psychiatrie und Nervenkrankheiten* 2 (1870): 626–628. Meynert criticized Griesinger for relying on the insufficiently physiological concept of the *Strebung* to explain which sensory and motor nerves were associated (628). Apart from Meynert's work, association plays a significant role in the work of Carl Wernicke, Paul Flechsig, August Forel, Sigmund Exner, and Otfrid Foerster.

94. N.B., however, that David Hartley had a physiological theory. Friedrich Herbart's approach in the German context was more of a mathematical theory. See Ingrid Kleeberg, "Poetik der nervösen Revolution: Psychophysiologie und das politisch Imaginäre, 1750–1860" (PhD diss., University of Constance, 2011). See also Barbara Bowen Oberg, "David Hartley and the Association of Ideas," *Journal of the History of Ideas* 37 (1976): 441–454. On the impact of association psychology on nineteenth-century brain science, see also Kenneth Levin, *Freud's Early Psychology of the Neuroses: A Historical Perspective* (Pittsburgh: University of Pittsburgh Press, 1978), 27.

95. Considering Meynert's use of the example in 1865 and 1884 respectively, we see that his views on the nature of *Vorstellungen* and their associations remained the same. Theodor Meynert, "Anatomie der Hirnrinde als Träger des Vorstellungslebens und ihrer Verbindungsbahnen mit den empfindenden Oberflächen und den bewegenden Massen," in *Lehrbuch der psychischen Krankheiten*, ed. Maximilian Leidesdorf (Erlangen: Enke, 1865), 45–73. Meynert is not named as author in Leidesdorf, but he makes the authorship clear in his *Psychiatry*, 153.

96. More specifically, the connection is from the eye to the originating cell (*Ursprungszelle*) of the optical nerve and from there, through the corona radiata, to the cortical cell. Meynert, "Anatomie," 52–53. The same was true for other sensory surfaces (*Sinnesoberflächen*), such as the ear.

97. Leidesdorf, *Lehrbuch der psychischen Krankheiten*, 52.

98. This was an induction process, as in the philosophy of John Stuart Mill. Cf. Meynert, *Psychiatry*, 153–154, 177–179. Meynert's view of consciousness corresponded to Theodor Fechner's notion of partial sleep, according to which the "Hemisphärenleistung" was always in a state of partial sleep. The "Funktionshöhe" of the different cortical territories varied; they were never all active at the same time, a process that was to Meynert regulated by cortical functional hyperemia. Meynert, *Psychiatrie*, 199.

99. Cf. Carl Wernicke's model of speech that constituted a similar extension of association principles to the sensory-motor sphere, in chapter 2.

100. Example presented in Meynert, *Psychiatry*, 160–161.

101. Meynert, "Skizzen über Umfang," 4; Meynert, "Ueber die Nothwendigkeit," 575.

102. Meynert, "Sprachstörung," 154.

103. See chapter 2 for a discussion of these terms.

104. Meynert, *Psychiatrie*, 198–203.

105. Theodor Meynert, "Ueber Fortschritte der Lehre von den psychiatrischen Krankheitsformen," *Psychiatrisches Centralblatt* 1 (1878): 9.

106. Meynert, *Psychiatrie*, 268.

107. Theodor Meynert, "Zum Verständniss der functionellen Nervenkrankheiten," *Wiener medizinische Blätter* 5 (1882): 484. For similar notions of inhibition, see also Meynert's explanation of delusion ("Wahn") in "Ueber Fortschritte der Lehre," 7–8.

108. E.g. Theodor Meynert, "Vom Gehirne der Säugethiere," in Salomon Stricker, ed., *Handbuch der Lehre von den Geweben des Menschen und der Thiere* (Leipzig: Engelmann, 1872), 1:695. In this emphasis on the one-to-one fiber connection between cortex and periphery, Meynert conformed to the tradition. See Müller, *Handbuch*, 1, 659.

109. True, it was a determination arising from the *connectivity* of sites on peripheries of the brain and the body; also, localization was not fixed, but dynamic and changing. But it was still static in the sense that the location of an individual *Vorstellung* was in a specific cell, connected to a specific peripheral site.

110. As Michael Hagner has perceptively noted, for Meynert, "there was no hierarchy and no center" within the association system, but only a "dynamic constellation of innumerable elements." Hagner, *Homo cerebralis*, 270.

111. Theodor Meynert, "Fragment aus den anatomischen Corollarien und der Physiologie des Vorderhirns," *Jahrbücher für Psychiatrie* 2 (1881): 87.

112. Theodor Meynert, "Ueber die Bedeutung der Stirnentwicklung," talk given at the Wissenschaftlicher Club in Vienna, 1886, in *Sammlung*, 105. Even when he studied brain centers or areas, it was generally motivated by his interest in fiber connections, e.g., in Meynert, "Sprachstörung."

113. Meynert, "Zum Verständniss," 483.

114. Meynert, *Psychiatry*, 214, my emphasis. It is also interesting that Meynert in his later writings chose the term *Beleidigung* (offense) over *lesion*, e.g., in Theodor Meynert, "Kraniologische Beiträge zur Lehre von der psychopathischen Veranlagung," *Jahrbücher für Psychiatrie* 1 (1879): 71.

115. Meynert, "Zusammenwirken." Meynert referred to similar social models elsewhere in his work, e.g. Theodor Meynert, *Klinische Vorlesungen über Psychiatrie auf wissenschaftlichen Grundlagen für Studirende und Ärzte, Juristen und Psychologen* (Vienna: Braumüller, 1890), iv, 41; Meynert, *Psychiatrie*, 158.

116. Meynert, "Zusammenwirken," 203.

117. Ibid., 204.

118. Ibid., 204–205.

119. Ibid., 205. Cf. Cornelius Borck's article "Fühlfäden und Fangarme: Metaphern des Organischen als Dispositiv der Hirnforschung," in *Ecce Cortex: Beiträge zur Geschichte des modernen Gehirns*, ed. Michael Hagner (Göttingen: Wallstein, 1999), 144–176.

120. Meynert, "Zusammenwirken," 205.

121. Meynert, *Psychiatry*, 168. See 168–178 for his full theory of individuality.

122. Ibid., 168.

123. Ibid., 169. Meynert refers to Wundt here.

124. Ibid., 169. We will see in chapter 3, on Freud, that the process of *Bahnung* was central for Meynert.

125. Ibid., 175.

126. Meynert, "Zusammenwirken," 205; Meynert, *Psychiatry*, 175.

127. Meynert, *Psychiatry*, 176.

128. Hirschmüller, *Freud's Begegnung*, 124.

Chapter Two

1. Today it is known by the Polish name, Wrocław.

2. Norman Davies and Roger Moorhouse, *Microcosm: Portrait of a Central European City* (London: Pimlico, 2003), 289.

3. Between 1871 and 1914, its population grew by more than 150 percent. Ibid., 317.

4. Ibid., 283, 282.

5. Ibid., 292.

6. Adolf Strümpell, *Aus dem Leben eines deutschen Klinikers: Erinnerungen und Beobachtungen* (Leipzig: Vogel, 1925), 230. See also Universitätsbund Breslau, *Aus dem Leben der Universität Breslau* (Breslau: Breslau Genossenschafts Buchdruckerei, 1936), 166.

7. Breslau also had the first chair of physiology in Germany, established in 1811, the year the University of Breslau was founded. It was held initially by Johann Purkinje. On Pavlov, see Todes, *Pavlov's Physiology Factory*; and Todes, *Ivan Pavlov*. Also see Thomas N. Bonner, *American Doctors and German Universities: A Chapter in International Intellectual Relations, 1870–1914* (Lincoln: University of Nebraska Press, 1963), 18.

8. Carl Wernicke, *Der aphasische Symptomencomplex: Eine psychologische Studie auf anatomischer Basis* (Breslau: Cohn and Weigert, 1874), reprinted in Carl Wernicke, *Gesammelte Aufsätze und kritische Referate zur Pathologie des Nervensystems* (Berlin: H. Kornfeld, 1893), 1–70.

9. Carl Wernicke, "Ueber den wissenschaftlichen Standpunkt in der Psychiatrie." *Wiener medizinische Wochenschrift* 42 (1880): 1149–1152; 43 (1880): 1179–1182.

10. Carl Wernicke, "Die Aufgaben der klinischen Psychiatrie," *Breslauer ärztliche Zeitschrift* 13 (1887), reprinted in Wernicke, *Gesammelte Aufsätze*, 146.

11. Karl Bonhoeffer, "Lebenserinnerungen von Karl Bonhoeffer—Geschrieben für die Familie," in *Karl Bonhoeffer Zum Hundersten Geburtstag am 31. März 1968*, ed. J. Zutt, E. Straus, and H. Scheller (Berlin: Springer, 1969), 39. Wernicke's *Lehrbuch* had in fact three volumes.

12. We can also see this change of heart in Wernicke's public statements. For example, in the first issue of *Monatsschrift für Psychiatrie und Neurologie*, which Wernicke founded in 1897 together with his colleague Theodor Ziehen at Jena, he harshly criticized the *Zentrenlehre* of their colleague Paul Flechsig. Carl Wernicke, "Tagesfragen," *Monatsschrift für Psychiatrie und Neurologie* 1 (1897): 5.

13. Engstrom, *Clinical Psychiatry*, 121–127.

14. In response to the great nosological heterogeneity that prevailed in psychiatry at the time, Neumann had suggested the concept of "unitary psychosis" (*Einheitspsychose*). The lack of schematism in Neumann's system could well have contributed to Wernicke's later rich psychopathological description. Arthur Leppmann, "Heinrich Neumann 1814–1884," in *Deutsche Irrenärzte: Einzelbilder ihres Lebens und Wirkens*, ed. T. Kirchhoff (Berlin: Springer, 1921), 1:261–265; cf. Engstrom, *Clinical Psychiatry*, 27.

15. E.g., by Henry Charlton Bastian (1869) and Johann Baptist Schmidt (1871).

16. Gertrude Eggert, *Wernicke's Work on Aphasia: A Sourcebook and Review* (The Hague: Mouton, 1977), 21, 43. See also Jürgen Tesak, *Geschichte der Aphasie*

(Idstein: Schulz-Kirchner, 2001); and Jacyna, *Lost Words*. Wernicke himself continued to work on aphasia throughout his life. See Eggert, *Wernicke's Work on Aphasia*, 20–45, for a discussion of the different phases.

17. Thanks to Dr. Frank Pillmann for drawing my attention to this point. The literature on Wernicke and aphasia does not usually note this shortage of evidence. E.g., Norman Geschwind, "Carl Wernicke, the Breslau school and the history of aphasia," in *Selected Papers on Language and the Brain*, ed. R. S. Cohen and M. W. Wartofsky (Dordrecht: D. Reidel, 1963), 42–61; Eggert, *Wernicke's Work on Aphasia.* On Wernicke's influence today, see Jürgen Tesak, *"Der aphasische Symptomencomplex" von Carl Wernicke* (Idstein: Schulz-Kirchner-Verlag, 2005); and Tesak, *Geschichte der Aphasie*. See also Jacyna, *Lost Words*, especially on critiques of the Wernicke model.

18. There was another case in the *Nachtrag* (addendum). Wernicke, *Symptomencomplex*, 69–70.

19. Eggert, *Wernicke's Work on Aphasia*, 21. On the other hand, in a later paper of 1895, Wernicke followed the more conventional structure of presenting the two cases first and then discussing them. Carl Wernicke, "Zwei Fälle von Rindenläsion," *Arbeiten aus der psychiatrischen Klinik in Breslau* 2 (1895): 33–52.

20. Wernicke never studied or worked with Griesinger, but his understanding of mental function, in particular his concept of the "psychic reflex," was strongly influenced by Griesinger's ideas. A similar form of the psychic reflex was also developed in England by Thomas Laycock (1812–1876). Although Griesinger's and Laycock's concepts show great parallels, there was no direct connection between the two, as Clarke and Jacyna emphasize in *Origins*, 127–128. On cerebral reflexes in the British tradition, see also Quick, "Techniques of life." The Russian physiologist Ivan Sechenov developed a similar concept of "cerebral reflexes" in 1863, independent of Griesinger and Laycock's earlier work, see *Origins*, 145–146. See there for literature on Sechenov. For a discussion of Sechenov's physiology, see Jeannerod, *Brain Machine*, 38–41.

21. Griesinger, "Reflexactionen"; Wernicke, *Symptomencomplex*, 7.

22. Wernicke, *Symptomencomplex*, 66, translation in Eggert, *Wernicke's Work on Aphasia*, 142.

23. Wernicke, "Symptomencomplex," 8, translation in Eggert, *Wernicke's Work on Aphasia*, 97.

24. For a different view, see Paul Eling, "Hatte Wernickes Sprachtheorie ihre Wurzeln in Berlin?" *Schriftenreihe der Deutschen Gesellschaft für Geschichte der Nervenheilkunde* 7 (2001): 33–38.

25. On a more anecdotal note, we know that Meynert's portrait was the only picture in the lecture hall of the *Krankenvorstellungen*. Cf. Karl Heilbronner, "Nekrolog C. Wernicke," *Allgemeine Zeitschrift für Psychiatrie* 62 (1905): 882; or Kurt Goldstein, "Carl Wernicke (1848–1904)," in *Founders of Neurology: One Hundred and Forty-Six Biographical Sketches by Eighty-Eight Authors*, ed. Webb Haymaker and Francis Schiller (Springfield, IL: Thomas, 1970), 531–535.

26. Wernicke spent six months working with Meynert in 1871, while an assistant to Neumann in Breslau; see below for more detail.

27. Wernicke, *Symptomencomplex*, 3, translation in Eggert, *Wernicke's Work on Aphasia*, 92.

28. In my translation of *Herdsymptome* (lesion symptoms), *Ausfallsymptome* (symptoms of loss), *Allgemeinsymptome* (general symptoms), *Herderkrankungen* (lesion diseases), and *Allgemeinerkrankungen* (general diseases), I depart from Eggert, *Wernicke's Work on Aphasia*, who translated *Herdsymptome* as "focal psychic diseases."

29. Carl Wernicke, *Grundriss der Psychiatrie in klinischen Vorlesungen*, 2nd ed. (Leipzig: Thieme, 1906), 109.

30. Wernicke, *Symptomencomplex*, 68, translation in Eggert, *Wernicke's Work on Aphasia*, 144.

31. Carl Wernicke, "Ein Fall von Ponserkrankung," *Archiv für Psychiatrie und Nervenkrankheiten* 7 (1877): 513–538.

32. Ibid., 522.

33. Eggert, *Wernicke's Work on Aphasia*, 6.

34. Carl Wernicke, "Erkrankungen der inneren Kapsel: Ein Beitrag zur Diagnose der Heerderkrankungen" (Breslau: Cohn and Weigert, 1875), in *Gesammelte Aufsätze*, 170. Cf. Wernicke's paper "Herabsetzung der electrischen Erregbarkeit bei cerebraler Lähmung," *Breslauer ärztliche Zeitschrift* 17 (1886), in *Gesammelte Aufsätze*, 260, where he mentioned his previous experience in the field. After a conflict with the Charité-Direktion, Wernicke was dismissed from his position at the Charité and settled down in private practice in Berlin, where he remained for the next seven years. See Archiwum Państwowe we Wrocławiu, Akta miasta Wrocławia 33972, 16, State Archive Breslau (hereafter SAB). On details of the case, see K.-J. Neumärker, "Carl Wernicke und Karl Kleist. Zwei Biographien—eine Richtung in ihrer Entwicklung," *Fundamenta Psychiatrica* 8 (1994): 178–184; and Eric Engstrom, "The birth of clinical psychiatry: Power, knowledge, and professionalization in Germany, 1867–1914" (PhD diss., University of North Carolina at Chapel Hill, 1997), 242n135.

35. In his work in Breslau, Wernicke continued to use electrotherapy and electrodiagnosis in his polyclinic. See Ludwig Mann, "Wernickes Stellung zur Elektrodiagnostik und Elektrotherapie," *Zeitschrift für Elektrotherapie und Elektrodiagnostik* 7 (1905): 311–317. Cf. Andreas Marneros and Frank Pillmann, *Das Wort Psychiatrie wurde in Halle geboren: Von den Anfängen der deutschen Psychiatrie* (Stuttgart: Schattauer, 2005), 124.

36. Akta miasta Wrocławia III/8783 Wernicke 46-51—December 13, 1889, medizinische Fakultät an Minister, SAB; Carl Wernicke, *Lehrbuch der Gehirnkrankheiten für Aerzte und Studirende* (Kassel: Fischer, 1881–83).

37. Davies and Moorhouse, *Microcosm*, 270; Wolfgang Mommsen, "Das Ringen um den nationalen Staat: Die Gründung und der innere Ausbau des Deutschen

Reiches unter Otto von Bismarck 1850–1980," in *Propyläen Geschichte Deutschland*, ed. Dieter Groh (Berlin, Propyläen, 1993), vol. 7, part 1, 652.

38. Bismarck did not build the social insurance system from scratch. There were several local sickness funds for which membership was mandatory. Henry Sigerist, "From Bismarck to Beveridge: Developments and trends in social security legislation," *Journal of Public Health Policy* 20 (1999): 474–496, originally published in *Bulletin of the History of Medicine* 13 (1943): 365–388. See also Werner Gerabek, "Der Weg zur Bismarckschen Invaliditäts- und Altersversicherung aus medizinhistorischer Sicht," *Würzburger medizinhistorische Mitteilungen* (1992).

39. The overall patient numbers increased from 76 to 173 between 1872 and 1883; the number of admissions to the Allerheiligenhospital and a newly opened branch at the Wenzel-Hancke Hospital increased from 290 to 722. Carl Wernicke, "Ueber die Irrenversorgung der Stadt Breslau," *Allgemeine Zeitschrift für Psychiatrie und psychisch-gerichtliche Medizin* 45 (1889): 434.

40. Alfred Buchwald, *Das Kranken-Hospital zu Allerheiligen in Breslau* (Breslau: Schletter, 1896), 1.

41. The figures are for the year 1880.

42. Richard Plüddemann (1846–1910) was commissioned to design several other buildings in Breslau, including the University Library, the Market Hall, and the Grunwaldzki Bridge. Johann Robert Mende (1824–1899) worked in Breslau and Berlin. The psychiatric hospital building unites elements of Northern German Gothic and Berlin Neo-Renaissance; cf. Andrzej Kiejna and Małgorzata Wójtowicz, *Z dziejów Kliniki Psychiatrycznej i Chorób Nerwowych we Wrocławiu: wybitni przedstawiciele i budowle—Zur Geschichte der psychiatrischen und Nervenklinik in Breslau: Bedeutende Vertreter und Bauwerke* (Wrocław: Fundacja Ochrony Zdrowia Psychicznego—Stiftung zur Förderung der psychischen Gesundheit, 1999), 36.

43. Buchwald, *Kranken-Hospital zu Allerheiligen*, 19.

44. Engstrom, *Clinical Psychiatry*, esp. chap. 3.

45. Griesinger, *Die Pathologie und Therapie der psychischen Krankheiten für Aerzte und Studirende*, 4th ed. (Braunschweig: Wreden, 1876), 9–11.

46. Carl Wernicke, "Stadtasyle und psychiatrische Kliniken," *Klinisches Jahrbuch* 2 (1890): 186.

47. Ibid., 187. On Griesinger's reform program, see Engstrom, *Clinical Psychiatry*, chap. 3.

48. Wernicke, "Stadtasyle," 187.

49. J. M. Charcot, *Poliklinische Vorträge*, vol. 1, *Schuljahr 1887–1888*, trans. Sigmund Freud (Leipzig: Deuticke, 1892). Cf. Didi-Huberman, *Invention of Hysteria: Charcot and the Photographic Iconography of the Salpêtrière* (Cambridge, MA: MIT Press, 2003); and the more recent Asti Hustvedt, *Medical Muses: Hysteria in Nineteenth-Century Paris* (New York: Norton, 2011). See also Christopher Goetz, Michel Bonduelle, and Toby Gelfand, *Charcot: Constructing Neurology*

(New York: Oxford University Press, 1995); J. Bogousslavsky, ed., *Following Charcot: A Forgotten History of Neurology and Psychiatry* (Basel: Karger, 2011); Engstrom, "Assembling professional selves; Rainer Herrn and Alexander Friedland, "Der demonstrierte Wahnsinn—Die Klinik als Bühne," *Berichte zur Wissenschaftsgeschichte* 37.4 (2014): 309–331.

50. Charcot was often criticized, however, for the theatrical nature of his presentations. According to Krafft-Ebing, clinical presentations in psychiatry had been performed since about the mid-eighteenth century. Richard v. Krafft-Ebing, *Der klinische Unterricht in der Psychiatrie. Eine Studie* (Stuttgart: Enke, 1890), 4. On notation in psychiatry, see also Sophie Ledebur, "Sehend schreiben, schreibend sehen: Vom Aufzeichnen psychischer Phänomene in der Psychiatrie," in *Krankheit schreiben: Aufschreibeverfahren in Medizin und Literatur*, ed. Yvonne Wübben and Carsten Zelle (Göttingen: Wallstein, 2013), 82–108.

51. As Engstrom has pointed out, though, they often clashed with middle-class sensibilities, because demonstrations were perceived as intrusions into the private sphere. "On psychiatric instruction," 132.

52. Engstrom, "Assembling professional selves," 137.

53. Carl Wernicke, *Krankenvorstellungen aus der Psychiatrischen Klinik in Breslau* (Breslau: Schletter, 1899–1900), 3:13.

54. Ibid., 3:57–58.

55. Wernicke, "Aphasie und Geisteskrankheit," 157–158.

56. Wernicke, *Grundriss* (1906), 16.

57. Carl Wernicke, "Ueber das Bewusstsein," *Allgemeine Zeitschrift für Psychiatrie* 35 (1879), in *Gesammelte Aufsätze*, 135.

58. Ibid.; Wernicke, *Grundriss* (1906), 62.

59. In fact, the patients were mostly male, and so were the assistants and medical students in the room. We know, however, that Breslau began to admit women to study medicine starting in 1892. Krysztof Popinski, *Studenten an der Universität Breslau 1871 bis 1921: Eine sozialgeschichtliche Untersuchung*, trans. Thorsten Möllenbeck, Historia Academica, vol. 26 (Würzburg: Studentengeschichtliche Vereinigung des Coburger Convents, 2009), 66.

60. Wernicke, *Krankenvorstellungen*, 2:50–51.

61. Ibid., 2:46. Although in *Grundriss* (1906), 46, Wernicke was clear that the somatopsyche was the bottom layer, covered by the allopsyche, in "Aufgaben der Psychiatrie," 151, he leaves the exact location of the layers open.

62. Wernicke, *Krankenvorstellungen*, 2:61.

63. The second sphere, the allopsyche, was "constructed" from the somatopsyche in the sense that the body was required for perception. Wernicke called the autopsyche a resultant (*die Resultante*) from the other two spheres. "Aufgaben der Psychiatrie," 148.

64. Wernicke compares his concept of the somatopsyche with Meynert's idea of the primary ego (*primäres Ich*). Meynert's primary ego is the most basic, infantile

ego, which is expanded to the secondary, adult ego through association processes. Cf. Meynert, *Klinische Vorlesungen*.

65. Wernicke, "Aufgaben der Psychiatrie," 147.

66. Wernicke, *Grundriss* (1906), 45.

67. Ibid., 47.

68. Wernicke, "Aufgaben der Psychiatrie," 149.

69. Wernicke, "Stadtasyle," 193.

70. Wernicke, *Krankenvorstellungen*. In the *Chronik der Königlichen Universität zu Breslau*, Wernicke mentions the publication of four volumes of *Krankenvorstellungen*, but there seem to be only three volumes. Universitas Vratislaviensis, *Chronik der Königlichen Universität zu Breslau (der Schlesischen Friedrich-Wilhelms-Universität zu Breslau)*, vols. 1–27 (Breslau: 1888–1913), vol. 13 (1898–1899), 78; vol. 14 (1899–1900), 67.

71. The selection probably took place the day before, when Wernicke examined all patients to be presented, which is indicated in the *Krankenvorstellungen*, e.g., 2:34. The selection was based on "didactic purposes," to introduce students to "psychiatric symptomatology" and, at the same time, present them with the most prevalent conditions. *Krankenvorstellungen*, 1:1.

72. Akta miasta Wrocławia, 33975, SAB.

73. On the theatricality of clinical demonstrations, see also Herrn and Friedland, "Der demonstrierte Wahnsinn."

74. Wernicke, *Krankenvorstellungen*, 1:16.

75. Ibid., 1:14.

76. The Liebigsche Lokal is mentioned four times in the *Krankenvorstellungen*, on 1:14–15 and 1:56–57.

77. Ibid., 1:14, 56.

78. There, as well, we find that patients take on active roles, for example by collaborating with the doctors and making themselves into "model patients." See the recent analysis in Hustvedt, *Medical Muses*. See also Didi-Huberman, *Invention of Hysteria*.

79. Wernicke, *Krankenvorstellungen*, 1:47, 56.

80. Ibid., 2:84, 1:87.

81. Ibid., 1:88, 14, 28.

82. Ibid., 1:14.

83. Ibid., 1:13.

84. Ibid., 1:16.

85. Ibid., 1:74.

86. Ibid., 2:82.

87. Guillaume-Benjamin Duchenne, "Exposition d'une nouvelle méthode de galvanisation, dite galvanisation localisée," *Archives générales de médecine* 4 (1850): 286–289.

88. Indeed it seems that Duchenne, in electrically stimulating individual facial muscles, adhered to the same principle. His concept of synergies applied, strictly

speaking, only to the muscles of the body. See Stéphanie Dupouy, "Künstliche Gesichter: Rodolphe Töpffer und Duchenne de Boulogne," in *Kunstmaschinen: Spielräume des Sehens zwischen Wissenschaft und Ästhetik*, ed. Andreas Mayer and Alexandre Métraux (Frankfurt: S. Fischer, 2005): 24–60.

89. Guillaume-Benjamin Duchenne, preface to *Physiologie des mouvements, démontrée à l'aide de l'expérimentation électrique et de l'observation clinique et applicable à l'étude des paralysies et des déformations* (Paris: Baillière, 1867), viii.

90. Ibid., viii (my emphasis), xi.

91. There is a growing literature on notation systems in the medicine of mind and brain. See, e.g., Cornelius Borck and Armin Schäfer, *Psychographien* (Berlin: Diaphanes, 2006); Sophie Ledebur, "Schreiben und Beschreiben: Zur epistemischen Funktion von psychiatrischen Krankenakten, ihrer Archivierung und deren Übersetzung in Fallgeschichten," *Berichte zur Wissenschaftsgeschichte* 34 (2011): 102–124; Volker Hess and Sophie Ledebur, "Taking and keeping: A note on the emergence and function of hospital patient records," *Journal of the Society of Archivists* 32 (2011): 21–32; Volker Hess and Andrew Mendelsohn, "Case and series: Medical knowledge and paper technology, 1600–1900," *History of Science* 48 (2010): 287–314. See also Yvonne Wübben's book on the function of language in psychiatry, *Verrückte Sprache: Psychiatrie und Dichter in der Anstalt des 19. Jahrhunderts* (Konstanz: Konstanz University Press, 2012).

92. Hugo Liepmann, "Carl Wernicke," in *Deutsche Irrenärzte: Einzelbilder ihres Lebens und Wirkens*, ed. Theodor Kirchhoff (Berlin: Julius Springer, 1921–24), 242. Cf. Wernicke's own description of the technique, *Krankenvorstellungen*, 1:1.

93. Wernicke, *Krankenvorstellungen*, 1:81.

94. Ibid., 3:14–16, also 55 and 72.

95. Karl Bonhoeffer, "Die Stellung Wernickes in der modernen Psychiatrie," *Berliner klinische Wochenschrift* 42 (1905): 927–928. For a similar statement, see Hugo Liepmann, "Wernickes Einfluss auf die klinische Psychiatrie," *Monatsschrift für Psychiatrie und Neurologie* 30 (1911): 28.

96. A comparison with the relatively rare occurrence of *Herdsymptome* in the *Krankenvorstellungen* is instructive here.

97. Wernicke, *Krankenvorstellungen*, 1:97.

98. Ibid., 1:52. Ludwig Windthorst (1812–1891) was the leader of the Catholic Center Party from 1874 to his death. Eduard Lasker (1829–1884) founded the National Liberal Party in 1867. Otto von Bismarck died shortly before the *Krankenvorstellungen*, in 1898. When directly asked about the year, the patient answered "1890," but his answers about politics indicate that he must have thought it was earlier (or held several beliefs at the same time). Emperor Wilhelm I, for example, had died in 1888, his wife Augusta of Saxe-Weimar early in 1890, and the National Liberal Eduard Lasker in 1884.

99. That he sympathized with them can be inferred from the remark that "he had talked knowledgably about them on the ward." Ibid., 1:53.

100. Ibid., 1:53.

101. Universitas Vratislaviensis, *Chronik Breslau*, 13:66–67.

102. To my knowledge, the "Golden Book" has not survived. The only existing clinical records of Wernicke's are from his time as director of the Psychiatric Hospital in Halle (Saale).

103. Heilbronner, "Nekrolog C. Wernicke," 889.

104. Ibid., 887. Another account of the same incident was given by Wernicke's assistant Karl Kleist in "Carl Wernicke," in *Große Nervenärzte*, ed. Kurt Kolle, 3 vols. (Stuttgart: Thieme, 1959), 2:106–128. It is interesting that Wernicke's loose collection of clinical cases has become the subject of these hagiographic accounts. Cf. a parallel passage by Thomas Huxley on Descartes: "He was an unwearied dissector and observer; and it is said, that, on a visitor once asking to see his library, Descartes led him into a room set aside for dissections, and full of specimens under examination. 'There,' said he, 'is my library.'" Thomas Huxley, "On the hypothesis that animals are automata and its history," *Fortnightly Review* 22 (1874): 201.

105. Carl Wernicke Patient Files (hereafter CWPF), Universitätsklinik und Poliklinik für Psychiatrie, Psychotherapie und Psychosomatik, Halle. File number 12348, admission date 3/19/1905.

106. As Karl Kleist, Wernicke's assistant in Halle, pointed out, he and the other assistants "read . . . until late at night section by section of his 'Grundriss der Psychiatrie' . . . under the impression of a splendid . . . presentation, which left the older psychiatry . . . far behind. Kleist "Carl Wernicke," 111.

107. File number 12348, admission date 3/19/1905, CWPF. For a quantitative analysis of the Wernicke patient files at Halle, see Pillmann and Marneros, *Das Wort Psychiatrie*; and Frank Pillmann et al., "An analysis of Wernicke's original case records: His contribution to the concept of cycloid psychoses," *History of Psychiatry* 11 (2000): 355–369.

108. In fact, as the analysis by Frank Pillmann suggests, the tripartite model did not figure so prominently in the overall diagnoses at Halle. Of the 723 admissions to Wernicke's clinic in 1904–5, the psychoses only constituted 11.1% of all diagnoses. The related motility psychoses constituted 9.2%. The most frequently made diagnosis was progressive paralysis. Overall, then, the diagnoses that Wernicke made, which were distinct from those drawing on his tripartite model, were relatively conventional. Marneros and Pillmann, *Das Wort Psychiatrie*, 132, 139.

109. Bonhoeffer, "Stellung Wernickes," 928.

110. Eight in vol. 2 and seven in vol. 3.

111. Two of them turned out not to have a *Herderkrankung* in the end. Wernicke, *Krankenvorstellungen*, vols. 1–3.

112. Wernicke, *Grundriss der Psychiatrie in klinischen Vorlesungen*, 1st ed. (Leipzig: Thieme, 1894), 1:3. For nonpsychiatric diseases, Wernicke did not give up on the lesion model, as his 1895 publication "Zwei Fälle von Rindenläsion" suggests. The cases (from 1886 and 1892) presented there were exceptionally clear,

showing (rarely corresponding) lesions in the "mid third of the post-central gyrus" (47) and the corresponding main symptom of "loss of *Vorstellungen* of touch in the right hand" (48). Wernicke used the same cases in his *Grundriss* (1906), 20–21. The paper is an exemplar of the pathological anatomical model, following the conventional structure of presenting two cases first and then discussing them.

113. Universitas Vratislaviensis, *Chronik Breslau*, vol. 9 (1894–1895), 60. Privatdozent Sachs also gave the neuroanatomical lecture. Akta miasta Wrocławia 33874.1, 48–52, Universitäts-Curator an Magistrat, SAB.

114. Bonhoeffer, "Lebenserinnerungen," 40.

115. Wernicke, *Grundriss* (1894), 1:9.

116. Wernicke, "Tagesfragen," 5.

117. Wernicke, "Zweck und Ziel der Psychiatrischen Kliniken," *Klinisches Jahrbuch* 1 (1889): 220.

118. Volker Roelcke, "Unterwegs zur Psychiatrie als Wissenschaft: Das Projekt einer 'Irrenstatistik' und Emil Kraepelins Neuformulierung der psychiatrischen Klassifikation," in Engstrom and Roelcke, *Psychiatrie im 19. Jahrhundert*, 183. The hope did not materialize, however; not until about two decades later did it become an official field of examination (*Prüfungsfach*). See also Engstrom, "Assembling professional selves."

119. See chapter 1.

120. This is most clearly seen in Wundtian psychology and psychophysics.

121. The dispute concerned the question whether or not the nervous system was made up of individual discrete elements. In Germany, the debate came to a head at the 1893 meeting of the Gesellschaft Deutscher Naturforscher und Ärzte.

122. For details, see Engstrom, *Clinical Psychiatry*, esp. 123–127. Younger psychiatrists such as Emil Kraepelin, Robert Sommer, Theodor Ziehen, Konrad Rieger, and Paul Möbius in particular, were attracted by the work by the psychologists Gustav Fechner and Wilhelm Wundt. As Engstrom has pointed out, the turn to the more recent science of experimental psychology and psychophysics was another attempt of academic psychiatrists to distinguish themselves from alienist psychiatry. Ibid., 125.

123. Ibid., 125.

124. Roelcke, "Unterwegs zur Psychiatrie als Wissenschaft."

125. Ibid., 184. This turn to psychology might have seemed like a turn to the subject; but, as Roelcke suggests, it actually eclipsed the subject, because an interest in the content of the patient's inner life—his delusions, his fixed ideas—was displaced by an interest in the form of the disease course (186).

126. See, e.g., Hugo Liepmann's work on aphasia and apraxia.

127. Cf. Bonhoeffer, who also published in the series Psychiatrische Abhandlungen (which included 15 volumes), in which the *Krankenvorstellungen* appeared. According to him, the "*Abhandlungen* . . . given the limited reach of the press appeared largely at the exclusion of the public." Bonhoeffer, *Lebenserinnerungen*, 47.

128. Arleen Tuchman, *Science, Medicine and the State in Germany: The Case of Baden, 1815–1871* (New York: Oxford University Press, 1993).

129. If Wernicke in a different context wanted his clinic to be more like the internal medicine or surgical departments, here he insisted on his right to be different, aiming for greater administrative powers like other asylum directors. Akta miasta Wrocławia 33874.1, 31–44, SAB.

130. Ibid.

131. Universitäts-Curator to Magistrat Breslau, July 17, 1888, Akta miasta Wrocławia 33874.1, 1–4, SAB.

132. This was, of course, also a problem for other doctors who held a dual position of *Primararzt* and professor, as Alfred Buchwald, *Primararzt* and professor of internal medicine, pointed out in 1896. This in part led university clinics to move out of the Allerheiligenhospital. Alfred Buchwald, *Kranken-Hospital zu Allerheiligen*, 19.

133. Akta miasta Wrocławia 33874.1, 59–64, SAB.

134. Akta miasta Wrocławia 33874.1, 65–88; 33929, 1–13, both in SAB.

135. Eduard Hitzig, "Neubau der psychiatrischen und Nervenklinik für die Universität Halle a. S.," *Klinisches Jahrbuch* 2 (1890): 383–405; Marneros and Pillmann, *Das Wort Psychiatrie*, esp. 94–103.

Chapter Three

1. Patients would have to pay for sessions that they missed. Freud compared this to music or language lessons in good society. Further, he stressed that payment, as for surgical treatment, was justified because of the effectiveness of his therapy. Sigmund Freud, "On beginning the treatment (further recommendations on the technique of psychoanalysis I)," *Standard Edition*, 12:125. For the Freud texts that have been translated into English, I rely on *The Standard Edition of the Complete Psychological Works of Sigmund Freud* (hereafter *SE*), trans. and ed. James Strachey, 24 vols. (London: Hogarth Press, 1953–1974); Mark Solms and Michael Saling, *A Moment of Transition: Two Neuroscientific Articles by Sigmund Freud* (London: Institute of Psycho-Analysis/Karnac, 1990); and E. Stengel's translation of the aphasia book. I have amended the translations where necessary.

2. Paul Roazen, *How Freud Worked: First-Hand Accounts of Patients* (Northvale, NJ: J. Aronson, 1995), xiv.

3. Edmund Engelman, *Berggasse 19: Sigmund Freud's Home and Offices, Vienna, 1938* (New York: Basic Books, 1976). For a detailed reconstruction of Freud's workplace, see Diana Fuss, "Freud's Ear," in *The Sense of an Interior: Four Writers and the Rooms That Shaped Them*, by Diana Fuss (New York: Routledge, 2004); Lydia Flem, *La vie quotidienne de Freud et de ses patients* (Paris: Hachette, 1986); Mayer, *Sites of the Unconscious*, chaps. 7 and 8. Notable patients of Freud, such as

Hilda Doolittle ("H.D.") and Joseph Wortis, have also described the setting and structure of their treatment with Freud.

4. Hilda Doolittle, *Tribute to Freud: Writing on the Wall: Advent* (1956; Boston: David R. Godine, 1974), 22.

5. As Freud pointed out, he did not like to be "stared at by other people for eight hours a day (or more)." Freud, "On beginning the treatment," 12:132.

6. E.g., Hirschmüller, *Freuds Begegnung*; Peter Amacher, *Freud's Neurological Education and Its Influence on Psychoanalytic Theory* (New York: International Universities Press, 1965); Giselher Guttmann and Inge Scholz-Strasser, eds., *Freud and the Neurosciences: From Brain Research to the Unconscious* (Vienna: Verlag der Österreichischen Akademie der Wissenschaften, 1998); Makari, *Revolution in Mind.*

7. The historical connection between Meynert and Freud, especially Freud's work in Meynert's anatomical laboratory and their gradual estrangement, has been researched in great detail by Hirschmüller, *Freuds Begegnung*. See also the work of Bernd Nitzschke, esp. "Warum wurde Freud nicht Psychiater?" in Bernd Nitzschke, *Aufbruch nach Inner-Afrika: Essays über Sigmund Freud und die Wurzeln der Psychoanalyse* (Göttingen: Vandenhoeck and Ruprecht, 1998), 197–208; Mayer, *Sites of the Unconscious*, 125–130; Makari, *Revolution in Mind*, 57–64. The debate between Meynert and Freud over male hysteria and hypnosis has also attracted the interest of scholars, for various reasons. See Mayer, *Sites of the Unconscious*, 125–131; Mark Micale, *Hysterical Men: The Hidden History of Male Nervous Illness* (Cambridge, MA: Harvard University Press, 2008), 237–343; Frank Sulloway, *Freud, Biologist of the Mind: Beyond the Psychoanalytic Legend* (Cambridge, MA: Harvard University Press, 1992), 49–50; Mai Wegener, *Neuronen und Neurosen: Der psychische Apparat bei Freud und Lacan: Ein historisch-theoretischer Versuch zu Freuds "Entwurf" von 1895* (Munich: Fink, 2004), 151–69.

8. In these accounts, Freud had to liberate himself from his early scientific work to develop his psychoanalysis, e.g., Ernest Jones, *The Life and Work of Sigmund Freud* (New York: Basic Books, 1953), 1:379; James Strachey, "Editor's introduction to J. Breuer and S. Freud (1893–95) *Studies on Hysteria*," *SE*, 2:ix–xxvii. Some scholars have argued that Freud, though not caught in it, still developed his psychology independently, e.g., Levin, *Freud's Early Psychology*. Others have emphasized the relevance of brain science for psychoanalysis, e.g., Maria Dorer, *Historische Grundlagen der Psychoanalyse* (Leipzig: Meiner, 1932); Amacher, *Freud's Neurological Education*. The debate has continued to resonate with scholars; see, among others, Mark Solms and Michael Saling, "On Psychoanalysis and Neuroscience: Freud's Attitude to the Localizationist Tradition," *International Journal of Psycho-Analysis* 67 (1986): 397–416; Guttmann and Scholz-Strasser, *Freud and the Neurosciences*; Alexandre Métraux, "Metamorphosen der Hirnwissenschaft: Warum Sigmund Freuds 'Entwurf einer Psychologie' aufgegeben wurde," in *Ecce Cortex: Beiträge zur Geschichte des modernen Gehirns*, ed. Michael Hagner (Göttingen: Wallstein, 1999), 75–109; Makari, *Revolution in Mind.*

9. In arguing for a continuity between two seemingly unconnected realms, I see my work in line with more recent historical scholarship on the history of psychoanalysis. For an overview of this and the older historiography, see Burnham, "'New Freud Studies'"; Lydia Marinelli and Andreas Mayer, *Forgetting Freud? For a New Historiography of Psychoanalysis, Science in Context*, special issue, 19.1 (2006). For two examples of this scholarship, see Makari, *Revolution in Mind*; and Mayer, *Sites of the Unconscious*.

10. Other scholars have emphasized different traditions in their contextualization, commenting on the influence of, e.g., the English neurologist John Hughlings Jackson (Solms and Saling, "Psychoanalysis and neuroscience," esp. 403–404; John Forrester, *Language and the Origins of Psychoanalysis* [London: Macmillan, 1980], esp. 18–21; Harrington, *Medicine, Mind, and the Double Brain*, chap. 8), or the German experimental physiologists Ernst Brücke and Sigmund Exner (Amacher, *Freud's Neurological Education*). See also Valerie Greenberg, *Freud and His Aphasia Book: Language and the Sources of Psychoanalysis* (Ithaca, NY: Cornell University Press, 1997); and Makari, *Revolution in Mind*, esp. 9–84.

11. In this sense I do not argue for absolute continuity between psychoanalysis and the biological sciences of Freud's day, as have other scholars such as Frank Sulloway, who reduced Freud's psychoanalysis to a "crypto-biology." Sulloway, *Freud, Biologist of the Mind*.

12. To Solms and Saling, "Psychoanalysis and neuroscience," Freud's *Aphasia* book marks his departure from German neurology. John Forrester has called *On Aphasia* the "*sine qua non* of the birth of psychoanalytic theory," in *Language*, 14. On the challenge of the new work on aphasia to the *Geisteswissenschaften*, see the important work by Stephen Jacyna and Michael Hagner: Jacyna, *Lost Words*; Hagner, *Homo cerebralis*. For an analysis of Freud's aphasia book as a work of rhetoric, see Greenberg, *Freud and His Aphasia Book*. On the historiography of aphasia, see Jacyna, *Lost Words*, 12–21.

13. Sigmund Freud, *On Aphasia: A Critical Study*, trans. E. Stengel (London: Imago, 1953), 55. In the following, I will quote from Stengel's English translation of Freud's aphasia book.

14. In this way, Freud's criticism resembles that of Jaspers presented at the beginning of chapter 1. But unlike Jaspers, who recommended a wholesale rejection of a method marked by "brain mythology," Freud engaged with and worked through the principles of the neuropsychiatric tradition.

15. The book has the following structure: mismatch between "psychic" clinical symptoms and the Wernicke-Lichtheim model of nervous architecture (sections 1–4); development of a new nervous architecture without drawing on psychological factors (section 5); and reassessment of the clinical (psychological) symptoms with the new model (section 6).

16. It is worthwhile noting that the neurologist Norman Geschwind was attracted to Wernicke's work precisely because it was not localizationist in a simple

sense. Geschwind, "Carl Wernicke," 42–61; Norman Geschwind, "Wernicke's contribution to the study of aphasia," *Cortex* 3 (1967): 449–463. See Kushner, "Norman Geschwind." Here he overlaps with Freud, who described, and productively drew upon, the tensions within Wernicke's system.

17. E.g., Wolfgang Leuschner, "Einleitung," in *Zur Auffassung der Aphasien: Eine kritische Studie*, by Sigmund Freud, ed. Paul Vogel (Frankfurt: Fischer, 2001), 7–31.

18. See Freud, *Aphasia*, 3. These two meanings of "functional" correspond broadly to what Levin calls "functional$_a$" and "functional$_b$" though I interpret their meaning and place within Freud's thought differently. Levin, *Freud's Early Psychology*, 76.

19. See Freud, *Aphasia*, 10–18, 19, 49, 53–54, 58, 68, 87, among others.

20. Ibid., 15, 29–31, 39–40, 43, 71, 83–84; Freud, *Project for a Scientific Psychology*, *SE*, 1:296–391, e.g., 296. Even in Freud's more psychologically oriented texts, function maintains this cellular and intercellular level. See "Some points for a comparative study of organic and hysterical motor paralyses," *SE*, 1:160–172. Readers should note moments when Freud seems to slip between the two meanings, e.g., *Aphasia*, 30, 87.

21. Cf. Meynert's notion of functional energy as a 'physiological force' in *Psychiatry*, 138–139, and "functional hyperaemia," 188, among others.

22. Sigmund Freud, "Kritische Einleitung in die Nervenpathologie," 1887: container 50, reel 1, Sigmund Freud Papers, Sigmund Freud Collection, Manuscript Division, Library of Congress (hereafter LOC). For a translation and critical edition, see Sigmund Freud, "Critical introduction to neuropathology," trans. Katja Guenther, *Psychoanalysis and History* 14.2 (2012): 151–202. See also Katja Guenther, "Recasting neuropsychiatry—Freud's 'critical introduction' and the convergence of French and German brain science," *Psychoanalysis and History* 14.2 (2012): 203–226.

23. Freud, "Critical introduction," 166. The text overlaps in parts with Freud's article "Gehirn," in *Handwörterbuch der gesamten Medizin*, ed. Albert Villaret (Stuttgart: Ferdinand Enke, 1888–91), 1:684–697, whose authorship is contested (cf. Solms and Saling, *Moment of Transition*, 7–12). See also Anneliese Menninger, *Sigmund Freud als Autor in Villarets Handwörterbuch der Gesamten Medizin von 1888–1891* (Hamburg: Dr. Kovač, 2011); and Anneliese Menninger, "Zu den Beiträgen Sigmund Freuds in Villarets Handwörterbuch der Gesamten Medizin (1888–91)" *Luzifer Amor* 49.1 (2012): 83–105.

24. Freud, "Critical Introduction," section titled "Physiological Methods." Note that contemporary neuroanatomical works, such as Schwalbe's, considered secondary degeneration as an anatomical-pathological, not a physiological method; they would use the method in a slightly different way, corresponding to traditional anatomy's interest in the connections between gray matters, stopping at the gray matter to which the damaged fiber was connected. Gustav Schwalbe, *Lehrbuch der Neurologie, zugleich des zweiten Bandes von Hoffmann's Lehrbuch der Anatomie des Menschen* (Erlangen: Besold, 1881), 321–327.

25. In the manuscript, Freud argued that Meynert's system, although itself relying on results gained through purely anatomical investigations—Meynert after all was the person who primarily promoted the method of cleavage—was a deeply physiological account of the nervous system, "a creation saturated [*durchtränkt*] with physiological ideas." Thus, any critique brought up against it would have to be based on "physiological aspects" as well (196). Freud, "Critical introduction."

26. Freud acknowledged that Meynert and Wernicke did not envision a simple "localization" of an elementary *Vorstellung* in individual cells, but rather its physiological correlate. Freud, *Aphasia*, 55.

27. Ibid., 55–56.

28. It has to be noted, however, that Freud defended Meynert and Wernicke at the same time as criticizing them (cf. *Aphasia*, 3, 64 and 103). Note also that in "Gehirn," as translated in Solms and Saling, *Moment of Transition*, 65, Freud seems to endorse the localization of elementary functions. In his correspondence with Ludwig Binswanger, Freud presented his critique of Wernicke more aggressively: "Wernicke always seemed to me an interesting example of the poverty of scientific thinking." But there, too, he qualified his criticism: "But in so judging him, I am measuring him by a high standard; I know very well that with others, whose names resound throughout the world, the question of scientific thought never arises at all." Freud to Binswanger, September 10, 1911, in *The Sigmund Freud-Ludwig Binswanger Correspondence 1908–1938*, by Sigmund Freud and Ludwig Binswanger, ed. Gerhard Fichtner, trans. Arnold J. Pomerans (New York: Other Press, 2003), 74. I thank Howard Kushner for drawing my attention to this letter. See also Valerie Greenberg, *Freud and His Aphasia Book*, 30.

29. Freud, *Aphasia*, 56.

30. Ibid., 57.

31. Most scholars, in contrast, have seen Freud's use of association as a strange introduction of psychological elements into his account: e.g., Otto Marx, "Freud and aphasia: An historical analysis," *American Journal of Psychiatry*, 124 (1967): esp. 822. John Forrester, in his brilliant book *Language and the Origins of Psychoanalysis*, recognizes that Freud was moving toward physiology. However, he remains suspicious of Freud's declarations to that effect, and rather suggests a linguistic structure. Forrester, *Language*, esp. 14–29. My argument builds off Forrester's but lends greater credence to Freud's own assertions that he was developing a purely physiological model.

32. Freud, *Aphasia*, 57, my emphasis. The postmortem finding that Freud refers to has been described by Heubner. See *Aphasia*, 23–24.

33. Ibid., 56–57.

34. Ibid., 50–52. In his own research in the 1870s and 1880s, Freud also showed how gray matter challenged the principle of fiber identity. Working on the anatomy of fiber tracts, he criticized the tendency in neuroanatomical research to look for "only one continuation [of a fiber] for each fiber bundle." Sigmund Freud

and L. Darkschewitsch, "Ueber die Beziehung des Strickkörpers zum Hinterstrang und Hinterstrangkern nebst Bemerkungen über zwei Felder der Oblongata," *Neurologisches Centralblatt* 5 (1886): 121–129. See also Sigmund Freud, "Über Spinalganglien und Rückenmark des Petromyzon," *Sitzungsberichte der Mathematisch-Naturwissenschaftlichen Classe der k. Akademie der Wissenschaften*, Vienna, III. Abtheilung, 78 (1878): 81–167. Indeed, as Freud pointed out in his paper on the anatomy of the acoustic nerve, it was impossible to do this. Fiber tracts would not travel beyond gray matter in the same way; instead, they changed their thickness and color. Sigmund Freud, "Ueber den Ursprung des N. acusticus," *Monatsschrift für Ohrenheilkunde* 20 (1886): 250. For example, from the nucleus of the acoustic nerve, two fiber tracts emerged. These had therefore to be considered only as "mediate continuations of the N. acusticus" (*mittelbare Acusticusfortsetzungen*). Ibid.

35. Freud, *Aphasia*, 53.

36. Ibid., 51.

37. Ibid., 63. "Speech territory" is the translation Stengel uses.

38. Cf. Freud's 1891 article "Lokalisation" in Villaret, *Handwörterbuch*, 2:231–233. See also Johann Reicheneder, "'Lokalisation': Ein bisher unbekannt gebliebener Beitrag Freuds zu Villarets Handwörterbuch der gesamten Medizin," *Jahrbuch der Psychoanalyse* 32 (1994): 155–182. Anneliese Menninger suggests that "Lokalisation" was authored by Johannes Gad, rather than Sigmund Freud. Menninger, "Zu den Beiträgen."

39. Freud, *Aphasia*, 63.

40. Ibid., 64.

41. Indeed, with the emphasis on thoroughfare, the notion of a center dissolves.

42. Freud, *Aphasia*, 64.

43. Of course we should remember that Wernicke's model did account for lesions on association fibers (sejunction), but this understanding was still structured by the idea of centers, which those association fibers connected.

44. Freud, *Aphasia*, 64.

45. Cornelius Borck has shown the progression of Freud's move away from anatomical explanations at the level of illustration, working out the performative aspects of Freud's diagrams. Cornelius Borck, "Visualizing nerve cells and psychic mechanisms: The rhetoric of Freud's illustrations," in Guttmann and Scholz-Strasser, *Freud and the Neurosciences*, 75–86. Cf. Jacyna, who calls Freud's diagram of the "speech association field" an "anti-diagram." Jacyna, *Lost Words*, 179. My work is complementary to this, by developing how the connections that the diagrams emphasized served to disrupt the attempt at localization.

46. In the English version of *Aphasia*, the word *lesion* translates two German words: *Läsion* and *Verletzung*. In *Zur Auffassung der Aphasien: Eine kritische Studie*, edited by Paul Vogel (Frankfurt: Fischer, 2001), Freud predominantly uses *Läsion*, except on two occasions when he uses *Verletzung*: once where he is trans-

lating from the English "injury" (66) and another time where he seems to be referring specifically to the event ("nach der Verletzung") (74).

47. Freud's theory of hysteria follows predominantly from Charcot. The Charcot tradition, however, accounts for only one aspect of the history of hysteria. See esp. Mark Micale, *Approaching Hysteria: Disease and Its Interpretations* (Princeton, NJ: Princeton University Press, 1994); Micale, *Hysterical Men*; and Sander Gilman, Helen King, Roy Porter, G. S. Rousseau, and Elaine Showalter, eds., *Hysteria beyond Freud* (Berkeley: University of California Press, 1993).

48. Cf. my discussion about the "physiological" and the "functional" above.

49. Freud, *Aphasia*, 29.

50. Freud, "Organic and hysterical paralyses," 1:160–72. Levin, in *Freud's Early Psychology*, places great emphasis on the psychological perspective that Freud takes here and suggests that the physiological developments in *Aphasia* are unimportant for "Organic and hysterical paralyses." As I argue here, we can only understand the form of Freud's psychological explanations by relating it to his physiological developments in *Aphasia*.

51. The scholarship on Charcot is contradictory on this point. Some hold that in his shift of interest from an anatomical to a physiological, "dynamic" model for hysteria, Charcot never abandoned materialist explanations. E.g., Goetz, Bonduelle, and Gelfand, *Charcot*, 206–207; J. Aguayo, "Charcot and Freud: Some implications of late 19th century French psychiatry and politics for the origins of psychoanalysis," *Psychoanalysis and Contemporary Thought* 9 (1986): 235. According to Levin, Charcot rejected an anatomical basis for hysteria. Levin, *Freud's Early Psychology*, esp. 6, 44–46. Freud himself, after introducing Charcot's notion of "purely dynamic or functional" cortical lesion, mentioned that "many who read M. Charcot's works believe that a dynamic lesion is indeed a [real organic] lesion." Freud, "Organic and hysterical paralyses," 168.

52. Freud, "Organic and hysterical paralyses," 168.

53. Ibid., 170.

54. E.g. Jean-Martin Charcot, *Clinical Lectures on Diseases of the Nervous System* (London: Sydenham Society, 1889), 3:281.

55. Freud, "Organic and hysterical paralyses," 169.

56. Ibid., 170–171.

57. Ibid., 170. Note that Meynert as well uses the expression "play of the associations" (*Spiel der Associationen*). Meynert, "Zusammenwirken," 223.

58. Freud, "Organic and hysterical paralyses," 170.

59. Sigmund Freud, "Hysterie," in Villaret, *Handwörterbuch*, 1:886–892, quotation on 886. Freud's link between the lack of visible changes in hysteria and the "neurosis in the strictest meaning of the term" (1:886) can be explained in view of the larger history of the neurosis concept. From the mid-1830s onward, hysteria and other neuroses were seen as physiological or functional, leaving no anatomical trace. See José María López Piñero, *Historical Origins of the Concept of*

Neurosis, trans. D. Berrios (Cambridge: Cambridge University Press, 1983), esp. 44–58.

60. Freud, "Hysterie," 1:886. The quote is generally used as evidence for Freud's moving away from materialist conceptions of the mind. But as I have shown, the move away from anatomical explanations represented for Freud a more radically materialist conception of the nervous system, rather than forgoing materialism.

61. Sigmund Freud, *The Interpretation of Dreams*, *SE*, 4:536. It should be noted, though, that the optical analogy points to a more radical conception than is presented in the *Project*. Even though the apparatus of the camera was material, a purely materialist way of explaining its workings was no longer helpful for understanding the science behind it.

62. Some scholars, e.g., James Strachey, John Forrester, and Jean Laplanche, have argued for its importance in its own right. Strachey, "Editor's introduction," 283–293, esp. 290–293; Forrester, *Language*, 223n40; Jean Laplanche, *Life and Death in Psychoanalysis*, trans. Jeffrey Mehlman (1970; Baltimore, MD: Johns Hopkins University Press, 1985). Other scholars have expressed doubt about the status of the *Project* for several reasons: first, the *Project* was a draft sent by Freud to Wilhelm Fliess that was never revised or published. Second, although key elements of the text reappear in chapter 7 of Freud's *Traumdeutung*, Freud distanced himself from the text and never asked Fliess to return it to him. Scholars have instead suggested that *On Aphasia* should be considered the more important text. Solms and Saling, "Psychoanalysis and neuroscience"; Borck, "Visualizing," 71n15. For other critical readings of the *Project*, see Métraux, "Metamorphosen"; Wegener, *Neuronen und Neurosen*; Erik Porath, "Vom Reflexbogen zum psychischen Apparat: Neurologie und Psychoanalyse um 1900," *Berichte zur Wissenschaftsgeschichte* 32 (2009): 53–69; Sandra Janßen, "Von der Dissoziation zum System: Das Konzept des Unbewussten als Abkömmling des Reflexparadigmas in der Theorie Freuds," *Berichte zur Wissenschaftsgeschichte* 32 (2009): 36–52.

63. As I argue below, Freud's criticism of Meynert's distinction between projection and association still remains.

64. Freud, *Project*, 299–300.

65. Ibid., 319.

66. The economic concept of cathexis was first used by Freud in 1895, in *Studies on Hysteria* and in the *Project*. J. Laplanche and J.-B. Pontalis, *The Language of Psycho-Analysis*, trans. Donald Nicholson-Smith, The International Psycho-Analytical Library, 94:1–497. (London: Hogarth Press and the Institute of Psycho-Analysis, 1973), 62.

67. Freud, *Project*, 319.

68. The opening sentence expresses this succinctly: "The intention is to furnish a psychology that shall be a natural science: that is, to represent psychical processes as quantitatively determinate states of specifiable material particles, thus making those processes perspicuous and free from contradiction," *Project*, 295.

That the notion of quality derives from psychological explanations can be seen in the section "The Problem of Quality," 307–310.

69. In the structure of the *Project*, Freud explains the construction of *Vorstellungen* before explaining how similar processes could associate them.

70. Ibid., 327. When Freud continued, saying, "So far we have neglected this feature; it is time to take it into account" (327), he probably was referring to his earlier simplification: "If the wished-for object is abundantly cathected, so that it is activated in a hallucinatory manner" (325), which (misleadingly) seemed to indicate that one object was localized (cathected) in an individual cell.

71. Ibid., 363. Throughout *Aphasia*, Freud used the terms *Vorstellung* and *Empfindung* interchangeably. In the *Project*, he identified *Vorstellung* (perception) with *Erinnerung* (memory), 325. While in *Aphasia* Freud just described the structure of *Vorstellungen/Empfindungen*, in the *Project* he was concerned with their genesis. Since *Vorstellungen* were produced from *Erinnerungsbildern* (memory images), this explains the shift.

72. Ibid., 299–305.

73. Ibid., 300.

74. Jacques Derrida, "Freud and the Scene of Writing" in *Writing and Difference*, by Jacques Derrida, trans. Alan Bass (Chicago: University of Chicago Press, 1978). As Edward Baring has shown, Derrida's paper emerged as an intervention in a debate between organicist and linguistic readings of Freud within French psychoanalysis, and Derrida appealed to writing and the trace as a way of mediating between the two. Edward Baring, *The Young Derrida and French Philosophy: 1945–68* (Cambridge: Cambridge University Press, 2011), 215.

75. Freud, *Project*, 298, 303.

76. Ibid., 304. Likewise, in "Gehirn," as translated in Solms and Saling, *Moment of Transition*, 64, Freud suggested that "the individual cortical elements . . . are differentiated . . . essentially by their connection with the different centripetal and centrifugal conductors of excitation."

77. The discussion of memory raises a whole series of problems and questions that I do not have the space to discuss here, but which have offered scholars different ways of characterizing and understanding the relationship between Freud's early work and his mature psychoanalysis. See, in particular, David Farrell Krell, *Of Memory, Reminiscence, and Writing. On the Verge* (Bloomington: Indiana University Press, 1990).

78. Freud, *Project*, 348.

79. Ibid., 350.

80. Ibid., 349. As we shall see, it is the idea that an association can be repressed, i.e., that it isn't available to consciousness and that an association can "pass through unconscious intermediate links until it comes to a conscious one" (ibid., 355), that marks Freud's distance from association *psychology*.

81. Ibid., 356. He also briefly mentions the term in "Organic and hysterical paralyses," 171–172. Freud first used the term "psychical trauma" in 1893 in a way that related to the "Organic and hysterical paralyses" definition of the hysterical lesion. See Sigmund Freud and Josef Breuer, *Studies on hysteria*, *SE*, 2:6. There is a history of psychological trauma before Freud. Most conspicuously, Charcot converted John Erichsen's "railway spine" into a psychological condition. See esp. Young, *Harmony of Illusions*, 12–42; and Ruth Leys, *Trauma: A Genealogy* (Chicago: University of Chicago Press, 2000), 3–4.

82. Marcel Gauchet, in *L'inconscient cérébral* (Paris: Seuil, 1992), does not spend much time on the (anatomical) complexity of the reflex—e.g., he lumps together Müller and Hall (42)—nor on association. Sandra Janßen, in "Dissoziation," offers a closer reading of Freud's texts to support her argument that the concept of the unconscious originated in the reflex paradigm. See also D. Smith, "Freud's neural unconscious," in *The Pre-Psychoanalytic Writings of Sigmund Freud*, ed. G. Van de Vijver and F. Geerardyn (London: Karnac, 2002), 155–164.

83. E.g., Freud, "The unconscious," *SE*, 14:167–168.

84. Freud suggested that the *Bahnungen* in the creation of *Vorstellungen* could be understood without appealing to consciousness. Freud, *Project*, 308.

85. Freud, and Breuer, *Studies on hysteria*, 12, 11.

86. As Andreas Mayer shows, resistance arose in the hypnotic context, but at that point the resistance was bypassed by the action of the doctor. In this way, resistance takes a central part in Freud's therapy only after 1900. This was also when transference transformed from a "disturbing factor" (*Störfaktor*) into a therapeutic tool. Mayer, *Mikroskopie der Psyche*, 181, 213–217, 221–224.

87. Freud and Breuer, *Studies on Hysteria*, 12, 15.

88. He was also an autodidact, learning electrotherapy directly from Erb's book.

89. Sigmund Freud to Martha Bernays, April 24, 1884, in *Briefe 1873–1939*, by Sigmund Freud, ed. Ernst Freud (Frankfurt am Main: S. Fischer, 1960). As a comparison, in the second half of the 1890s, when he was more established, Freud asked for an average fee of 4 gulden (or florin) for a one-hour psychoanalytic consultation at his office. Record book, 1896–1899 ("Kassen-Protokoll"), box 50, Sigmund Freud Papers, Sigmund Freud Collection, Manuscript Division, LOC.

90. Sander Gilman, "Sigmund Freud, electrotherapy, and the voice," in *Diseases and Diagnoses: the Second Age of Biology* (New Brunswick, NJ: Transaction, 2010), 174. Cf. Hannah S. Decker, *Freud, Dora, and Vienna 1900* (New York: Free Press, 1991), 8. Decker presents the adoption of the cathartic method as a break with physical methods, but from Freud's writings of the period, especially from his letters to Martha, it rather seems that he made use of both at the same time.

91. Freud, "Hysterie," 892.

92. Freud, "The aetiology of hysteria," *SE*, 3:217. See also Margarete Vöhringer and Yvonne Wübben, "Phantome im Labor: Die Verbreitung der Reflexe in Hirn-

forschung, Kunst und Technik," *Berichte zur Wissenschaftsgeschichte*, special issue, 32 (2009); esp. Porath, "Vom Reflexbogen zum psychischen Apparat"; and Janßen, "Dissoziation."

93. Freud, *Interpretation of Dreams*, 4:538.

94. Freud's comparison of psychoanalysis with the analysis of the chemist points in this direction as well. He wrote that it "is not true that something in the patient has been divided into its components and is now quietly waiting for us to put it somehow together again." Rather, the makeup of mental life changes in the process. This "psycho-synthesis" happened on its own, without the doctor's intervention. Freud, "Lines of advance in psycho-analytic therapy" (1919), *SE*, 17:161.

95. The case was published as "Bemerkungen über einen Fall von Zwangsneurose" ("Notes upon a case of obsessional neurosis") (1909), *SE*, vol. 10. On Freud and the case history, see John Forrester, "If p, then what? Thinking in cases," *History of the Human Sciences* 9 (1996): 1–25. See also Anne Sealey, "The strange case of the Freudian case history: The role of long case histories in the development of psychoanalysis," *History of the Human Sciences* 24 (2011): 36–50. For a book-length discussion of the case, see Patrick Mahony, *Freud and the Rat Man* (New Haven, CT: Yale University Press, 1986). Mahony complements his reading of the case by consulting other sources, including minutes of the discussions in the Vienna Psychoanalytic Society, Freud's commentaries, and correspondence about the case.

96. Freud, "Case of obsessional neurosis," 158. The case was first published in 1909.

97. Strachey, "Freud's psycho-analytic procedure" ([1903] 1904), *SE*, 7:249.

98. Freud, "Case of obsessional neurosis," 174.

99. Freud, "Recommendations to physicians practising psycho-analysis," *SE*, vol. 12, 114–115.

100. Freud, "Case of obsessional neurosis," 159.

101. According to Laplanche and Pontalis, the rule emerged somewhere between 1892 and 1898. See also Freud, "On beginning the treatment," 134.

102. Freud, "Case of obsessional neurosis," 174.

103. Ibid., 166.

104. Freud, "Two encyclopaedia articles," *SE*, 18:247. See also Freud, "On the history of the psycho-analytic movement," *SE*, 14:16.

105. Freud, "Case of obsessional neurosis," 205.

106. Ibid., 205.

107. Freud used this term in the original record of the case, not in the published case. Sigmund Freud, "Addendum: Original record of the case," *SE*, 10:292.

108. Freud, "Case of obsessional neurosis," 166.

109. Freud, "On psychotherapy," *SE*, 7:266.

110. It did not even count as an insight, unless the patient was ready to accept and confirm it. We find the concept of "working-through" (*Durcharbeiten*) as early

as 1895 in Freud's work, but it was only later more fully formalized, e.g., in "Remembering, repeating and working-through (further recommendations on the technique of psycho-analysis II)" (1914), *SE*, 12:145–156. Cf. J. Laplanche and J.B. Pontalis, "Working-through," in Nicholson-Smith, *Language of Psycho-Analysis*, 488–489.

111. Freud, "Case of obsessional neurosis," 199.

112. Ibid., 176.

113. Ibid., 391.

114. Freud, "'Wild' Psycho-Analysis," *SE*, 11:226. In addition, transference had an epistemological function, allowing the doctor to make "the patient's hidden and forgotten erotic impulses immediate and manifest." Freud, "The dynamics of transference," *SE*, 12:108.

115. Freud, "Case of obsessional neurosis," 206n1.

116. On Freud's "turn to fantasy" in the etiology of the neuroses, see Laplanche and Pontalis, *Language of Psycho-Analysis*, 314–319.

117. As Lydia Marinelli and Andreas Mayer have suggested, Freud's *Interpretation of Dreams* was the first institution. See their *Dreaming by the Book: Freud's Interpretation of Dreams and the History of the Psychoanalytic Movement*, trans. Susan Fairfield (New York: Other Press, 2003), on the intertwined histories of the book and the psychoanalytic movement.

118. Sándor Ferenczi, "On the organization of the psycho-analytic movement" (1911), in *Final Contributions to the Problems and Methods of Psychoanalysis*, by Sándor Ferenczi (New York: Brunner/Mazel, 1980), 305.

119. For the history of psychoanalysis and the creation of the psychoanalytic profession, see Makari, *Revolution in Mind*. On Freud and psychoanalysis in America, see John C. Burnham, *After Freud Left: A Century of Psychoanalysis in America* (Chicago: University of Chicago Press, 2012); Nathan Hale, *The Rise and Crisis of Psychoanalysis in America: Freud and the Americans, 1917–1985* (New York: Oxford University Press, 1995); and Elizabeth Lunbeck, *The Americanization of Narcissism* (Cambridge, MA: Harvard University Press, 2014).

120. Carl Gustav Jung, "Vorbemerkung der Redaktion," in *Jahrbuch für psychoanalytische und psychopathologische Forschungen*, vol. 1, ed. Eugen Bleuler, Sigmund Freud, and C. G. Jung (Leipzig: Deuticke, 1909).

Chapter Four

1. Günter Hesse, "Patient Lenin: Ein Übermensch?" *Deutsches Ärzteblatt* 10 (1975): 682–686, 755–760, 835–839, 3205–3207. See also Leonard Crome, "The medical history of V.I. Lenin," *History of Medicine* 4 (1972): 3–9, 20–22.

2. Crome, "Medical history of Lenin," 7.

3. The team included Otfrid Foerster, Oswald Bumke, Max Nonne, Adolf v. Strümpell, and Oskar Minkowski from Germany; Solomon Henschen from Sweden;

and the Russian doctors Vladimir Bekhterev and V. P. Osipov. Manuscript of unpublished biography of Foerster, probably written by Carlos Gutiérrez-Mahoney, based on materials collected and drafts written by Foerster's daughter Ilse, chapter titled "Foerster in Russia with Lenin," 6, William Gutiérrez-Mahoney Papers, Department of Medical Ethics and History of Medicine (hereafter DMEHM).

4. Foerster to Oswald Bumke, June 8, 1923, Gutiérrez-Mahoney Papers, DMEHM. On the history of neurological gymnastics, see Katja Guenther, "Exercises in therapy—neurological gymnastics between *Kurort* and hospital medicine, 1880–1945," *Bulletin of the History of Medicine* 88.1 (2014): 102–131.

5. In contrast to the history of psychiatry, the history of neurology is relatively underexplored. As Toby Gelfand has noted, "the historiography of psychiatry has overwhelmed that of neurology in scope and sheer volume." Toby Gelfand, "Neurologist or psychiatrist? The public and private domains of Jean-Martin Charcot," *Journal of the History of the Behavioral Sciences* 36 (2000): 216. Historical work on neurology has tended to focus on its scientific, rather than therapeutic, aspects. See, for example, Fielding Garrison's classic *History of Neurology*, ed. Lawrence C. McHenry Jr. (Springfield, IL: Thomas, 1969), which mainly focuses on individuals and their neurological discoveries. In recent years, the trend has been toward inclusion of neurological therapies: e.g., see Stanley Finger, François Boller, and Kenneth L. Tyler, *History of Neurology* (Edinburgh: Elsevier, 2010), part 6, "Treatments and recovery"; the chapter "Babinski as therapist" in Jacques Philippon and Jacques Poirier, *Joseph Babinski: A Biography* (Oxford: Oxford University Press, 2009). Clifford Rose's *History of British Neurology* (London: Imperial College Press, 2012), too, has a chapter on neurosurgery and a section on spinal-cord rehabilitation. An early example of this development is Kolle's *Große Nervenärzte*, whose second volume includes a chapter "Die Therapeuten." As Delia Gavrus has shown, the de-emphasis of treatment in neurology in the American case stems in part from the efforts of mid-twentieth-century American neurologists to define themselves as a social group. "Men of dreams and men of action: Neurologists, neurosurgeons, and the performance of professional identity, 1920–1950," *Bulletin of the History of Medicine* 85 (2011): 57–92.

6. For a parallel account that emphasizes the formation of specialties, see Andreas Killen's discussion of electrotherapy, *Berlin Electropolis: Shock, Nerves and German Modernity* (Berkeley: University of California Press, 2006).

7. *Plasticity* was a term used by Foerster, but not by Freud, although he made use of a concept of malleability.

8. Despite Foerster's prominence, there is very little historical work on him so far. Apart from obituaries and shorter notices, mostly in medical journals, there is only one book-length account (mostly a compilation of his papers), written by the neurologist Klaus-Joachim Zülch, a student of Foerster's. *Otfrid Foerster, Arzt und Naturforscher, 9.11.1873–15.6.1941* (Berlin: Springer, 1966); in English: *Otfrid Foerster, Physician and Naturalist, November 9, 1873–June 15, 1941*, trans. Adolf Rosenauer and Joseph P. Evans (Berlin: Springer, 1969).

9. Otfrid Foerster, "Demonstration: Fall von einem eigenthümlichen Zwangsphänomen," *Allgemeine Zeitschrift für Psychiatrie* 57 (1901): 411–414; Otfrid Foerster, "Ein Fall von elementarer allgemeiner Somatopsychose (Afunktion der Somatopsyche): Ein Beitrag zur Frage der Bedeutung der Somatopsyche für das Wahrnehmungsvermögen," *Monatsschrift für Psychiatrie und Neurologie* 14 (1903): 189–205; Otfrid Foerster, "Vergleichende Betrachtungen über Motilitätspsychosen und über Erkrankungen des Projektionssystems," Antrittsvorlesung, Habilitation als Privatdozent, 1903.

10. Otfrid Foerster, *Beiträge zur Kenntnis der Mitbewegungen* (Jena: G. Fischer, 1903).

11. The three volumes of the atlas were published between 1897 and 1903. The work departed from the more conventional project of brain mapping by emphasizing, often at great expense, the fiber makeup of the brain. Carl Wernicke, *Atlas des Gehirns: Schnitte durch das menschliche Gehirn in photographischen Originalien herausgegeben mit Unterstützung der Königlichen Akademie der Wissenschaften in Berlin*, 3 vols. (Breslau: Franck and Weigert, 1897–1903).

12. Hans Kuhlendahl, "Otfrid Foerster (9.11.1873–14.6.1941)," International Congress Series No. 320, Recent Progress in Neurological Surgery, Proceedings of the Symposia of the First International Congress of Neurological Surgery, Tokyo, October 7–13, 1973, *Excerpta Medica*, Amsterdam: 207. Similarly, William Gutiérrez-Mahoney makes the following point about Foerster's contribution to the *Handbuch* on muscle function: "There is no more comprehensive source for the knowledge of individual muscle function than this." Gutiérrez-Mahoney Papers, DMEHM.

13. He went there at the suggestion of Wernicke. Zülch, *Foerster, Arzt und Naturforscher*, 3. At the time of Foerster's visit, Dejerine, a former student of Charcot's, ran the second neurological unit at the Salpêtrière hospital in Paris. Philippon and Poirier, *Babinski*, 325. On Dejerine's relationship with Charcot, see Goetz, Bonduelle, and Gelfand, *Charcot*, 309–310, 319–320.

14. See chapter 2. As we saw there, Duchenne's work helped Wernicke move away from a simple lesion model in his account of the three spheres of consciousness.

15. As noted in *Monatsschrift fuer Psychiatrie und Neurologie* 27.1 (1910): 94; Eulner, *Entwicklung*, 267. Cf. Fritz Ringer, *The Decline of the German Mandarins* (Middletown, CT: Wesleyan University Press, 1990), 35–36. However, in the *Vorlesungsverzeichnis* of the University of Breslau, after 1909, Foerster was still listed under *Privatdozenten*.

16. Salaried professors thus had a double income. This income varied depending on the popularity of their lectures, ranging from 6,000 to 40,000 marks, with an average of 12,000. Ringer, *Decline*, 37–38. At Breslau, an *ordentlicher* professor was required to give one lecture for free every semester, and an *außerordentlicher* professor was to give one free lecture every other semester. For *Privatdozenten*

there was no such requirement. *Reglement für die medicinische Facultät der königlichen Universität zu Breslau* (1840; reprint, Breslau, 1890), 22. Not until 1917, at age forty-four, did Foerster obtain his first salaried teaching position—he was two years younger than the national average. Ringer, *Decline*, 54. In early-twentieth-century Prussia, the *Habilitation* was still the necessary precondition for a professorship. For the changing status of the *Privatdozent*, see Alexander Busch, *Die Geschichte des Privatdozenten* (Stuttgart: Enke, 1959).

17. Johannes Pantel, "Streitfall Nervenheilkunde—eine Studie zur disziplinären Genese der klinischen Neurologie in Deutschland," *Fortschritte der Neurologie und Psychiatrie* 61 (1993): 148. See also Johannes Pantel, "Neurologie, Psychiatrie und Innere Medizin: Verlauf und Dynamik eines historischen Streites," *Würzburger medizinhistorische Mitteilungen* 11 (1993): 77–99. Oppenheim deplored precisely this point—that young German neurologists had to go abroad for their training. See Pantel, "Streitfall Nervenheilkunde," 148. In France, neurology found early institutional recognition; the creation of Charcot's chair in 1882 is usually presented as the starting point. See Goetz, Bonduelle, and Gelfand, *Charcot*, esp. chaps. 3 and 7; Philippon and Pourier, *Babinski*, esp. chapter 16.

18. According to Pantel, Griesinger was the "Prototyp des Internistenpsychiaters." "Streitfall Nervenheilkunde," 145.

19. Quoted in ibid. As Volker Hess and Eric Engstrom have shown for the case of the Charité in Berlin, psychiatry and internal medicine battled over the material of neurological patients, which each claimed for its own discipline. Volker Hess and Eric Engstrom, "Neurologie an der Charité zwischen medizinischer und psychiatrischer Klinik," in *Geschichte der Neurologie in Berlin*, ed. Bernd Holdorff and Rolf Winau (Berlin: De Gruyter, 2001), 100–101.

20. For details, see Pantel, "Streitfall Nervenheilkunde," 147–148.

21. Pantel, "Streitfall Nervenheilkunde."

22. This is somewhat ironic because psychiatry was itself not a particularly therapeutic specialty. For a recent overview of the history of psychiatry, see Schott and Tölle, *Geschichte der Psychiatrie*.

23. Karl Bonhoeffer, "Psychiatrie und Neurologie," *Monatsschrift für Psychiatrie und Neurologie* 37 (1915): 95.

24. Ibid., 100. Pantel, "Streitfall Nervenheilkunde," 150. Bonhoeffer's skepticism went so far that he even claimed that the neurologists could not be authentically therapeutic but would always run the risk of exploiting the situation to run neurological experiments on the patient. "Psychiatrie und Neurologie," 100.

25. On the history of the relationship between syphilis and neuropsychiatric diseases, see Gayle Davis, *"The Cruel Madness of Love": Sex, Syphilis and Psychiatry in Scotland, 1880–1930*, Clio Medica 85 (Amsterdam: Rodopi, 2008).

26. Otfrid Foerster, "Ueber einige seltenere Formen von Krisen bei der Tabes dorsalis, sowie über die tabischen Krisen im Allgemeinen," *Monatsschrift für Psychiatrie und Neurologie* 11 (1902): 259–283, quotation on 259.

27. Apart from sensory and motor disturbances, there were also disturbances in secretion, that is, hypersecretion. Ibid., 267.

28. Ibid., 267.

29. Foerster left it open whether the reflex mechanism happened at the level of the spinal cord or above. Ibid., 268. Apart from the vagus nerves, the rami communicantes of the N. splanchicus major that originated from the dorsal roots of the spinal cord were connected to the stomach. Otfrid Foerster, "Ueber operative Behandlung gastrischer Krisen durch Resektion der 7.–10. hinteren Dorsalwurzel," *Beiträge zur klinischen Chirurgie* 63 (1909): 245–256, 250.

30. Foerster, "Krisen bei der Tabes dorsalis," 269.

31. The other symptoms, pain and secretion, were caused by a similar reflex mechanism and were thus equally indirect.

32. Alexander Tietze, "Die Technik der Foersterschen Operation," *Mitteilungen aus den Grenzgebieten der Chirurgie und Medizin* 20 (1909): 559.

33. Foerster, "Ueber eine neue operative Methode der Behandlung spastischer Lähmungen mittels Resektion hinterer Rückenmarkswurzeln," *Zeitschrift für orthopaedische Chirurgie* 22 (1908): 204.

34. Otfrid Foerster, "Die Symptomatologie und Therapie der Kriegsverletzungen der peripheren Nerven," *Deutsche Zeitschrift für Nervenheilkunde* 59 (1918): 32–172, image on 72.

35. Foerster was not alone in this reconceptualization. For an analysis of how the concept of inhibition was invoked in the period, see Smith, *Inhibition*. For Foerster's engagement with Hughlings Jackson, see esp. Otfrid Foerster, "The motor cortex in man in light of Hughlings Jackson's doctrines," *Brain* 59 (1936): 135–159.

36. See, e.g., Max Nonne, "Über Wert und Bedeutung der modernen Syphilistherapie für die Behandlung von Erkrankungen des Nervensystems," *Deutsche Zeitschrift für Nervenheilkunde* (1911–12): 166–250. Treatment of tabes was on Foerster's mind even before he developed his surgical intervention. In a 1907 paper with the Breslau dermatologist Harttung, he wrote in favor of calomel (mercury) treatment, not in the more common form of *Schmierkur* (application of mercury ointment to large areas of the skin) but as intramuscular injections. Otfrid Foerster and Harttung, "Erfahrungen über die Behandlung von Störungen des Nervensystems auf syphilitischer Grundlage," *Archiv für Dermatologie und Syphilologie* 86 (1907): 3–44.

37. Foerster, "Ueber operative Behandlung."

38. Ibid., 253.

39. Ibid., 255.

40. Otfrid Foerster, "Die Indikation und Erfolge der Resektion hinterer Rückenmarkswurzeln," *Wiener klinische Wochenschrift* 25 (1912): 951.

41. Julius Wagner-Jauregg, "Über die lanzinierenden Schmerzen der Tabetiker," *Wiener klinische Wochenschrift* 40 (1924): 187–204, esp. 201–202. In 1909 Paul Ehrlich developed salvarsan, which quickly became the treatment of choice. But even

though great hopes were placed in Ehrlich's new drug, it proved to have little effect on the progress of the disease. See Allan Brandt, *No Magic Bullet: A Social History of Venereal Disease in the United States since 1880* (New York: Oxford University Press, 1987). See also Arthur Schlesinger, "Die Foerster'sche Operation, Sammelreferat," *Neurologisches Centralblatt* 29 (1910): 970–978.

42. Sigmund Freud, "Die infantile Cerebrallähmung," in *Specielle Pathologie und Therapie*, by Hermann Nothnagel, 24 vols. (1897), vol. 9, part 2, 310.

43. Otfrid Foerster, "Surgical Treatment of Neurogenic Contractures," *Surgery, Gynecology, and Obstetrics* 52 (1931): 360–361. Note that Foerster also developed an operation to treat pain: chordotomy.

44. Otfrid Foerster, "Ueber die Beeinflussung spastischer Lähmungen durch die Resektion hinterer Rückenmarkswurzeln," *Deutsche Zeitschrift für Nervenheilkunde* 41 (1911): 159.

45. A copy of the film is in my possession. The original is kept at the Belgian Royal Film Archives Brussels (RFAB). Gehuchten published an almost identical paper in *Le Névraxe* 11.2–3 (1910): 245–92. For more information on van Gehuchten's role in the history of medical film, see Geneviève Aubert, "Arthur van Gehuchten takes neurology to the movies," *Neurology* 59 (2002): 1612–1618.

46. See, for example, S. J. Hunkin, "Experience with Foerster's operation," *Journal of Orthopaedic Surgery* s2–11 (1913): 207–214.

47. "Hatten Sie eine Ahnung?" *Zeit am Montag* 21 (May 23, 1932). The girl was transported by airplane against medical advice. In his correspondence with the city of Breslau, which considered suing the *Zeit am Montag*, Foerster wrote: "I would like to point out that I warned the father by letter not to execute his plan." Acta miasta Wrocławia, 3377, 378 (2), SAB.

48. In line with their associationist model, Meynert's and Wernicke's brain maps were characterized by an emphasis on fiber connections. See, e.g., Wernicke's *Atlas des Gehirns*. On Meynert's anatomical preparations, see Borck, "Fühlfäden und Fangarme," 144–176. The mapping projects of the era that supported the notion of "centers" most were those by Fritsch and Hitzig, and the psychiatrist Karl Kleist, which were attacked by the holistic tradition. On holism in German culture, see Anne Harrington, *Reenchanted Science: Holism in German Culture from Wilhelm II to Hitler* (Princeton, NJ: Princeton University Press, 1996); Ash, *Gestalt Psychology*.

49. Foerster aimed to fully map out the dermatomes in man, building upon the work by Head and Sherrington in animals. See also Steven Greenberg, "The history of dermatome mapping," *Archives of Neurology* 60 (2003): 126–131.

50. Otfrid Foerster, "Methoden der Dermatombestimmung beim Menschen," *Archiv für Psychiatrie und Nervenkrankheiten* 77 (1926): 652.

51. For the best study of Henry Head, see Stephen Jacyna, *Medicine and Modernism: A Biography of Sir Henry Head* (London: Pickering and Chatto, 2008).

52. Tietze, "Die Technik," 560.

53. A. Borchard and W. von Brunn, eds., *Deutscher Chirurgenkalender: Verzeichnis der deutschen Chirurgen und Orthopäden mit Biographien und bibliographischen Skizzen*, 2nd ed. (Leipzig: Barth, 1926), 329.

54. For an image of the device used to drive the nail into the bone, see Tietze, "Die Technik," 560.

55. Otfrid Foerster, "Methoden der Dermatombestimmung," 657.

56. Electrical stimulation was not the only method Foerster used to determine dermatomes. Often, he relied on the "method of remaining sensibilities" that Charles Sherrington had developed for the study of monkeys. Cf. Otfrid Foerster, "Dermatomes in man," *Brain* 56 (1933): 2.

57. While aphasia has been a guiding thread in the first part of this book, epilepsy will take over that role in the second, and I will also address it in the final chapter. For the history of epilepsy, see especially the work by Oswei Temkin, *The Falling Sickness: A History of Epilepsy from the Greeks to the Beginnings of Modern Neurology* (Baltimore: Johns Hopkins Press, 1971); and by Ellen Dwyer, e.g., "Stories of Epilepsy, 1880–1930," in *Framing Disease. Studies in Cultural History*, ed. Charles Rosenberg and Janet Golden (New Brunswick, NJ: Rutgers University Press, 1992), 248–272; or "Neurological patients as experimental subjects: Epilepsy studies in the United States," in *The Neurological Patient in History*, ed. L. Stephen Jacyna and Stephen Casper (Rochester, NY: University of Rochester Press, 2012), 44–60.

58. There were other forms of late-onset traumatic epilepsy, not related to war trauma, for example those in which a tumor or brain lesion such as a birth defect was the "primary focus" from which the seizures originated. Foerster reported on some of these cases, but he did not begin to publish extensively on epilepsy until 1924, after the increased incidence of war-trauma-related epilepsy.

59. The effect of such a pull after brain injury, visible as the "wandering" of the ventricles (*Ventrikelwanderung*), the liquor-filled cavities of the brain, was first described by Foerster in 1925 in "Encephalographische Erfahrungen," *Zeitschrift für die gesamte Neurologie und Psychiatrie* 94 (1925): 539–584. Again we can see here how pathology resulted not merely from the damage but from how that damage affected the broader nervous structure.

60. The therapy consisted in the excision of scarred brain tissue, which in most cases permanently cured the patient of his epilepsy. It was this mechanism of scarring that Foerster hoped to learn from Wilder Penfield in their exchange. Penfield had acquired this knowledge on an earlier trip to Europe in 1924, when he worked with Río Hortega in Madrid, to learn the "Spanish techniques" of neuronal staining. These techniques, developed for the study of neuroglia, the support cells of the brain, shed light on the process of scarring in traumatic epilepsy. For an account of this episode, see A. S. Gill, "Wilder Penfield, Pío Del Río-Hortega, and the Discovery of Oligodendroglia," *Neurosurgery* 60 (2007): 940–948; E. García-Albea, "Wilder G. Penfield en la Residencia de Estudiantes (Madrid, 1924)," *Re-*

vista neurologia 39 (2004): 872–878; Penfield's autobiography, Wilder Penfield, *No Man Alone: A Neurosurgeon's Life* (Boston: Little, Brown, 1977), chap. 5.

61. Foerster was not the only one to perform this operation. The German neurosurgeon Fedor Krause operated on ninety-six patients for the treatment of epilepsy between 1893 and 1912. Like Foerster, he used electrical stimulation to map out the motor strip and to identify the epileptogenic areas that he would then remove. William Feindel, Richard Leblanc, and Jean-Guy Villemure, "History of the Surgical Treatment of Epilepsy," in *A History of Neurosurgery in Its Scientific and Professional Contexts*, ed. Samuel Greenblatt (Park Ridge, IL: American Association of Neurological Surgeons, 1997), esp. 468–471.

62. Otfrid Foerster and Wilder Penfield, "Der Narbenzug am und im Gehirn bei traumatischer Epilepsie in seiner Bedeutung für das Zustandekommen der Anfälle und für die therapeutische Bekämpfung derselben," *Zeitschrift für die gesamte Neurologie und Psychiatrie* 125 (1930): 481.

63. Foerster, "Die Pathogenese des epileptischen Krampfanfalles," *Tagung der Gesellschaft deutscher Nervenärzte*, Düsseldorf, 94 (1926): 42.

64. On Vogt, see Michael Hagner, *Geniale Gehirne: Zur Geschichte der Elitegehirnforschung* (Munich: dtv, 2007); Michael Hagner, "*Der Geist bei der Arbeit*: Die visuelle Repräsentation cerebraler Prozesse," in Hagner, *Der Geist bei der Arbeit*, 164–194; Helga Satzinger, *Die Geschichte der genetisch orientierten Hirnforschung von Cécile und Oskar Vogt in der Zeit von 1895 bis ca. 1927* (Stuttgart: Deutscher Apotheker, 1998).

65. The map had six as opposed to the seven arrows in figure 4.3. Cécile Vogt and Oskar Vogt, "Die vergleichend-architektonische und die vergleichend-reizphysiologische Felderung der Großhirnrinde unter besonderer Berücksichtigung der menschlichen," *Die Naturwissenschaften* 14.50–51 (1926): 1193.

66. He therefore called his research method "architectonics" rather than "histology."

67. Foerster to Vogt, September 7, no year given, but probably 1926. Cécile and Oskar Vogt Archive (COVA).

68. Ibid. Although Vogt faithfully reproduced Foerster's epilepsy arrows in his 1926 paper (in which he compared his and Foerster's results), he did not comment on them. In his publication, then, Foerster's map had undergone a transformation from the dynamic epileptogenic map into a static neuroanatomical map à la Vogt.

69. Otfrid Foerster, "Sensible corticale Felder," in *Handbuch der Neurologie*, ed. Otfrid Foerster and Oswald Bumke (Leipzig: Springer, 1936), 6:326.

70. Otfrid Foerster, "Motor cortex in man," 137.

71. This is not to say that there was a direct reception of Meynert's views by Foerster. Rather, I would like to draw attention to a family resemblance of their views. It could well be that Foerster, in framing his work in the language of the social *and* of work (*Arbeitsgemeinschaft*), was influenced by a reevaluation of the liberal economic model after the Wall Street crash.

72. Otfrid Foerster, "Motor cortex in man," 152. Not until the 1930s did Foerster begin to use social language to describe the workings of the nervous system, although his earlier views were compatible with the social perspective. In his English publications, Foerster uses (nervous or reflex) "society" to translate *Nervengemeinschaft* or *Reflexgemeinschaft*. I have used "reflex community" in this book.

73. Otfrid Foerster, "Die Symptomatologie der Erkrankungen des Rückenmarks und seiner Wurzeln," in Foerster and Bumke, *Handbuch der Neurologie*, 5:88. Note that these are reflex arcs in a narrow sense, not "mental reflexes" as in Wernicke's system. See also Otfrid Foerster, "Über die Bedeutung und Reichweite des Lokalisationsprinzips," *Verhandlungen der Deutschen Gesellschaft für Innere Medizin* 46 (1934): esp. 140.

74. Foerster, "Bedeutung und Reichweite des Lokalisationsprinzips," 141, 147.

75. Cf. Klaus-Joachim Zülch, who called the localization of function in the definition of Hughlings Jackson as "Beziehung verschiedener Teile des zentralen Nervensystems zu anderen und zu Teilen des Körpers," the "gemeinsamer Nenner für Foerster's Denken." Klaus-Joachim Zülch, "Otfrid Foerster und die Lokalisationslehre," *Leopoldina* 9 (1973): 168.

76. Otfrid Foerster, "Uebungstherapie," in Foerster and Bumke, *Handbuch der Neurologie*, 8:316.

77. Foerster, "Bedeutung und Reichweite des Lokalisationsprinzips," 149.

78. Ibid., 159.

79. Ibid., 160.

80. Ibid., 161.

81. Otfrid Foerster, *Die Physiologie und Pathologie der Coordination* (Jena: Fischer, 1902), 308. We know that Foerster practiced and taught *Übungstherapie* for most of his academic life. He offered a seminar on *Übungstherapie*, under changing titles, every year between the summer semester of 1904 and the summer semester of 1921. From the winter semester of 1921–22 onward, he offered a course "Klinik der Nervenkrankheiten" (clinic of nervous disease), which might have included *Übungstherapie*. From 1925–26, he added a course "Neurochirurgie" (neurosurgery). *Verzeichniss der Vorlesungen an der Königlichen Universität Breslau* (Breslau), SS 1903–SS 1935. 1904–05 to 1910 and 1923 are missing from the analysis.

82. Penzholz, Helmut. "Otfrid Foerster zum 100. Geburtstag vor Deutschen Neurochirurgen," in Gutiérrez-Mahoney Papers, DMEHM.

83. Otfrid Foerster, "Uebungstherapie bei Tabes dorsalis," *Deutsche Ärzte-Zeitung* (1901): 101.

84. Ibid., 102.

85. Foerster, "Uebungstherapie," 316.

86. In his late synthetic paper on *Übungstherapie* in the Bumke-Foerster *Handbuch*, Foerster made extensive use of social language.

87. Foerster, "Uebungstherapie bei Tabes dorsalis," 101.

88. This seems all the more remarkable because the majority of patients needing *Übungstherapie* were syphilitics, who, as sufferers of venereal disease, were traditionally given little respect in the hospital environment. According to Joel Braslow, psychiatric and neurological symptoms were associated with syphilis since the 1820s. Joel Braslow, *Mental Ills and Bodily Cures: Psychiatric Treatment in the First Half of the Twentieth Century* (Berkeley: University of California Press, 1997), 72.

89. In his book Braslow suggests another possibility: the ability to help patients made doctors change their attitude toward them; and syphilitics, he argues, were treated much better after the discovery of malaria therapy. Braslow, *Mental Ills*. Conversely, it mattered to the patients that their doctors could *do* something, a point that Charles Rosenberg made in "The therapeutic revolution: Medicine, meaning, and social change in 19th-century America," *Perspectives in Biology and Medicine* 20 (1977): 485–506.

90. Foerster, "Uebungstherapie," 329.

91. Ibid., 372.

92. Ibid., 373.

93. It should be noted that this reading offers a new dimension to the image conferred in accounts of Foerster, especially by his holistically oriented colleagues. Foerster's successor to the chair of neurology at Breslau, Victor von Weizsäcker, for instance, described Foerster as a "neurological virtuoso" who played his patients "like a cello." Victor von Weizsäcker, "Otfrid Foerster 1873–1941," *Deutsche Zeitschrift für Nervenheilkunde* 153 (1942): 7. Similarly, in the discussion of Foerster and Goldstein's papers given at the annual meeting of the Gesellschaft deutscher Nervenärzte in Dresden in 1930, while some participants saw their accounts as essentially compatible (e.g., the neurologist Walter from Bremen), Weizsäcker emphasized their difference. Kurt Goldstein, Viktor v. Weizsäcker, and Otfrid Foerster, "Aussprache zum Bericht und Schlussworte. 20. Jahresversammlung der Gesellschaft Deutscher Nervenärzte in Dresden," *Deutsche Zeitschrift für Nervenheilkunde* 116 (1930): 42 (Walter), 29 (Weizsäcker). At other times, however, Weizsäcker appeared less wedded to their opposition. E.g., Weizsäcker, "Otfrid Foerster." See also Bernd Holdorff, "Die Lokalisationsdiskussion vor 60 Jahren und heute (O. Foerster, K. Goldstein, V. v. Weizsäcker)," *Schriftenreihe der Deutschen Gesellschaft für Geschichte der Nervenheilkunde* 1 (1996): 139–141; Wilhelm Rimpau, "Die Krise der Neurologie in erkenntnistheoretischer Weise: Kontroverse zwischen Viktor von Weizsäcker, Kurt Goldstein und Otfrid Foerster zum Lokalisationsprinzip 1930," *Nervenarzt* 80 (2009): 970–974; Cornelius Borck, "Concrete existence: Goldstein's variations on the nervous body," article manuscript, in author's collection.

94. Ilse Foerster, "Die Arbeitsstätte von Foerster," typed manuscript, Gutiérrez-Mahoney Papers, DMEHM.

95. In his 1932 speech at the Neurological Congress, Foerster suggested that "at least in the larger hospitals separate neurological wards [should be] established."

Otfrid Foerster, "Eröffnungsansprache," *Deutsche Zeitschrift für Nervenheilkunde* 129 (1932): 175–184, quotation on 181.

96. Personalbogen Otfrid Foerster, Wrocław University Archive (WUA).

97. In Max Lewandowsky, ed., *Handbuch der Neurologie*, Ergänzungsband, vol. 2 (Berlin: Julius Springer, 1924), 976 (peripheral nerves); 1721 (spinal cord).

98. Acta miasta Wrocławia 33831, 3, June 7, 1924, SAB. The Wenzel-Hancke hospital in Breslau was first opened on January 1, 1878. It was subsequently expanded, esp. in 1890, 1895, and 1909. "Denkschrift über Entstehung, Bau, Einrichtung und Betrieb des Wenzel-Hancke-Krankenhauses," Breslau, 1912, Acta miasta Wrocławia, SAB.

99. E.g. June 7, 1924, 33831, 1; October 5, 1936, 33835, 199; August 8, 1927, 33850, 192. Foerster states the problem of "Platzmangel" on January 28, 1929, 33835, 3. All in Acta miasta Wrocławia, SAB.

100. Foerster to Kuratorium des Wenzel-Hancke-Krankenhauses: Breslau, October 23, 1921, 33773, 201, Acta miasta Wrocławia, SAB.

101. June 7, 1924, 33831, 3, Acta miasta Wrocławia, SAB.

102. August 11, 1921, 33849, 272 (2), Acta miasta Wrocławia, SAB.

103. June 7, 1924, 33831, 3, Acta miasta Wrocławia, SAB.

104. Davies and Moorhouse, *Microcosm*, 335.

105. Daniel O'Brien's diary, December 3, 1931, Rockefeller Foundation Archives (hereafter RFA), Record Group 1.1, series 717A, box 11, folder 76.

106. "Erläuterungsbericht zum Vorentwurfe C.2 für den Neubau eines Forschungslaboratoriums bei der neurologischen Abteilung im Wenzel-Hanckeschen Krankenhause zu Breslau," RFA, Record Group 1.1, series 717A, box 11, folder 76.

107. Otfrid Foerster, "Eröffnungsansprache," *Deutsche Zeitschrift für Nervenheilkunde* 110 (1929): 213–214.

108. Ibid., 129.

109. *Proceedings of the First International Neurological Congress, Berne (Switzerland), August 31st to September 4th, 1931*, 374. See also Otfrid Foerster, "Eröffnungsansprache" (1932), 178 and M. Minkowski, "Die Stellung der Neurologie im medizinischen Unterricht," *Schweizer Archiv für Neurologie und Psychiatrie* 30 (1933): 165, 168.

110. Foerster, "Eröffnungsansprache" (1932), 178–179. From Foerster's correspondence with Harvey Cushing and John Fulton, it seems that the congress was mostly remembered for social reasons: Cushing wrote to Foerster on November 14, 1931, saying "what fun" he had had in Bern. He did not mention Foerster's address or the resolution. Similarly, John Fulton wrote to Foerster on September 14, 1931, "The Bern meeting remains a very happy memory, and for me one of the pleasantest recollections of the week was seeing you there and hearing your paper and the beautiful address at the dinner. I sincerely hope you will find time to put the address down in permanent form, not alone for me, but for your many friends who will wish to have it." It appears that Fulton emphasized the address (in which,

after all, his teacher Harvey Cushing was singled out) less for its value for neurology. This perspective corresponds to the presentation of his report of the congress. John Fulton, "Arnold Klebs and Harvey Cushing at the First International Neurological Congress at Berne in 1931," *International Journal of Neurology* 14 (1980): 103–115, reprinted in *Bulletin of the History of Medicine*. From an American (neurosurgical) perspective, the question of an independent neurology must have appeared less pressing. A copy of the Harvey Cushing–Otfrid Foerster correspondence and a copy of the John Fulton–Otfrid Foerster correspondence are in the Gutiérrez-Mahoney Papers, DMEHM.

111. Foerster, "Eröffnungsansprache," *Deutsche Zeitschrift für Nervenheilkunde* 115 (1930):152.

112. Ernst Rüdin, "Eröffnungsansprache auf der 1. Jahresversammlung der Gesellschaft deutscher Neurologen und Psychiater," *Deutsche Zeitschrift für Nervenheilkunde* 139 (1936): 9–10.

113. Foerster was not alone in his repudiation. The neurologist Heinrich Pette, who was elected the head of the society's neurological section, expressed his regrets that "in einer Zeit wo die Neurologie in fast allen Ländern der Welt verselbständigt werde, ihr in Deutschland, dem Mutterland der Neurologie, die Türen verschlossen werden sollen." Quoted in Pantel, "Streitfall Nervenheilkunde," 152.

114. Percival Bailey, *Up from Little Egypt* (Chicago: Buckskin Press, 1969), 153.

115. Otfrid Foerster, "Operativ-experimentelle Erfahrungen beim Menschen über den Einfluss des Nervensystems auf den Kreislauf," *Zeitschrift für die gesamte Neurologie und Psychiatrie* 167 (1939): 461. Shorter references to the "Führer" can be found elsewhere, e.g., in Otfrid Foerster, "Über die Wechselbeziehungen von Herdsymptomen und Allgemeinsymptomen," *Verhandlungen der deutschen Gesellschaft für innere Medizin* 50 (1938): 458–485, quotation on 485; and Foerster's speech at the opening of Foerster institute in 1934, "Das Neurologische Institut eröffnet: Das erste in der ganzen Welt," January 31, 1934, *Breslauer 8-Uhr Abendblatt*, article clipping held in Gutiérrez-Mahoney Papers DMEHM.

116. This cannot be presented as de rigueur. Indeed, other contributors to the conference did not contain anything so political. See, for instance, the contribution by Wilhelm Tönnis in the same issue, "Zirkulationsstörungen bei krankhaftem Schädelinnendruck," *Zeitschrift für die gesamte Neurologie und Psychiatrie* 167 (1939): 462–465. This has probably contributed to the decline of Foerster's name in international neurology. Angelika Foerster, Otfrid's sister, wrote to Gutiérrez-Mahoney in 1948 that, to her sorrow, Pervical Bailey had written that "America now only respected Dr Nonne and had quite turned against Otfrid." February 3, 1948, Gutiérrez-Mahoney Papers, DMEHM.

117. Wilder Penfield, "Orientation of scientific research to war," *American Scientist* 30.2 (1941): 116. See also Wilder Penfield, "The electrode, the brain and the mind," *Journal of Neurology* 201 (1972): 301.

118. Lena Kaletowa, "Der Breslauer Arzt Lenins," 3 parts, *Gazeta Robotnicza*, April 6, 13, 20, 1984, translation of Polish article into German, held in the Gutiérrez-Mahoney Papers, DMEHM.

119. On the history of neurology post-World War II, see Pantel, "Streitfall Nervenheilkunde," 153. See also Klaus-Joachim Zülch, "The place of neurology in medicine and its future," *Handbook of Clinical Neurology* 1 (1969): 1–44. Zülch sees an important reason for the postwar success of neurology in Germany in the development of effective therapies, including anticonvulsants for epilepsy, steroids to treat inflammatory and degenerative conditions, treatments of ischemic conditions of the brain and spinal cord, treatment of metabolic disorders, and rehabilitation (34).

120. See Goldstein's paper "Die Restitution bei Schädigungen der Hirnrinde," *Deutsche Zeitschrift für Nervenheilkunde* 116 (1930): 2–26, esp. 24–25. See also his famous critique of the "atomistic method" within reflex theory in his 1934 *The Organism: A Holistic Approach to Biology Derived from Pathological Data in Man* (New York: Zone Books, 1995), esp. 69–93.

121. Goldstein, Weizsäcker, and Foerster, "Aussprache zum Bericht und Schlussworte," 40. Note that another "holist," Viktor von Weizsäcker, picked up on the opposition between their views. Cf. note 93.

Chapter Five

1. Schilder to Freud, December 21, 1935, Lauretta Bender Papers, IV, series 9, Brooklyn College Archives and Special Collections (hereafter BCA).

2. Schilder to Karolinska Institute, December 21, 1935, Lauretta Bender Papers, IV, series 9, BCA. Freud never got the Nobel Prize. The 1936 prize went to Sir Henry Dale and Otto Loewi instead, for their research on the chemical transmission of nerve impulses.

3. Schilder to Educational Committee, December 21, 1935, Lauretta Bender Papers, IV, series 9, BCA.

4. The literature on Paul Schilder is scarce. For a biography, see Dieter Langer, "Paul Ferdinand Schilder: Leben und Werk" (PhD diss., Mainz University, 1979). See also Donald Shaskan and William Roller, eds., *Paul Schilder: Mind Explorer* (New York: Human Sciences Press, 1985).

5. Paul Schilder and K. Kassowitz, "Einige Versuche über die Feinheit der Empfindungen bei bewegter Tastfläche," *Pflügers Archiv für die gesamte Physiologie der Menschen und Tiere* 122 (1908): 119–129.

6. E.g., Paul Schilder, "Über die amyloide Entartung der Haut," *Frankfurter Zeitschrift für Pathologie* 3 (1909): 782–794; with W. Haberfeld, "Die Tetanie der Kaninchen," *Mitteilungen aus dem Grenzgebiet der Medizin und Chirurgie* 20 (1909): 728–756. There are no publications from his time (three months) with Heinrich Obersteiner.

7. Lauretta Bender Papers, IV, series 9, BCA. The condition has come to be known as "Schilder's disease."

8. From December 27, 1909, to March 15, 1912. Lauretta Bender Papers, IV, series 9, BCA.

9. Paul Schilder, "Vita," in Shaskan and Roller, *Paul Schilder*, 267.

10. Ibid.

11. Despite Wernicke's own reticence in using such models, Anton extended the social language to describe Wernicke's ideas in his inaugural lecture in Halle. Gabriel Anton, *Über den Wiederersatz der Funktion bei Erkrankungen des Gehirnes* (Berlin: Karger, 1906). The social language and readings of Meynert and Wernicke had been especially useful for Anton in the area in which he had made his name: the study of brain damage and the compensatory processes of the nervous system. Writing about the "self help" (7) of the central nervous system and the "division of labor of the individual nerve junctions" (15), Anton affirmed that the principle of compensation was visible in Wernicke's work: "The symptoms of the language disturbances, to the knowledge of which Wernicke exerted an epoch-making influence, has shown that the individual sensory categories could stand up for each other and compensate each other" (28). Schilder also spent time in Leipzig, as *Assistenzarzt* to Paul Flechsig, but he did not single out Leipzig as an influential stage for his professional or intellectual development in his "Vita," as he did with Halle.

12. Certificate by Otto Poetzl, Lauretta Bender Papers, IV, series 9, BCA.

13. Otto Poetzl, "Julius Wagner von Jauregg," *Wiener klinische Wochenschrift* 53 (1940): 1–4, on 2, my emphasis. Quoted and translated in Lesky, *Vienna Medical School*, 361n115.

14. Elisabeth Young-Bruehl, *Anna Freud: A Biography* (New York: Summit Books, 1988).

15. Fritz Wittels, "Paul Schilder—1886–1940," *Psychoanalytic Quarterly* 10 (1941): 132.

16. Magda Whitrow, *Julius Wagner-Jauregg (1857–1940)* (London: Smith-Gordon, 1993), 102.

17. Wagner-Jauregg to Freud, March 9, 1932, extract from copies of letters between Sigmund Freud and Julius Wagner-Jauregg (supplied to Ernest Jones by Ernst Freud, c. 1955), Society and Institute Records, CFE/F13/004, Archives of the British Psychoanalytic Society (hereafter ABPS).

18. Wagner-Jauregg helped Schilder's career at several important moments, e.g., by recommending him as *Privatdozent*. Langer, "Paul Ferdinand Schilder," 30. He also supported his promotion to be extraordinary university professor. Lauretta Bender Papers, IV, series 9, BCA. In the end, he did not make him his *Adjunkt* (permanent assistant), so Schilder saw himself forced to move on. See Erwin Stransky, "Aus einem Gelehrtenleben um die Jahrhundertwende: Rückschau, Ausblick, Gedanken," typescript, 2065, History of Medicine Collections, Institute for

the History of Medicine, Vienna University (HMC). Schilder noted in his letters to Adolf Meyer, whose clinic later became his first point of contact with American medicine, that he was "not quite satisfied with my present position here." Schilder to Meyer, June 14, 1929, Individual correspondence Paul Schilder, I/3422/2, Alan Mason Chesney Medical Archives of the Johns Hopkins Medical Institutions (hereafter AMC). He also believed he could "work better in your [Meyer's] clinic than here in Austria." Schilder to Meyer, not dated, probably summer 1929, I/3422/2, Individual correspondence Paul Schilder, AMC.

19. Wernicke, *Grundriss* (1894), 1:36–47.

20. Ibid., 1:38.

21. Ibid., 1:45.

22. Sigmund Freud, *The Ego and the Id* (1923), in Freud, *Standard Edition*, 19:1–66, on 26. There are also (allusive) references to the concept in Freud's 1914 "On narcissism: An introduction," in Freud, *Standard Edition*, 14:67–102.

23. Freud, "Ego and id," 26–27.

24. Paul Schilder, "Probleme der klinischen Psychiatrie," *Medizinische Klinik* 21 (1925): 79.

25. Henry Head and Gordon Holmes, "Sensory disturbances from cerebral lesions," *Brain* 34 (2–3) (1911): 102–254; Paul Schilder, *Das Körperschema: Ein Beitrag zur Lehre vom Bewusstsein des eigenen Körpers* (Berlin: Springer, 1923), 2.

26. Schilder, *Körperschema*, 2–3.

27. Ibid., 5–6.

28. Ibid., 7.

29. Schilder alternates between the expressions optical or tactile "part" of "the" body schema and "optical body schema" or "tactile body schema."

30. Schilder, *Körperschema*, 12.

31. Ibid., 86.

32. Ibid., 42, 48, 45.

33. Ibid., 37.

34. Hugo Liepmann, *Über Störungen des Handelns bei Gehirnkranken* (Berlin: Karger, 1905), 45.

35. Schilder, *Körperschema*, 84.

36. Ibid., 35.

37. The two columns appear in half of the case histories. The others were written as continuous text, and it is likely that Schilder used a similar method there.

38. Schilder, *Körperschema*, 18.

39. Ibid., 35.

40. Ibid., 33, 51.

41. Ibid., 19.

42. Ibid., 57.

43. This attempt becomes especially clear in Schilder's work on the psychoses, e.g., Paul Schilder, F. Paul, and H. Weidner, "Zur Kenntnis symbolähnlicher Bildungen

im Rahmen der Schizophrenie," *Zeitschrift für die gesamte Neurologie und Psychiatrie* 26 (1914): 201–244; Paul Schilder, "Gesichtspunkte der allgemeinen Psychiatrie," *Archiv für Psychiatrie und Nervenkrankheiten* 59 (1918): 699–713; Paul Schilder, "Entwurf zu einer Psychologie der Schizophrenie und Paraphrenie," *Deutsche medizinische Wochenschrift* 49 (1923): 1433–1435; Paul Schilder, *Wahn und Erkenntnis* (Berlin, Springer: 1918); Paul Schilder, "Bemerkungen ueber die Psychologie des paralytischen Größenwahns," *Zeitschrift für die gesamte Neurologie und Psychiatrie* 74 (1922): 1–14; Paul Schilder "Die Psychotherapie der Psychosen," *Zentralblatt für die gesamte Neurologie und Psychiatrie* 44 (1926): 182; Paul Schilder, "Psychologie der Schizophrenie vom psychoanalytischen Standpunkt," *Zeitschrift für die gesamte Neurologie und Psychiatrie* 112 (1928): 279–282; Paul Schilder, "Scope of Psychiatry in Schizophrenia," *American Journal of Psychiatry* 11 (1932), 1181–1187. In this work, although noting certain differences as well, Schilder emphasized the continuity between the normal and the schizophrenic mind, suggesting that "schizophrenia . . . was accessible to empathetic understanding." Schilder, "Gesichtspunkte der allgemeinen Psychiatrie," 702. Similarly, in his work on the body schema–body image, especially in the second book, Schilder used examples from the realm of the normal to explain the workings of the body image, e.g., the "well-known phenomenon that the head appears to be larger after alcoholic intoxication" or the Japanese illusion in which hands and fingers were doubly crossed and doubly intertwined so that their "optic impression . . . becomes so complicated that the optic gnosia is not sufficient to disentangle the picture." Only through tactile or kinesthetic intervention (moving the fingers or touching the fingers while the subject is looking) did the body image fall into place again. Paul Schilder, *The Image and Appearance of the Human Body: Studies in the Constructive Energies of the Psyche* (New York: International Universities Press, 1950), 23–24. But despite these complications, Schilder's rejection of the unconscious remained in place.

44. Schilder, *Körperschema*, 18.

45. Ibid., 18–19.

46. It is possible, of course, that patients used this language themselves.

47. Schilder, *Körperschema*, 14.

48. Ibid., 17, 33.

49. Schilder's clinical reliance on a self-transparent and self-expressive subject made the problem of speech a particularly interesting one to him. For an exemplary paper on his approach to speech disturbances, which demonstrates his attempts to grapple with the problem of aphasia, see Frank Curran and Paul Schilder, "Paraphasic signs in diffuse lesions of the brain," *Journal of Nervous and Mental Disease* 82 (1935): 613–636.

50. Schilder, *Körperschema*, 27.

51. Ibid., 20—see Jacyna's account of this tour-de-force of psychological introspection in *Medicine and Modernism*, 125–137.

52. Schilder, "Gesichtspunkte zur allgemeinen Psychiatrie," 707.

53. Paul Schilder, "Das Unbewusste," *Zeitschrift für die gesamte Neurologie und Psychiatrie* 80 (1923): 99–100.

54. In line with this development, we see Schilder reassert the identity of the psychic and awareness, which Freud had broken. E.g., ibid., 96.

55. Stransky, "Aus einem Gelehrtenleben um die Jahrhundertwende." On scientific emigration, see Michael Hubenstorf, "Österreichische Ärzte-Emigration," in *Vertriebene Vernunft I: Emigration und Exil österreichischer Wissenschaft 1930–1940*, ed. Friedrich Stadler (Vienna: Jugend und Volk, 1987), 359–415; Mitchell Ash and Alfons Söllner, *Forced Migration and Scientific Change: Emigré German-Speaking Scientists after 1933* (Cambridge: Cambridge University Press, 1996).

56. For the only monograph on Meyer so far, see Susan D. Lamb, *Pathologist of the Mind: Adolf Meyer and the Origins of American Psychiatry* (Baltimore: Johns Hopkins University Press, 2014).

57. Ruth Leys, "Meyer's dealings with Jones: A chapter in the history of the American response to psychoanalysis," *Journal of the History of the Behavioral Sciences* 17 (1981): 445–465; Ruth Leys, "Adolf Meyer's Life Chart and the Representation of Individuality," *Representations* 34 (1991): 1–28; Hale, *Rise and Crisis*.

58. Meyer was present to receive one of several honorary degrees, awarded on the occasion of the institution's twentieth anniversary. For a critical account of Freud's 1909 visit and its role in the history of American psychoanalysis, see Burnham, *After Freud Left*.

59. Leys, "Meyer's dealings," 446. See also Hale, *Rise and Crisis*; and John C. Burnham, *Psychoanalysis and American Medicine: 1894–1918; Medicine, Science, and Culture* (New York: International Universities Press, 1967).

60. On the relationship between Meyer and psychoanalysis, see Ruth Leys, "Meyer, Jung, and the Limits of Association," *Bulletin of the History of Medicine* 59.3 (1985): 345–360. For a recent evaluation of the relationship, see Lamb, *Pathologist of the Mind*. See also Mical Raz, "Between the Ego and the Icepick: Psychosurgery, Psychoanalysis, and Psychiatric Discourse," *Bulletin of the History of Medicine* 82 (2008): 387–420, on the relations between somatic and dynamic approaches to mental illness in American medicine.

61. Adolf Meyer, "The dynamic interpretation of dementia praecox," lecture delivered at the celebration of the twentieth anniversary of Clark University, September 1909, *American Journal of Psychology* 21 (1910): 385–403, in *The Collected Papers of Adolf Meyer*, ed. Eunice E. Winters (Baltimore: Johns Hopkins Press, 1950–), vol. 2.

62. Ibid., 443, quoted in Leys, "Meyer's dealings," 462n53.

63. Adolf Meyer, "Psychobiology in the first year of medical school," *Journal of the Association of American Medical Colleges* 10 (1935): 365–72, in *Collected Papers*, 3:97.

64. Adolf Meyer, "A discussion of some fundamental issues in Freud's psychoanalysis," *State Hospitals Bulletin*, n.s., 2 (1909–10): 827–848, in *Collected Papers*, 2:615.

65. Adolf Meyer, "Interpretation of obsessions," *Psychological Bulletin* 3 (1906): 280–283, in *Collected Papers*, 2:633. Meyer must have read Freud in German. In fact, the first translation of any work by Freud into English was in 1909, by A. A. Brill, who translated Freud and Breuer's *Studies in Hysteria*. At the time, Freud himself showed some terminological looseness with respect to the unconscious, referring to the "subconscious" (*Unterbewusstes*) once, which he used in the same way. Cf. Laplanche and Pontalis, *Language of Psycho-Analysis*, 430–431. Putnam, whose *Recent Experiences in the Study and Treatment of Hysteria* Meyer reviewed in the same article, referred to this Freudian "subconscious" (using the term *unconscious* for describing the state of mind of a patient who had fallen down a staircase), which Meyer used as well in his article when directly referring to Putnam. In his later writings, Meyer adopted the term *unconscious* when English translations of Freudian texts became more widely available. Similarly, Meyer sometimes avoided the term *repressed*, referring to "displaced" or "converted" instead, e.g., in Meyer, "Interpretation of obsessions," in *Collected Papers*, 2:633.

66. E.g., *Collected Papers*, 3:238, 283–284, 407, 431.

67. Adolf Meyer, "Dealing with mental disease," *Modern Hospital* 51 (1938): 87–89, in *Collected Papers*, 4:102.

68. Adolf Meyer, "Preparation for psychiatry," *Archives of Neurology and Psychiatry* 30 (1933): 1111–1125, in *Collected Papers*, 3:78.

69. Adolf Meyer, "Outlines of examinations" (1923), privately printed, 1918, in *Collected Papers*, 3:244–245. This is one of the few passages in which Meyer did not place "the unconscious" in quotation marks.

70. Leys, "Meyer's dealings"; Hale, *Rise and Crisis*, 170; Lamb, *Pathologist of the Mind*.

71. Adolf Meyer, "Paul Ferdinand Schilder, M.D., 1886–1940," *Journal of Nervous and Mental Disease* 93 (1941): 812–814, in *Collected Papers*, 3:548.

72. He stayed from mid-October 1928 to January 1929. His lectures were published in 1931 in Paul Schilder, *Brain and Personality: Studies in the Psychological Aspects of Cerebral Neuropathology and the Neuropsychiatric Aspects of the Motility of Schizophrenics* (New York: Nervous and Mental Disease, 1931), part 2. The volume also contains the lectures Schilder delivered to the Washington Psychoanalytic Association during his stay (part 1). For Schilder's psychoanalytic practice at the Phipps, see, e.g., his discussion of a patient he treated over several months during his stay, in "Problems in the Technique of Psychoanalysis," *Psychoanalytic Review* 17 (1930): 1–19.

73. Meyer, "Paul Ferdinand Schilder," *Collected Papers*, 3:549.

74. Menas Gregory to Schilder, August 13, 1929, Lauretta Bender Papers, IV, series 9, BCA.

75. An English reprint was published in 1999 by Routledge and Kegan Paul (Kegan Paul, Trench, Trübner and Company had joined with Routledge in 1912, although the publishers continued to publish separately), in the International Li-

brary of Psychology, Philosophy and Scientific series (edited by C. K. Ogden), which was later renamed the International Library of Psychology. It published a total of 204 volumes on work by figures such as C. G. Jung, Sigmund Freud, Jean Piaget, Otto Rank, and James Hillman. Schilder's *The Image and Appearance of the Human Body* appeared in the minisubset "Physiological Psychology in 10 volumes." In addition to psychoanalysis, the series responded to the rise of analytic philosophy and Viennese logical positivism. Cf. M. Saunders, "'Science and futurology in the to-day and to-morrow series': Matter, consciousness, time and language," *Interdisciplinary Science Reviews* 34.1 (2009): 68–78.

76. Schilder, *Image and Appearance*, henceforth *The Body Image*.

77. Ibid., 249, 250.

78. Ibid., 11.

79. Ibid., 250. At the very least, we can say that the body image, or parts thereof, were considered to be unconscious (253). Note the slippage in terminology: schema/image—unconscious/conscious.

80. E.g., on 22, 31, 248, 291 and 296 in Schilder, *The Body Image*. Sometimes, e.g., on 31, he put the term in quotation marks the first or second time he used it, and when he used it again in the same matter, he left out the quotation marks. Whenever Schilder did not put the term in quotation marks, he referred not to "the unconscious" but rather to cases in which psychic processes never left the background of awareness, as in this example: "the muscular pull which tries to restore the previous position and its effect remain unconscious" (76). Other examples of the use of the term are on 81, 83, 250–251.—The clearest statement of Schilder's attitude toward the Freudian unconscious is in his paper "Der gegenwärtige Stand der Neurosenlehre," *Klinische Wochenschrift* 6.2. (1927): 50. There, Schilder drew on the works by Oswald Külpe, Narziss Ach, and Karl Bühler, proponents of the Würzburg school of thought psychology who studied processes through the method of experimental introspection. Cf. Edwin Boring, *A History of Experimental Psychology*, 2nd ed. (New York: Appleton-Century-Croft, 1950), esp. 401–410.

81. Schilder, *The Body Image*, 249.

82. Paul Schilder, *Mind: Perception and Thought in Their Constructive Aspects* (New York: Columbia University Press, 1942), 261–262.

83. Schilder, *The Body Image*, 296.

84. Ibid., 250. Similarly, when discussing the sociology of the body image, and more specifically the concept of identification, Schilder pointed out that we possessed a "collection of various images of bodies," which, as he conceded, "in the terminology of psychoanalysis . . . would be called unconscious identification" (274). But Schilder "prefer[red] to talk of the background of the consciousness or to use the term 'sphere'" (274–275).

85. Ibid., 11.

86. Ibid., 171.

87. Ibid., 182.

88. Ibid., 169.

89. Ibid., 132–133.

90. Ibid., 202.

91. Freud, "Character and Anal Erotism" (1908), in Freud, *Standard Edition*, 9:167–176.

92. Schilder, *The Body Image*, 202.

93. Ibid., 135.

94. Ibid., 131.

95. Hale, *Rise and Crisis*, 233–244. See also Makari, *Revolution in Mind*, 449–455.

96. In a letter to A. A. Brill, Schilder emphasized that he was "a member with full rights of the Viennese Psycho-analytic Society." He continued: "If the Society intended not to give me full rights it would have had the duty to inform me before I became a member. I would then certainly not have given up my membership of the Vienniese [*sic*] Society and would not have entered the New York Psychoanalytic Society." Schilder to Brill, November 19, 1933, Lauretta Bender Papers, IV, series 9, BCA.

97. Hale, *Rise and Crisis*, 121–124.

98. See also the epilogue to George Makari's impressive *Revolution in Mind*, esp. 475–485.

99. Hale, *Rise and Crisis*, 103.

100. Ibid., 105.

101. Cf., e.g., the Flexner Report, published in 1910. Abraham Flexner, *Medical Education in the United States and Canada: A Report to the Carnegie Foundation for the Advancement of Teaching*, Bulletin Number 4 (New York: Carnegie Foundation for the Advancement of Teaching, 1910).

102. Hale, *Rise and Crisis*, 30.

103. "The 1949 annual meeting," *Bulletin of the American Psychoanalytic Association*, 5C: 31–75.

104. A. A. Brill, *Psychoanalysis: Its Theories and Practical Application*, 3rd ed. (Philadelphia: Saunders, 1922), 4–5.

105. Ibid., 11.

106. Meyer, "Preparation," in *Collected Papers*, 3:78.

107. Schilder to Freud, November 1, 1935, cited in Hale, *Rise and Crisis*, 123.

108. Schilder to George E. Daniels, June 4, 1935, Lauretta Bender Papers, IV, series 9, BCA.

109. Brill to Jones, May 1, 1935, Society and Institute Records, CBD/F04/040, ABPS.

110. Brill to Jones, May 17, 1935, Society and Institute Records, CBD/F04/041, ABPS.

111. Schilder to Lewin, October 23, 1935, Lauretta Bender Papers, IV, series 9, BCA.

112. Daniels to Schilder, May 24, 1935, Lauretta Bender Papers, IV, series 9, BCA.

113. Brill to Jones, May 1, 1935. Brill feared that Schilder might take further steps to harm the society: "If Schilder should bring it before the County Medical Society, which he threatens to do, Kubie will be censured." Brill to Jones, May 17, 1935. Schilder, in fact, was already forming his own psychotherapeutic society, which Brill feared "will do us no good." Brill to Jones, November 7, 1934, Society and Institute Records, CBD/F04/037, ABPS.

114. The following psychiatrists were present at the first meeting of the society on January 17, 1935: George S. Amsden, Lauretta Bender, Abraham Arden Brill, Clarence O. Cheney, Frank Curran, Phyllis Greenacre, Beatrice Hinkle, Karen Horney, Robert B. McGraw, Clarence Paul Oberndorf, Howard W. Potter, Paul Schilder, Louis Wender, and Fritz Wittels. Schilder Society Records, 1935–1981, series 5, box 2, New York Academy of Medicine, Rare Book Room (hereafter NYAM).

115. "Outline for a Society for Psychotherapy and Psychopathology," Schilder Society Records, 1935–1981, series 1, box 1, NYAM.

116. Anna Freud, in a letter to Jones, summarized the content of her father's letter. Society and Institute Records, CFA/F02/061, ABPS.

117. Freud to Schilder, quoted in Hale, *Rise and Crisis*, 123.

Chapter Six

1. The 3-D homunculus was not produced by Penfield himself. He did, however, draw several detached versions of the figure (see fig. 6.3).

2. The plaster cast in figure 6.1 is a motor homunculus; it has a sensory sibling, with slightly different proportions.

3. Wilder Penfield, *The Mystery of the Mind: A Critical Study of Consciousness and the Human Brain* (Princeton, NJ: Princeton University Press, 1975), 46.

4. Eric Kandel, James Schwartz, Thomas Jessel, and Steven Siegelbaum, *Principles of Neural Science*, 5th ed. (New York: McGraw-Hill, 2012), 364.

5. Sandra Blakeslee and Matthew Blakeslee, *The Body Has a Mind of Its Own: How Body Maps in Your Brain Help You Do (Almost) Everything Better* (New York: Random House, 2007), 20.

6. Richard Dawkins and Yan Wong, *The Ancestor's Tale: A Pilgrimage to the Dawn of Evolution* (Boston: Houghton Mifflin, 2004), 245. Dawkins and Wong also criticize Penfield's homunculus, pointing out that "it was based on very scanty data" (237).

7. Blakeslee and Blakeslee, *Body Has a Mind of Its Own*, 27.

8. E.g., Wilder Penfield and Edwin Boldrey, "Somatic motor and sensory representation in the cerebral cortex of man as studied by electrical stimulation," *Brain* 60.4 (1937): 389–443; see also Kandel et al., *Principles*, 344; or Bryan Kolb and Ian Whishaw, *An Introduction to Brain and Behavior* (New York: Worth, 2009), 363–

364. Cf. Michael Hagner, quoting Penfield, in Hagner, "*Der Geist bei der Arbeit*," 183, reprinted from *Anatomien medizinischen Wissens: Medizin, Macht, Moleküle*, by Cornelius Borck (Frankfurt am Main: Fischer, 1996), 259–286.

9. G. D. Schott, "Penfield's homunculus: A note on cerebral cartography," *Journal of Neurology, Neurosurgery, and Psychiatry* 56 (1993): 329–333. Within the MNI, the homunculus has become the subject of ridicule as well, but also of community building, for example in the song by John Kershman: "The sad reflections of a research fellow at the MNI on first reading the 'cerebral cortex of man'" of 1951, sung to the tune of "Goodnight Irene." W/U 120: "John Kershman," A/N 31-1/2, Wilder Penfield Fonds, Osler Library of the History of Medicine (hereafter OLHM). Similarly, hand drawings of the homunculus made it to the cover page of the 1961 MNI Fellows Society Annual Newsletter. A/N 31-4, Wilder Penfield Fonds, OLHM.

10. Penfield referred to his exchange (an "amusing correspondence") with a "distinguished neurologist" in a footnote in his book *The Cerebral Cortex of Man*. Wilder Penfield and Theodore Rasmussen, *The Cerebral Cortex of Man: A Clinical Study of Localization of Function* (New York: Macmillan, 1950), 26.

11. Walshe was referring to the 1937 edition of the homunculus. F.M.R. Walshe to Penfield, Correspondence Wilder Penfield with Sir Francis Walshe, C/D 20, Wilder Penfield Fonds, OLHM. The letter is dated April 25, 1943, which is presumably a mistake on Walshe's part, since the rest of the correspondence took place between May and December 1946.

12. Sir Francis Walshe to Wilder Penfield, August 6, 1946, Correspondence Wilder Penfield with Sir Francis Walshe, C/D 20, Wilder Penfield Fonds, OLHM.

13. After 1954, most notably in his Sherrington lecture of 1958, Penfield ceased to include the figure in his publications, replacing pictorial maps by maps that designated body parts by words. Wilder Penfield, *The Excitable Cortex in Conscious Man* (Springfield, IL: Thomas, 1958).

14. Penfield in fact engaged with the criticism in two separate letters, one from May 3, 1946, and one from August 20, 1946. Correspondence Wilder Penfield with Sir Francis Walshe, C/D 20, Wilder Penfield Fonds, OLHM.

15. Penfield and Boldrey, "Somatic motor and sensory representation," 431–432. See also Wilder Penfield and Theodore Erickson, *Epilepsy and Cerebral Localization: A Study of the Mechanism, Treatment and Prevention of Epileptic Seizures* (Springfield, IL: Thomas, 1941), 48; or Penfield and Rasmussen, *Cerebral Cortex of Man*, 25.

16. For the historical scholarship on Penfield's work and the figure of the homunculus, see Winter, *Memory*, 75–192; Hagner, "*Der Geist bei der Arbeit*," esp. 182–187; Claudio Pogliano, "Penfield's homunculus and other grotesque creatures from the land of if," *Annali di Storia della Scienza* 27.1 (2012): 141–162; Peter J. Snyder and Harry A. Whitaker, "Neurologic heuristics and artistic whimsy: The cerebral cartography of Wilder Penfield," *Journal of the History of the Neurosciences*

22.3 (2013): 277–291; Cathy Gere, "Nature's Experiment": Epilepsy, Localization of Brain Function and the Emergence of the Cerebral Subject," in *Neurocultures: Glimpses into an Expanding Universe*, ed. Francisco Ortega and Fernando Vidal (Frankfurt: Peter Lang, 2011): 235–247.

17. For a detailed summary of his education, see Penfield, *No Man Alone*. On experimental physiology as an epistemological basis for neurosurgery, see Thomas Schlich, "'Physiological surgery': Laboratory science as the epistemic basis of modern surgery (and neurosurgery)," in *Technique in the History of the Brain and Mind Sciences*, ed. Stephen Casper and Delia Gavrus (under review).

18. For example, he had performed a study on the responses to experimentally induced wounds to rabbit brains. P. del Río Hortega and Wilder Penfield, "Cerebral cicatrix: The reaction of neuroglia and microglia to brain wounds," *Bulletin of the Johns Hopkins Hospital* 41 (1927): 278–303.

19. Penfield published more than a dozen papers between 1924 and 1930 on the formation of microglia and oligodendroglia and cicatrical contraction, using the Spanish methods of histopathology on animal and human brains. For Penfield's account of his work with Río Hortega in Madrid in 1924, see the chapter in his autobiography, *No Man Alone*.

20. Ibid., 157.

21. The case has been described in detail in Wilder Penfield, "The radical treatment of traumatic epilepsy and its rationale," *Canadian Medical Association Journal* 23 (1930): 193–194.

22. For accounts of Cushing and the history of neurosurgery, see the work of Samuel Greenblatt, e.g., "The historiography of neurosurgery: Organizing themes and methodological issues," in *A History of Neurosurgery in Its Scientific and Professional Contexts*, ed. Samuel Greenblatt (Park Ridge, IL: American Association of Neurological Surgeons, 1997), 3–10. See also the article by Dale Smith, "The Evolution of Modern Surgery: A Brief Overview" in the same volume, 11–26. The most recent biography of Harvey Cushing is Michael Bliss, *Harvey Cushing: A Life in Surgery* (New York: Oxford University Press, 2005). See also Gavrus, "'Men of strong opinions.'"

23. Wilder Penfield, "Impressions of neurology, neurosurgery, and neurohistology in Central Europe," 1928 Report to the Rockefeller Foundation, 6, W/U 17, Wilder Penfield Fonds, OLHM.

24. Ibid., p. 17.

25. Penfield, *No Man Alone*, 164.

26. In his autobiography, Penfield suggested that Foerster's clinic lacked appropriate expertise in neuropathology. He implied that an earlier cooperation between Foerster and the Munich neuropathologist Walter Spielmeyer had not been satisfactory. Ibid., 164.

27. Wilder Penfield, Operations Notebook, W/Z 9, see esp. 13, 24, and 25 (patients Z., B., and L.), Wilder Penfield Fonds, OLHM.

28. Penfield to Archibald, July 17, 1927, Correspondence Wilder Penfield with Edward Archibald, A/M 11 1/1, 1, Wilder Penfield Fonds, OLHM.

29. Penfield to Archibald, January 18, 1929, cited in *No Man Alone*, 223. The final building had eight floors. See Wilder Penfield, *The Significance of the Montreal Neurological Institute*, McGill University (London: Oxford University Press, 1936): 40. This was a collection of neurological biographies and addresses, a foundation volume published for the staff, to commemorate the opening of the Montreal Neurological Institute.

30. For an account of Penfield's relationship to the Rockefeller Foundation, see Wilder Penfield, *The Difficult Art of Giving: The Epic of Alan Gregg* (Boston: Little, Brown, 1967); Penfield, *No Man Alone*; see also William Feindel's numerous publications, e.g., "The Montreal Neurological Institute," *Journal of Neurosurgery* 75 (1991): 821–822.

31. Principal and vice-chancellor of McGill University to Alan Gregg, November 27, 1931, C/G 5-3/1, Wilder Penfield Fonds, OLHM. See also Penfield, *No Man Alone*, 301; and Penfield, *Difficult Art*, 262–263.

32. Penfield, *Significance*, 47.

33. Feindel, Leblanc, and Villemure, "History of the surgical treatment of epilepsy," 475.

34. The numbers did not improve before the opening of the MNI. From November 1, 1928, to August 1, 1930, Penfield operated on 21 cases of epilepsy. In comparison, the number of skull fractures was 33; the number of brain tumors operated on was still high, with a total of 78 cases. We should add to these 24 operations for spinal cord tumors. Report, "The neurosurgical clinic," A/N 27-1/1, Wilder Penfield Fonds, OLHM.

35. See, for instance, his organization of research colloquia. "Report on neurological surgery at the Royal Victoria Hospital from September 9th to December 31st 1928," A/N 27-1/1, Wilder Penfield Fonds, OLHM. The colloquia were a great success; they continued until the 1970s, when the Montreal Neurological Society was put in their place. Delia Gavrus has analyzed Penfield's self-fashioning as a neurosurgeon vis-à-vis the weak, yet older and more established, specialty of neurology. She suggests that Penfield, by presenting himself as "the embodiment of a new kind of neuro-surgeon . . . could claim authority over the whole of clinical neurology." In initiating a public debate about the advantages of neurosurgery over other fields and, in particular, its therapeutic superiority over clinical neurology, Penfield managed to effectively displace the clinical neurologists in Montreal from their position of authority, first by creating the joint Department of Neurology and Neurosurgery, headed by him, and ultimately by the foundation of the MNI. Gavrus, "Men of dreams," 65. See also Gavrus, "'Men of strong opinions." Penfield touched upon the somewhat sensitive displacement of Colin Russel at the foundation of the joint department in October 1930 in a letter to his mother. Cf. Penfield, *No Man Alone*, 273.

36. Penfield and Boldrey, "Somatic motor and sensory representation," 420.

37. The importance of the anesthetist to the success of Penfield's project was underlined by his frequent acknowledgement of his or her work throughout his writings, e.g., ibid., 397; or Wilder Penfield and Herbert Jasper, *Epilepsy and the Functional Anatomy of the Human Brain* (Boston: Little, Brown, 1954), 752.

38. After Herbert Jasper, one of the first to take up electroencephalography in North America, joined Penfield at the MNI in 1938, EEGs were run regularly. Cf. the chapter "Electroencephagraphy" in Penfield and Jasper, *Epilepsy*. On Jasper, see Borck, *Hirnströme*, esp. 193–199.

39. The last patient I could find on whom Penfield used galvanic and faradic stimulation was J.G., 2/10/34, Stimulation Reports held at the Montreal Neurological Institute (hereafter SRMNI). The first patient on whom the thyratron stimulator was used was M.J, 1/16/35, SRMNI. I refer to the stimulation reports according to their admission date and initials.

40. The thyratron stimulator uses both types of current. Penfield and Boldrey, "Somatic motor and sensory representation," 397.

41. Penfield and Erickson, *Epilepsy*, 531.

42. Penfield and Boldrey, "Somatic sensory and motor representation," 425–427.

43. Penfield and Erickson, *Epilepsy*, 532.

44. A.B., 2/18/31, SRMNI. Exceptions are E.W, 1/31/31; and B.W., 5/28/32, both SRMNI, which list the results of electrical exploration under separate headings.

45. In the end, their title has often been left out, which further suggests that they had become the main part of the operation record.

46. Listed under "Objective findings": J.R., 5/2/33; E.S., 6/23/53; M.R., 4/20/34; F.F., 11/6/34; J.M., 3/19/35. Under "Procedure": A.M., 12/27/33; J.D., 12/28/33; J.G., 2/10/34; N.N., 1/9/35; M.J., 1/16/35. Their own section: G.P, 3/18/33; D.S., 4/15/33; M.D., 5/26/33; D.H., 6/29/33; H.M., 4/6/35; F.S., 4/23/35; H.M., 6/20/35; M.A., 9/21/35; T.P., 9/25/35; F.R., 10/30/35; H.T., 11/12/35; M.G., 12/5/35; L.M., 12/7/35. Not part of the typed protocol: L.P., 1/17/33; J.F., 1/20/33, all SRMNI.

47. E.g., to the late William Feindel, who kindly gave me access to them.

48. All cases in this group have drawings of blood vessels; only two of them were not labeled (V.R., 11/11/31; and B.W., 5/28/32, SRMNI). In the first years after his arrival, Penfield worked on the blood vessels in the brain, among other things. As he wrote to Dean Martin in his annual report dated February 2, 1931: "Most important perhaps is the fact that we have demonstrated that the blood vessels within the substance of the brain are equipped with perivascular nerve fibers like the blood vessels elsewhere in the body, a fact which up to the present has been denied." While initially Penfield's work was on vascular changes more generally, he increasingly related this research to the problem of epilepsy. Of his four papers on this topic, the first two were on the general physiology of blood vessels (although one states briefly that epilepsy was a motivation for study); the second two examine vascular changes in epilepsy. In Wilder Penfield, "Intracerebral vas-

cular nerves," *Archives of Neurology and Psychiatry* 27 (1932): 34, Penfield studied blood vessels in the normal subject, although he wrote that epilepsy was his motivation for his study of cerebral blood vessels. In Wilder Penfield and J. Chorobski, "Cerebral vasodilator nerves and their pathway from the medulla oblongata with observations on the pial and intracerebral vascular plexus," *Archives of Neurology and Psychiatry* 28 (1932): 1257–1289, Penfield did not mention epilepsy. The other two papers are Wilder Penfield, "The evidence for a cerebral vascular mechanism in epilepsy," *Annals of Internal Medicine* 7.3 (1933): 303–310; and Wilder Penfield, "Les effets des spasmes vasculaires dans l'épilepsie," *Union Médicale du Canada* 63 (1934): 1275–1282.

49. Penfield and Boldrey, "Somatic motor and sensory representation," 425. See also Penfield and Rasmussen, *Cerebral Cortex of Man*, 9.

50. In an earlier iteration, Penfield also relied on Vogt's cytoarchitectonics, attempting to align his results with the cytoarchitectonical work of Oskar and Cécile Vogt and Korbinian Brodmann. The different areas of his maps were labeled with architectonic fields such as 6aα or 8γ. Penfield increasingly came to emphasize the limitations of this attempt (e.g., Penfield and Boldrey, "Somatic motor and sensory representation," 425), however; finally this dropped out in favor of the fissures in his later work.

51. The fold dividing the precentral and postcentral, or motor and sensory, strips from each other is called the "Rolandic fissure," or "central sulcus" (sometimes "central fissure"). For the sake of simplicity, I will use only the term *central sulcus*.

52. Of the eleven cases between 1930 and 1932 that have been considered, the central sulcus was either not drawn at all (five times), drawn and crossed out (once), labeled with a question mark (once), or drawn and labeled (four times).

53. For example, in the case of patient J.D. (1933), when stimulating with galvanic current, he received seven sensory results, lying on one gyrus. J.D., 12/28/33, SRMNI. He thus assumed that he was dealing with the precentral gyrus. When he stimulated point 8, on the other side of the visible fold, and the response was motor, he concluded that he had found the line and noted in his operation report that "the central fissure was made out" (that is, the Rolandic fissure). J.D., 12/28/33, SRMNI. Similar cases are A.M., 12/27/33; H.M., 4/6/35, both SRMNI.

54. W.O., 1/22/37, SRMNI.

55. There might have been a confusion with another fold lying more toward the occipital part of the brain.

56. This did not get him out of trouble. Of the nine points that he moved (in addition to point 11), four remained on the sensory side (points 4, 5, 6 and 7). Points 1, 2, 3, and 10 crossed the line; of those, however, points 2, 3, and 10 were ambiguous: 2 was both sensory and motor, and points 3 and 10 caused the feeling of imminent attack. Only points 1 and 8 were purely sensory. Penfield's hesitation to move those two points over is visible in the fact that for them, the arrows are subtly shorter than the errors for point 11 and points 4 to 7.

57. D.S., 4/15/33, SRMNI.

58. E.g. Penfield and Jasper, *Epilepsy*, 763.

59. D.S., 4/15/33, SRMNI.

60. That the stimulation key and outlining of the central fissure were written down before the generalized seizure is suggested by the fact that in the stimulation key, Penfield wrote down for point II: "Sensation in right jaw and chewing movements," corresponding to his expectations. In the operation report, however, Penfield wrote that at first he did not recognize that he had caused generalized seizure; he only heard about it from the anesthetist, which was presumably after he had noted down the response. "There was no evidence of the attack at first until I was told that he was going into a seizure," after which Penfield also saw (and described) the vascular and tissue changes on the brain.

61. D.S., 4/15/33, SRMNI.

62. E.g., Penfield and Boldrey, "Somatic motor and sensory representation," 398.

63. J.F., 1/20/33; and W.G., 4/6/33, both SRMNI.

64. See, for example, the case of J.G. in 1934. J.G., 2/10/34, SRMNI.

65. Penfield and Boldrey, "Somatic motor and sensory representation," 392–393. Other explanations he gave for unexpected results included "epileptic spread" (ibid., 402) and nervous connections between pre- and postcentral areas (Penfield and Erickson, *Epilepsy*, 50).

66. E.g. in Penfield and Rasmussen, *Cerebral Cortex of Man*, 21, 46.

67. Penfield, *Mystery of the Mind*, 53.

68. Ibid., 7.

69. General anesthesia was administered only in the final stages of the operation, when patients no longer needed to cooperate; in children under ten years of age and in uncooperative patients, general anesthesia was also given in the early stages of the operation. Penfield and Jasper, *Epilepsy*, 753.

70. Though the brain has no pain receptors, the operator still needed to take precautions not to hurt the patient at the site of incision and at the opening of the skull.

71. Penfield and Jasper, *Epilepsy*, 748.

72. Penfield, *Mystery of the Mind*, 13.

73. Penfield and Jasper, *Epilepsy*, 748. For a concern with the voices of epileptic patients, see Ellen Dwyer's work: "Stories of epilepsy," 248–272; and "Neurological patients," 44–60.

74. Penfield, *Significance*, 52.

75. See Penfield, *Difficult Art*, 237.

76. However, he published a prominent paper with Penfield based on his studies at the MNI: Hebb and Penfield, "Human behavior." This did not mean that psychology played no role in the MNI in the 1930s. M.R. Harrower-Erickson contributed a chapter to Penfield and Erickson's first epilepsy book, and various papers by Penfield included psychological studies of patients after the removal of consid-

erable portions of the brain, most notably the frontal lobes. See the chapter "Psychological studies of patients with epileptic seizures," in Penfield and Erickson, *Epilepsy*. The papers include Wilder Penfield and J. Evans, "The frontal lobe in man: A clinical study of maximum removals," *Brain* 58.1 (1935): 115–133; and Wilder Penfield and Donald Hebb, "Human behavior after extensive bilateral removal from the frontal lobes," *Archives of Neurology and Psychiatry* 44 (1940): 421–438. But these evaluations were either more focused on personality structures (Harrower-Erickson) or were largely anecdotal, e.g., drawing on Penfield's own experience with his sister, who stated that "her own home provided in some ways a better background for study than the consulting room of the psychologist" (Penfield and Evans, "The frontal lobe in man," 131), or, in the same paper, quoting from a letter by Colin Russel, who discussed literature with her to test her cognitive function (118). Case 2 of the same paper, on the other hand, has a more formal psychological examination. See also Penfield and Hebb, "Human behavior," for a more formal and more extensive psychological evaluation.

77. Wilder Penfield, "Clinical notes from a trip to Great Britain," *Archives of Neurology and Psychiatry* 47 (1942): 1030–1036, quotations on 1035. On the other hand, Penfield responded to the needs of treating the psychoneuroses at home in the MNI, where, in addition to the physiological studies, research also included investigations into the psychoneuroses. See Penfield, "Some problems of wartime neurology," *Archives of Neurology and Psychiatry* 47 (1942): 840.

78. Despite making general recommendations on the organization of military hospitals in England, Penfield, during the war, was less involved in work abroad; he only went to Europe once, on a two-month trip in the summer of 1941, when he compiled a report to the National Research Council on the surgical treatment of war wounds. On the same trip, he also paid a visit to the Canadian No. 1 Neurological Hospital, at Hackwood Park near Basingstoke in England, where his close colleagues Lt.-Col. Cone and Lt.-Col. Russel worked. Jefferson Lewis, *Something Hidden: A Biography of Wilder Penfield* (Toronto: Doubleday, 1981), 166–170.

79. Wilder Penfield, "Overcrowding of the Montreal Neurological Institute and the care of patients from the combatant services, February 8th, 1943," A/N 10-2, 4, Wilder Penfield Fonds, OLHM.

80. Wilder Penfield, "Organization and administration—military annex," A/N 10-3; see also "Celebration programme: Transfer of the Military Annex to the Neurological Institute, January 17, 1947," A/N 10-3, both in Wilder Penfield Fonds, OLHM.

81. William Feindel, "Brain physiology at the Montreal Neurological Institute," *Journal of Clinical Neurophysiology* 9.2 (1992): 180–181. See also Penfield, "Some problems of wartime neurology," 840.

82. The average number of epilepsy operations was 22 per year between 1935 and 1938. From September 10, 1939, until the end of 1940, the number was 30. From 1941 on, the average number was 11 (1941, 12; 1942, 14; 1943, 4, 1944, 13),

with 1943 a bit of an outlier. Note that the numbers during the war are somewhat higher because Penfield's co-worker Cone did not operate, and only Penfield's operations were counted for the analysis.

83. See, e.g., Wilder Penfield and Herbert Jasper, "Electroencephalograms in post-traumatic epilepsy: Pre-operative and post-operative studies," *American Journal of Psychiatry* 100 (1943): 365–377; Wilder Penfield, "Posttraumatic epilepsy," *American Journal of Psychiatry* 100 (1944): 750–751.

84. Numbers by Robert Morison, quoted in Theodore Brown, "Alan Gregg and the Rockefeller Foundation's support of Franz Alexander's psychosomatic research," *Bulletin of the History of Medicine* 61 (1987): 156.

85. The grant for Cobb was a renewal grant, given in 1935. Alexander's institute was supported from 1935 to 1943. Other funding projects for which Gregg was responsible included allocations to the Munich Institute for Psychiatric Research in 1931 and to the Berlin Institute for Brain Research in 1932. See ibid.

86. Penfield, *Difficult Art*.

87. The hippocampus is part of the temporal lobe.

88. The collaboration between Penfield and psychologists Milner and Hebb was ironic because, as Winter has observed, they came from very different directions scientifically. Whereas Penfield's approach, with certain limitations that have been outlined in this chapter, was largely localizationist, Milner, at least by her training, leaned toward the connective models of the nervous system proposed by her Cambridge mentor Bartlett. But perhaps it is evidence of the flexibility of the reflex model that Penfield modified by cutting the reflex and, as this section has suggested, the related interdisciplinarity of the MNI, that integration between these different approaches was possible. Winter, *Memory*, 82–87.

89. Gordon Shepherd, *Creating Modern Neuroscience: The Revolutionary 1950s* (Oxford: Oxford University Press, 2010).

90. What Edward Shorter has called the "second biological psychiatry," in Shorter, *A History of Psychiatry: From the Era of the Asylum to the Age of Prozac* (New York: Wiley, 1997), 612; Heinz-Peter Schmiedebach, "Siegeszug des somatischen Positivismus," in *Psychiatrie und Psychologie im Widerstreit: Die Auseinandersetzung in der Berliner medicinisch-psychologischen Gesellschaft, 1867–1899* (Husum: Matthiesen, 1986), 241.

91. "McGill University, Department of Psychiatry, 50th Anniversary," *Journal of Psychiatry and Neuroscience* 18.4 (1993): 142. According to investigative journalist Anne Collins, Cameron used chlorpromazine for his brainwashing experiments. See her book also on the relationship between Cameron and Penfield. Anne Collins, *In the Sleep Room: The Story of the CIA Brainwashing Experiments in Canada* (Toronto: Lester and Orpen Dennys, 1998).

92. Case H.M., 4/6/35, SRMNI.

93. For left-handed people, it could be found on either the right or the left hemisphere.

94. Wilder Penfield, "The cerebral cortex in man. I. The cerebral cortex and consciousness," *Archives of Neurology and Psychiatry* 40 (1938): 421. See also Penfield and Rasmussen, *Cerebral Cortex of Man*, 88. In contrast to these findings, Sherrington was unable to produce vocalization in anthropoids.

95. Penfield used the term "ideational speech" in his 1959 book with Roberts. Wilder Penfield and Lamar Roberts, *Speech and Brain Mechanisms* (Princeton, NJ: Princeton University Press, 1959), e.g., 200. In his late publications, Penfield preferred the term "psychical speech." See Penfield, *Mystery of the Mind*, xii.

96. Penfield and Roberts, *Speech*, 36.

97. Ibid., 36.

98. Penfield, *Mystery of the Mind*, 55.

99. Penfield and Rasmussen, *Cerebral Cortex of Man*, 107.

100. Stimulation of precentral motor gyrus exclusively to elicit motor speech: one case in 1932, two in 1933, two in 1935, one in 1936, one in 1937, two in 1938, one in 1939, one in 1940, and one in 1945.

101. Other speech areas stimulated in one case in 1935, two in 1946, one in 1947, one in 1948, two in 1949, one in 1950, and one in 1951.

102. The superior frontal seems to have been stimulated: once in 1935, once in 1938, once in 1939, once in 1946, and once in 1947. The superior frontal has clearly been stimulated: once in 1958 (3 responses), once in 1948, and once in 1951.

103. It seems that Penfield neglected to stimulate the area once in 1938, once in 1939, and once in 1947. He clearly neglected to stimulate the area twice in 1948, twice in 1949, twice in 1950, once in 1952, and once in 1959.

104. One in 1946, one in 1946 (counting backward), one in 1948, three in 1949, two in 1950, and one in 1952.

105. Parietal aphasic arrest area: one in 1946, one in 1948, two in 1949, one in 1950, one in 1951, and one in 1959. Temporal aphasia arrest area: one in 1948, one in 1949, three in 1950, and one in 1951. Note the difference between speech arrest and aphasic arrest, the first being part of "motor" the second part of "ideational" speech; cf. figure 6.9.

106. Penfield, *Mystery of the Mind*, 27.

107. Penfield and Roberts, *Speech*, 42. Penfield insisted on the restriction of psychical responses to the temporal lobe on numerous occasions. See, e.g., Penfield and Jasper, *Epilepsy*, 20, 45; Wilder Penfield and P. Perot, "The brain's record of auditory and visual experience: A final summary and discussion," *Brain* 86 (1963): 596, 597, 601 (2x); Penfield and Roberts, *Speech*, 35, 40, 42; Wilder Penfield, "Some observations on the cerebral cortex of man," Ferrier Lecture, *Proceedings of the Royal Society of London* B 134 (1947): 340.

108. Y.N., 2/6/51, SRMNI.

109. Penfield claimed that he had included all the cases of psychical responses he had ever encountered in the summary paper, but I have found exceptions.

110. Penfield and Perot, "Brain's record," 626.

111. Penfield and Jasper, *Epilepsy*, 23. A third category was automatism, located in a separate area of the brain.

112. Ibid., 23.

113. Other accounts have suggested that Penfield's turn to dualism was part of broader cultural trends, such as an interwar turn to holism, or the engagement with the philosophy of Charles Hendel. Delia Gavrus, "Mind over Matter: Sherrington, Penfield, Eccles, Walshe and the dualist movement in neuroscience," MCIS Briefings, Comparative Program on Health and Society, Lupina Foundation Working Papers Series 2005–6, ed. Jillian Clare Cohen and Lisa Forman, University of Toronto, 2006, 51–75; Nicholas Schiff, "Wilder Graves Penfield: Philosophy, physiology, and the mystery of mind" (Senior thesis, Stanford University, Stanford, CA, April 1987).

114. The foldable version (fig. 6.11, bottom) was introduced in 1951, after Penfield had drawn the area of the island of Reil in a separate pencil sketch (by hand) besides the chart from 1949 onward (e.g., R.W., A.J., J.M., E.O., D.H.). However, even beyond the introduction of the foldable chart, the two representations coexisted. Another way of representing deep stimulations was through a double numbering system, e.g., 19_1, 19_2, 19_3, 19_4, going deeper with the electrode each time. This method was used from 1950 on (e.g., E.L.).

115. This hypothesis was further supported by the observation that removal of cortical areas did not bring about any changes in consciousness, whereas removal of the higher brain stem did. Penfield suggested that he first made these claims in 1938, but he refers predominantly to papers from 1952. Penfield, *Mystery of the Mind*, 18.

116. In his 1938 paper "Cerebral cortex in man," which Penfield retrospectively (in his 1954 book) referred to as his first encounter with "psychical" responses (see above), he referred to Jackson's "mental diplopia," or doubling of consciousness (431).

117. E.g., Penfield, *Mystery of the Mind*, 55.

118. Ibid., 5.

119. For the standard narrative, see, e.g., Kandel et al., *Principles*, 13.

Epilogue

1. Wilder Penfield, "Memory Mechanisms," *Archives of Neurology and Psychiatry* 67 (1952): 191.

2. Ibid., 193.

3. Alison Winter discusses Kubie's work at the MNI in considerable detail in *Memory*, 81. As Winter points out, Penfield's work caught the interest of other psychoanalysts, including Sydney Margolin and John Lilly (90). See also Lawrence Kubie, "Some implications for psychoanalysis of modern concepts of the organization of the brain," *Psychoanalytic Quarterly* 22 (1953): 22.

4. For the most detailed account of psychoanalysis in America, see Hale, *Rise and Crisis*. See also Burnham, *After Freud Left*.

5. The two were not always in opposition, however, as Mical Raz has shown. Raz, "Between the ego and the ice pick." On the coexistence of psychoanalysis and biological psychiatry in midcentury America, see also Metzl, *Prozac on the Couch*; and David Herzberg, *Happy Pills in America: From Miltown to Prozac* (Baltimore: Johns Hopkins University Press, 2009), 38.

6. Herbert Jasper, Cosimo Ajmone-Marson and Julius Stoll, "Corticofugal projections to the brain stem," *Archives of Neurology and Psychiatry* 67 (1952): 155–171.

7. Gavrus, "'Men of strong opinions,'" esp. part 3. See also Shepherd, *Creating Modern Neuroscience*, chap. 14.

8. For example, he discussed at length Houston Merritt's suggestion that his electrode should have elicited memories of smell, since it was placed near the olfactory area of the brain, and he responded to Karl Lashley's skeptical remarks about his postulated location of a system responsible for processing consciousness, the centrencephalic system. This is not to say that Penfield did not think his work was relevant for psychology. But he admitted to not being an expert himself, thus leaving it to others to analyze the psychological implications of his work. See e.g. Penfield, "Cerebral cortex in man," 417. See also Winter, *Memory*, 81.

9. For example, the brand of neuropsychology that Brenda Milner brought to the MNI.

10. Nima Bassiri, "Freud and the matter of the brain: On the rearrangements of neuropsychoanalysis," *Critical Inquiry* 40 (Autumn 2013): 6.

11. See, e.g., Leon Wieseltier's recent response to Steven Pinker, "Science is not your enemy: An impassioned plea to neglected novelists, embattled professors, and tenure-less historians," *New Republic*, August 6, 2013; Leon Wieseltier, "Crimes against humanities: Now science wants to invade the liberal arts: Don't let it happen," *New Republic*, September 3, 2013.

12. In thinking through this relationship, we should be attentive to the differences between the psychoanalytic and neuroscientific embrace of connectivity. Freud gave up his project of "brain architecture" because of the insufficiency of existing tools for such a complex task. The technological conditions of such a project are very different today. Moreover, though Freud used what I have called his "detour through physiology" in order to recast the language of psychology, which had been directly imported into Meynert's model, it has become a long-standing criticism that modern neuroscience has yet to do more than simply transfer psychological terms into "neuro-speak." See, e.g., Hagner in *"Der Geist bei der Arbeit,"* 165n1.

13. For critical discussions of the role of current neuroscientific ideas in broader culture, and their effects on subjectivity, see Vidal and Ortega, *Neurocultures*; Nikolas Rose, *The Politics of Life Itself: Biomedicine, Power and Subjectivity in the Twenty-first Century* (Princeton, NJ: Princeton University Press, 2007); Nikolas Rose and Joelle Abi-Rached, *Neuro: The New Brain Sciences and the Man-*

agement of the Mind (Princeton, NJ: Princeton University Press, 2013); Choudhury and Slaby, *Critical Neuroscience*; Littlefield and Johnson, *Neuroscientific Turn*; Roger Cooter, "The end? History-writing in the age of biomedicine (and before)," in *Writing History in the Age of Biomedicine*, ed. Roger Cooter and Claudia Stein (New Haven, CT: Yale University Press, 2013), 1–40; Isis Focus Section "Neurohistory and History of Science" (with papers from Steve Fuller, Daniel Lord Smail, Stephen Casper, Max Stadler, and Roger Cooter), *Isis* 105.1 (2014): 1154; Cathy Gere and Charlie Gere, "The brain in a vat," special issue, ed. Cathy Gere and Charlie Gere, *Studies in History and Philosophy of Science Part C: Studies in History and Philosophy of Biological and Biomedical Sciences* 35.2 (2004). As a way of overcoming this personhood-defining dominance of the neurosciences, scholars drawing on the phenomenological tradition have emphasized an "enactive understanding" of the human condition, e.g., Alva Noë, *Action in Perception* (Cambridge, MA: MIT Press, 2004); Francisco Varela, Evan Thompson, and Eleanor Rosch, *The Embodied Mind: Cognitive Science and Human Experience* (Cambridge, MA: MIT Press, 1991). For the history of the neurosciences in the twentieth century, see also the special issue of *Science in Context* 14.4 (2001); and Michael Hagner and Cornelius Borck, "Mindful practices: On the neurosciences in the twentieth century," *Science in Context* 14.4 (2001): 507–510.

14. See Martha Nussbaum, *Not for Profit: Why Democracy Needs the Humanities* (Princeton, NJ: Princeton University Press, 2012). Note that Nussbaum does not frame her critique within the two-cultures paradigm; instead, she thinks "science, rightly pursued, is a friend of the humanities rather than their enemy" (8).

Bibliography

Abbott, Alison. "Solving the brain." *Nature* 499 (July 18, 2013): 272–274.

Ackerknecht, Erwin. *Medicine at the Paris Hospital, 1794–1848*. Baltimore: Johns Hopkins Press, 1967.

Aguayo, J. "Charcot and Freud: Some implications of late 19th century French psychiatry and politics for the origins of psychoanalysis." *Psychoanalysis and Contemporary Thought* 9 (1986): 223–260.

Amacher, Peter. *Freud's Neurological Education and Its Influence on Psychoanalytic Theory*. New York: International Universities Press, 1965.

Anton, Gabriel. *Über den Wiederersatz der Funktion bei Erkrankungen des Gehirnes*. Berlin: Karger, 1906.

Ash, Mitchell. *Gestalt Psychology in German Culture 1890–1967: Holism and the Quest for Objectivity*. Cambridge: Cambridge University Press, 1995.

Ash, Mitchell, and Alfons Söllner. *Forced Migration and Scientific Change: Emigré German-Speaking Scientists after 1933*. Cambridge: Cambridge University Press, 1996.

Aubert, Geneviève. "Arthur van Gehuchten takes neurology to the movies." *Neurology* 59 (2002): 1612–1618.

Bailey, Percival. *Up from Little Egypt*. Chicago: Buckskin Press, 1969.

Ballenger, Jesse. *Self, Senility and Alzheimer's Disease in Modern America: A History*. Baltimore: Johns Hopkins University Press, 2006.

Baring, Edward. *The Young Derrida and French Philosophy: 1945–68*. Cambridge: Cambridge University Press, 2011.

Bassiri, Nima. "Freud and the matter of the brain: On the rearrangements of neuropsychoanalysis." *Critical Inquiry* 40 (Autumn 2013): 1–26.

Bastian, Henry Charlton. "On the various forms of loss of speech in cerebral disease." *British and Foreign Medical and Chirurgical Review* 43 (1869): 209–236, 470–492.

Bauer, Axel. "Die Formierung der Pathologischen Anatomie als naturwissenschaftliche Disziplin und ihre Institutionalisierung an den deutschsprachigen Universitäten." *Würzburger medizinhistorische Mitteilungen* 10 (1992): 315–330.

Beaulieu, Anne. *The Space Inside the Skull: Digital Representations, Brain Mapping and Cognitive Neuroscience in the Decade of the Brain*. Amsterdam: University of Amsterdam, 2000.

Beller, Steven. *Rethinking Vienna 1900*. New York: Berghahn Books, 2001.

Bennett, M. R., and P. M. S. Hacker. *Philosophical Foundations of Neuroscience*. Malden, MA: Blackwell, 2003.

Blakeslee, Sandra, and Matthew Blakeslee. *The Body Has a Mind of Its Own: How Body Maps in Your Brain Help You Do (Almost) Everything Better*. New York: Random House, 2007.

Bliss, Michael. *Harvey Cushing: A Life in Surgery*. New York: Oxford University Press, 2005.

Bonhoeffer, Karl. "Die Stellung Wernickes in der modernen Psychiatrie." *Berliner klinische Wochenschrift* 42 (1905): 893–894, 927–928.

———. "Lebenserinnerungen von Karl Bonhoeffer—Geschrieben für die Familie." In *Karl Bonhoeffer Zum Hundersten Geburtstag am 31. März 1968*, edited by J. Zutt, E. Straus, and H. Scheller, 8–107. Berlin: Springer, 1969.

———. "Psychiatrie und Neurologie." *Monatsschrift für Psychiatrie und Neurologie* 37 (1915): 94–104.

Bonner, Thomas N. *American Doctors and German Universities: A Chapter in International Intellectual Relations, 1870–1914*. Lincoln: University of Nebraska Press, 1963.

Borchard, A., and W. von Brunn, eds. *Deutscher Chirurgenkalender: Verzeichnis der deutschen Chirurgen und Orthopäden mit Biographien und bibliographischen Skizzen*. 2nd ed. Leipzig: Barth, 1926.

Borck, Cornelius. *Anatomien medizinischen Wissens: Medizin, Macht, Moleküle*. Frankfurt am Main: Fischer, 1996.

———. "Concrete existence: Goldstein's variations on the nervous body." Article manuscript, in author's collection.

———. "Fühlfäden und Fangarme: Metaphern des Organischen als Dispositiv der Hirnforschung." In *Ecce Cortex: Beiträge zur Geschichte des modernen Gehirns*, edited by Michael Hagner, 144–176. Göttingen: Wallstein, 1999.

———. *Hirnströme: Eine Kulturgeschichte der Elektroenzaphalographie*. Göttingen: Wallstein, 2005.

———. "Recording the brain at work: The visible, the readable, and the invisible in electroencephalography." *Journal of the History of the Neurosciences* 17 (2008): 367–379.

———. "Visualizing nerve cells and psychic mechanisms: The rhetoric of Freud's illustrations." In Guttmann and Scholz-Strasser, *Freud and the Neurosciences*, 57–86.

Borck, Cornelius, and Armin Schäfer. *Psychographien*. Berlin: Diaphanes, 2006.

Boring, Edwin. *A History of Experimental Psychology*. 2nd ed. New York: Appleton-Century-Croft, 1950.

Brandt, Allan. *No Magic Bullet: A Social History of Venereal Disease in the United States since 1880*. New York: Oxford University Press, 1987.

Braslow, Joel. *Mental Ills and Bodily Cures: Psychiatric Treatment in the First Half of the Twentieth Century*. Berkeley: University of California Press, 1997.

Breidbach, Olaf. *Die Materialisierung des Ichs: Zur Geschichte der Hirnforschung im 19. und 20. Jahrhundert*. Frankfurt: Suhrkamp, 1997.

Brill, A. A. *Psychoanalysis: Its Theories and Practical Application*. 3rd ed. Philadelphia, London: Saunders, 1922.

Broca, Paul. "Remarques sur le siège de la faculté du langage articulé, suivies d'une observation d'aphémie (perte de la parole)." *Bulletin de la Société de l'Anatomie de Paris* 36 (1861): 330–357.

Brown, Theodore. "Alan Gregg and the Rockefeller Foundation's support of Franz Alexander's psychosomatic research." *Bulletin of the History of Medicine* 61 (1987): 155–182.

Buchwald, Alfred. *Das Kranken-Hospital zu Allerheiligen in Breslau*. Breslau: Schletter, 1896.

Burnham, John C. *After Freud Left: A Century of Psychoanalysis in America*. Chicago: University of Chicago Press, 2012.

———. "The 'New Freud Studies': A historiographical shift." *Journal of the Historical Society* 6.2 (2006): 213–233.

———. *Psychoanalysis and American Medicine: 1894–1918; Medicine, Science, and Culture*. New York: International Universities Press, 1967.

Busch, Alexander. *Die Geschichte des Privatdozenten*. Stuttgart: Enke, 1959.

Canguilhem, Georges. *Die Herausbildung des Reflexbegriffs im 17. und 18. Jahrhundert*. Translated by Henning Schmidgen. Paderborn: Fink, 2008.

———. *La formation du concept de réflexe au XVIIe et XVIIIe siècles*. Paris: Presses Universitaires de France, 1955.

Casper, Stephen. "Anglo-American neurology: Lewis H. Weed and Johns Hopkins neurology, 1917–1942." *Bulletin of the History of Medicine* 82 (2008): 646–671.

———. *The Neurologists: A History of a Medical Specialty in Modern Britain, c. 1789–2000*. Manchester, UK: Manchester University Press, 2014.

Casper, Stephen, and Delia Gavrus. *Technique in the History of the Mind and Brain Sciences*, under review.

Charcot, Jean-Martin. *Clinical Lectures on Diseases of the Nervous System*. Vol. 3. London: Sydenham Society, 1889.

Choudhury, Suparna, and Jan Slaby, eds. *Critical Neuroscience: A Handbook of the Social and Cultural Contexts of Neuroscience*. Chichester, West Sussex, UK: Wiley-Blackwell, 2012.

Clarke, Edwin, and Stephen Jacyna. *Nineteenth-Century Origins of Neuroscientific Concepts*. Berkeley: University of California Press, 1987.

Clarke, Edwin, and C. D. O'Malley. *The Human Brain and Spinal Cord: A Historical Study Illustrated by Writings from Antiquity to the Twentieth Century*. San Francisco: Norman, 1996.

Collins, Anne. *In the Sleep Room: The Story of the CIA Brainwashing Experiments in Canada*. Toronto: Lester and Orpen Dennys, 1998.

Cooter, Roger. *The Cultural Meaning of Popular Science: Phrenology and the Organization of Consent in Nineteenth-Century Britain*. Cambridge: Cambridge University Press, 1984.

———. "The end? History-writing in the age of biomedicine (and before)." In *Writing History in the Age of Biomedicine*, edited by Roger Cooter and Claudia Stein, 1–40. New Haven, CT: Yale University Press, 2013.

Crome, Leonard. "The medical history of V.I. Lenin." *History of Medicine* 4 (1972): 3–9, 20–22.

Curran, Frank, and Paul Schilder. "Paraphasic signs in diffuse lesions of the brain." *Journal of Nervous and Mental Disease* 82 (1935): 613–636.

D'Angelo, Egidio. "Toward the connectomic era." *Functional Neurology* 27.2 (2012): 77.

Danziger, Kurt. *Constructing the Subject: Historical Origins of Psychological Research*. Cambridge: Cambridge University Press, 1990.

———. *Marking the Mind: A History of Memory*. Cambridge: Cambridge University Press, 2008.

"Das Neurologische Institut eröffnet: Das erste in der ganzen Welt." *Breslauer 8-Uhr Abendblatt*, January 31, 1934.

Davies, Norman, and Roger Moorhouse. *Microcosm: Portrait of a Central European City*. London: Pimlico, 2003.

Davis, Gayle. *"The Cruel Madness of Love": Sex, Syphilis and Psychiatry in Scotland, 1880–1930*. Clio Medica 85. Amsterdam: Rodopi, 2008.

Dawkins, Richard, and Yan Wong. *The Ancestor's Tale: A Pilgrimage to the Dawn of Evolution*. Boston: Houghton Mifflin, 2004.

Decker, Hannah S. *Freud, Dora, and Vienna 1900*. New York: Free Press, 1991.

Derrida, Jacques. "Freud and the scene of writing." In *Writing and Difference*, by Jacques Derrida. Translated by Alan Bass. Chicago: University of Chicago Press, 1978.

Doolittle, Hilda. *Tribute to Freud: Writing on the Wall: Advent*. 1956. Reprint, Boston: David R. Godine, 1974. Page references are to the reprint edition.

Dorer, Maria. *Historische Grundlagen der Psychoanalyse*. Leipzig: Meiner, 1932.

Duchenne, Guillaume-Benjamin. "Exposition d'une nouvelle méthode de galvanisation, dite galvanisation localisée." *Archives générales de médecine* 4 (1850): 286–289.

———. *Physiologie des mouvements, démontrée à l'aide de l'expérimentation électrique et de l'observation clinique et applicable à l'étude des paralysies et des déformations*. Paris: Baillière, 1867.

Dumit, Joseph. *Picturing Personhood: Brain Scans and Biomedical Identity*. Princeton, NJ: Princeton University Press, 2004.

Dupouy, Stéphanie. "Künstliche Gesichter: Rodolphe Töpffer und Duchenne de Boulogne." In *Kunstmaschinen: Spielräume des Sehens zwischen Wissenschaft und Ästhetik*, edited by Andreas Mayer and Alexandre Métraux, 24–60. Frankfurt: S. Fischer, 2005.

Dwyer, Ellen. "Neurological patients as experimental subjects: Epilepsy studies in the United States." In *The Neurological Patient in History*, edited by L. Stephen Jacyna and Stephen Casper, 44–60. Rochester, NY: University of Rochester Press, 2012.

———. "Stories of epilepsy, 1880–1930." In *Framing Disease: Studies in Cultural History*, edited by Charles Rosenberg and Janet Golden, 248–272. New Brunswick, NJ: Rutgers University Press, 1992.

Eggert, Gertrude. *Wernicke's Work on Aphasia: A Sourcebook and Review*. The Hague: Mouton, 1977.

Eling, Paul. "Hatte Wernickes Sprachtheorie ihre Wurzeln in Berlin?" *Schriftenreihe der Deutschen Gesellschaft für Geschichte der Nervenheilkunde* 7 (2001): 33–38.

Engelman, Edmund. *Berggasse 19: Sigmund Freud's Home and Offices, Vienna, 1938*. New York: Basic Books, 1976.

Engstrom, Eric. "Assembling professional selves: On clinical instruction in German academic psychiatry." In Engstrom and Roelcke, *Psychiatrie im 19. Jahrhundert*, 117–152.

———. "The birth of clinical psychiatry: Power, knowledge, and professionalization in Germany, 1867–1914." PhD diss., University of North Carolina at Chapel Hill, 1997.

———. *Clinical Psychiatry in Imperial Germany: A History of Psychiatric Practice*. Ithaca, NY: Cornell University Press, 2003.

———. "Neurowissenschaften und Gehirnforschung." In *Geschichte der Universität Unter den Linden, 1810–2010*, edited by Heinz-Elmar Tenorth, 777–797. Berlin: Akademie, 2010.

Engstrom, Eric, and Volker Roelcke, eds. *Psychiatrie im 19. Jahrhundert: Forschungen zur Geschichte von psychiatrischen Institutionen, Debatten und Praktiken im deutschen Sprachraum*. Mainz: Schwabe, 2003.

Erb, Wilhelm H. "Ueber Sehenreflexe bei Gesunden und bei Rückenmarkskranken." *Archiv für Psychiatrie und Nervenkrankheiten* 5 (1875): 792–802.

Eulner, Hans-Heinz. *Die Entwicklung der medizinischen Spezialfächer an den Universitäten des deutschen Sprachgebietes*. Stuttgart: Ferdinand Enke, 1970.

Fearing, Franklin. *Reflex Action: A Study in the History of Physiological Psychology*. New York: Hafner, 1930.

Feindel, William. "Brain physiology at the Montreal Neurological Institute." *Journal of Clinical Neurophysiology* 9.2 (1992): 176–194.

———. "The Montreal Neurological Institute." *Journal of Neurosurgery* 75 (1991): 821–822.

Feindel, William, Richard Leblanc, and Jean-Guy Villemure. "History of the surgical treatment of epilepsy." In *A History of Neurosurgery in Its Scientific and Professional Contexts*, edited by Samuel Greenblatt, 465–488. Park Ridge, IL: American Association of Neurological Surgeons, 1997.

Ferenczi, Sándor. "On the organization of the psycho-analytic movement" (1911). In *Final Contributions to the Problems and Methods of Psychoanalysis*, by Sándor Ferenczi, 299–307. New York: Brunner/Mazel, 1980.

Ferrier, David. "Experimental research in cerebral physiology and pathology." *West Riding Lunatic Asylum Medical Reports* 3 (1873): 30–96.

Finger, Stanley, François Boller, and Kenneth L. Tyler. *History of Neurology*. Edinburgh: Elsevier, 2010.

Flem, Lydia. *La vie quotidienne de Freud et de ses patients*. Paris: Hachette, 1986.

Flexner, Abraham. *Medical Education in the United States and Canada: A Report to the Carnegie Foundation for the Advancement of Teaching*. Bulletin Number 4. New York: Carnegie Foundation for the Advancement of Teaching, 1910.

Foerster, Otfrid. *Beiträge zur Kenntnis der Mitbewegungen*. Jena: G. Fischer, 1903.

———. "Demonstration: Fall von einem eigenthümlichen Zwangsphänomen." *Allgemeine Zeitschrift für Psychiatrie* 57 (1901): 411–414.

———. "Dermatomes in man." *Brain* 56 (1933): 1–39.

———. "Die Indikation und Erfolge der Resektion hinterer Rückenmarkswurzeln." *Wiener klinische Wochenschrift* 25 (1912): 950–954.

———. "Die Pathogenese des epileptischen Krampfanfalles." *Tagung der Gesellschaft deutscher Nervenärzte*, Düsseldorf, 94 (1926): 15–53.

———. *Die Physiologie und Pathologie der Coordination*. Jena: Fischer, 1902.

———. "Die Symptomatologie der Erkrankungen des Rückenmarks und seiner Wurzeln." In *Handbuch der Neurologie*, edited by Otfrid Foerster and Oswald Bumke, 5:1–403. Leipzig: Springer, 1936.

———. "Die Symptomatologie und Therapie der Kriegsverletzungen der peripheren Nerven." *Deutsche Zeitschrift für Nervenheilkunde* 59 (1918): 32–172.

———. "Ein Fall von elementarer allgemeiner Somatopsychose (Afunktion der Somatopsyche): Ein Beitrag zur Frage der Bedeutung der Somatopsyche für das Wahrnehmungsvermögen." *Monatsschrift für Psychiatrie und Neurologie* 14 (1903): 189–205.

———. "Encephalographische Erfahrungen." *Zeitschrift für die gesamte Neurologie und Psychiatrie* 94 (1925): 512–584.

———. "Eröffnungsansprache." *Deutsche Zeitschrift für Nervenheilkunde* 110 (1929): 208–330.

———. "Eröffnungsansprache." *Deutsche Zeitschrift für Nervenheilkunde* 115 (1930): 147–159.

———. "Eröffnungsansprache." *Deutsche Zeitschrift für Nervenheilkunde* 129 (1932): 175–184.

———. "Methoden der Dermatombestimmung beim Menschen." *Archiv für Psychiatrie und Nervenkrankheiten* 77 (1926): 652–658.

———. "The motor cortex in man in light of Hughlings Jackson's doctrines." *Brain* 59 (1936): 135–159.

———. "Operativ-experimentelle Erfahrungen beim Menschen über den Einfluss des Nervensystems auf den Kreislauf." *Zeitschrift für die gesamte Neurologie und Psychiatrie* 167 (1939): 439–461.

———. "Sensible corticale Felder." In *Handbuch der Neurologie*, edited by Otfrid Foerster and Oswald Bumke. Vol. 6. Leipzig: Springer, 1936.

———. "Surgical treatment of neurogenic contractures." *Surgery, Gynecology, and Obstetrics* 52 (1931): 360–366.

———. "Über die Bedeutung und Reichweite des Lokalisationsprinzips." *Verhandlungen der Deutschen Gesellschaft für Innere Medizin* 46 (1934): 117–211.

———. "Ueber die Beeinflussung spastischer Lähmungen durch die Resektion hinterer Rückenmarkswurzeln." *Deutsche Zeitschrift für Nervenheilkunde* 41 (1911): 146–169.

———. "Über die Wechselbeziehungen von Herdsymptomen und Allgemeinsymptomen." *Verhandlungen der deutschen Gesellschaft für innere Medizin* 50 (1938): 458–485.

———. "Ueber eine neue operative Methode der Behandlung spastischer Lähmungen mittels Resektion hinterer Rückenmarkswurzeln." *Zeitschrift für orthopaedische Chirurgie* 22 (1908): 203–223.

———. "Ueber einige seltenere Formen von Krisen bei der Tabes dorsalis, sowie über die tabischen Krisen im Allgemeinen." *Monatsschrift für Psychiatrie und Neurologie* 11 (1902): 259–283.

———. "Ueber operative Behandlung gastrischer Krisen durch Resektion der 7.–10. hinteren Dorsalwurzel." *Beiträge zur klinischen Chirurgie* 63 (1909): 245–256.

———. "Uebungstherapie." In *Handbuch der Neurologie*, edited by Otfrid Foerster and Oswald Bumke, 8:316–414. Leipzig: Springer, 1936.

———. "Uebungstherapie bei Tabes dorsalis." *Deutsche Ärzte-Zeitung* (1901): 100–104, 128–131.

———. "Vergleichende Betrachtungen über Motilitätspsychosen und über Erkrankungen des Projektionssystems." Inaugural lecture after qualification to teach as *Privatdozent*. Jena: Gustav Fischer, 1903.

Foerster, Otfrid, and Harttung. "Erfahrungen über die Behandlung von Störungen des Nervensystems auf syphilitischer Grundlage." *Archiv für Dermatologie und Syphilologie* 86 (1907): 3–44.

Foerster, Otfrid, and Wilder Penfield. "Der Narbenzug am und im Gehirn bei traumatischer Epilepsie in seiner Bedeutung für das Zustandekommen der Anfälle und für die therapeutische Bekämpfung derselben." *Zeitschrift für die gesamte Neurologie und Psychiatrie* 125 (1930): 475–572.

Forrester, John. "If p, then what? Thinking in cases." *History of the Human Sciences* 9 (1996): 1–25.

———. *Language and the Origins of Psychoanalysis*. London: Macmillan, 1980.

Foucault, Michel. *The Birth of the Clinic: An Archeology of Medical Perception*. Translated by A. M. Sheridan Smith. New York: Pantheon Books, 1973.

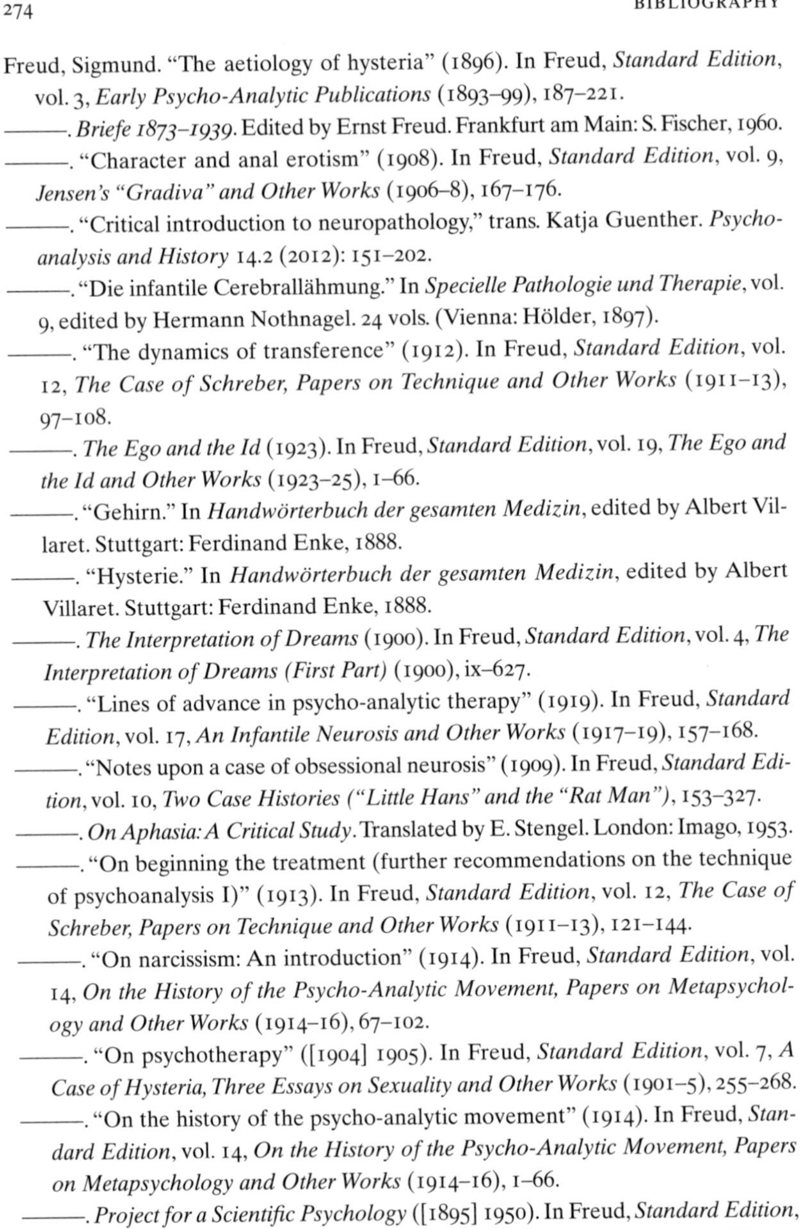

Freud, Sigmund. "The aetiology of hysteria" (1896). In Freud, *Standard Edition*, vol. 3, *Early Psycho-Analytic Publications* (1893–99), 187–221.

———. *Briefe 1873–1939*. Edited by Ernst Freud. Frankfurt am Main: S. Fischer, 1960.

———. "Character and anal erotism" (1908). In Freud, *Standard Edition*, vol. 9, *Jensen's "Gradiva" and Other Works* (1906–8), 167–176.

———. "Critical introduction to neuropathology," trans. Katja Guenther. *Psychoanalysis and History* 14.2 (2012): 151–202.

———. "Die infantile Cerebrallähmung." In *Specielle Pathologie und Therapie*, vol. 9, edited by Hermann Nothnagel. 24 vols. (Vienna: Hölder, 1897).

———. "The dynamics of transference" (1912). In Freud, *Standard Edition*, vol. 12, *The Case of Schreber, Papers on Technique and Other Works* (1911–13), 97–108.

———. *The Ego and the Id* (1923). In Freud, *Standard Edition*, vol. 19, *The Ego and the Id and Other Works* (1923–25), 1–66.

———. "Gehirn." In *Handwörterbuch der gesamten Medizin*, edited by Albert Villaret. Stuttgart: Ferdinand Enke, 1888.

———. "Hysterie." In *Handwörterbuch der gesamten Medizin*, edited by Albert Villaret. Stuttgart: Ferdinand Enke, 1888.

———. *The Interpretation of Dreams* (1900). In Freud, *Standard Edition*, vol. 4, *The Interpretation of Dreams (First Part)* (1900), ix–627.

———. "Lines of advance in psycho-analytic therapy" (1919). In Freud, *Standard Edition*, vol. 17, *An Infantile Neurosis and Other Works* (1917–19), 157–168.

———. "Notes upon a case of obsessional neurosis" (1909). In Freud, *Standard Edition*, vol. 10, *Two Case Histories ("Little Hans" and the "Rat Man")*, 153–327.

———. *On Aphasia: A Critical Study*. Translated by E. Stengel. London: Imago, 1953.

———. "On beginning the treatment (further recommendations on the technique of psychoanalysis I)" (1913). In Freud, *Standard Edition*, vol. 12, *The Case of Schreber, Papers on Technique and Other Works* (1911–13), 121–144.

———. "On narcissism: An introduction" (1914). In Freud, *Standard Edition*, vol. 14, *On the History of the Psycho-Analytic Movement, Papers on Metapsychology and Other Works* (1914–16), 67–102.

———. "On psychotherapy" ([1904] 1905). In Freud, *Standard Edition*, vol. 7, *A Case of Hysteria, Three Essays on Sexuality and Other Works* (1901–5), 255–268.

———. "On the history of the psycho-analytic movement" (1914). In Freud, *Standard Edition*, vol. 14, *On the History of the Psycho-Analytic Movement, Papers on Metapsychology and Other Works* (1914–16), 1–66.

———. *Project for a Scientific Psychology* ([1895] 1950). In Freud, *Standard Edition*, vol. 1, *Pre-Psycho-Analytic Publications and Unpublished Drafts* (1886–99), 281–391.

———. "Recommendations to Physicians Practising Psycho-Analysis" (1912). In Freud, *Standard Edition*, vol. 12, *The Case of Schreber, Papers on Technique and Other Works* (1911–13), 109–120.

———. "Remembering, repeating and working-through (further recommendations on the technique of psycho-analysis II)" (1914). In Freud, *Standard Edition*, vol. 12, *The Case of Schreber, Papers on Technique and Other Works* (1911–13), 145–156.

———. "Some points for a comparative study of organic and hysterical motor paralyses" (1888). In Freud, *Standard Edition*, vol. 1, *Pre-Psycho-Analytic Publications and Unpublished Drafts* (1886–99), 157–172.

———. *The Standard Edition of the Complete Psychological Works of Sigmund Freud*, translated and edited by James Strachey. 24 vols. London: Hogarth Press, 1953–74.

———. "Two encyclopaedia articles" (1923). In Freud, *Standard Edition*, vol. 18, *Beyond the Pleasure Principle, Group Psychology and Other Works* (1920–22), 233–260.

———. "Über Spinalganglien und Rückenmark des Petromyzon." *Sitzungsberichte der Mathematisch-Naturwissenschaftlichen Classe der k. Akademie der Wissenschaften*, Vienna, III. Abtheilung 78 (1878): 81–167.

———. "Ueber den Ursprung des N. acusticus." *Monatsschrift für Ohrenheilkunde* 20 (1886): 245–282.

———. "The unconscious" (1915). In Freud, *Standard Edition*, vol. 14, *On the History of the Psycho-Analytic Movement, Papers on Metapsychology and Other Works* (1914–16), 159–215.

———. "'Wild' psycho-analysis." In Freud, *Standard Edition*, vol. 11, *Five Lectures on Psychoanalysis, Leonardo da Vinci and Other Works* (1910), 219–228.

Freud, Sigmund, and Ludwig Binswanger. *The Sigmund Freud-Ludwig Binswanger Correspondence, 1908–1938*. Edited by Gerhard Fichtner. Translated by Arnold J. Pomerans. New York: Other Press, 2003.

Freud, Sigmund, and Josef Breuer. *Studies on Hysteria* (1893–95). In Freud, *Standard Edition*, vol. 2, *Studies on Hysteria* (1893–95).

Freud, Sigmund, and L. Darkschewitsch. "Ueber die Beziehung des Strickkörpers zum Hinterstrang und Hinterstrangkern nebst Bemerkungen über zwei Felder der Oblongata." *Neurologisches Centralblatt* 5 (1886): 121–129.

Fritsch, Gustav, and Eduard Hitzig. "Ueber die elektrische Erregbarkeit des Grosshirns." *Archiv für Anatomie, Physiologie und wissenschaftliche Medizin* 37 (1870): 300–332.

Fuller, Steve, Daniel Lord Smail, Stephen Casper, Max Stadler, and Roger Cooter. "Neurohistory and history of science." Isis Focus Section. *Isis* 105.1 (2014): 110–154.

Fulton, John. "Arnold Klebs and Harvey Cushing at the First International Neurological Congress at Berne in 1931." *International Journal of Neurology* 14 (1980): 103–115.

Fuss, Diana. "Freud's Ear." In *The Sense of an Interior: Four Writers and the Rooms That Shaped Them*, by Diana Fuss, 71–105. New York: Routledge, 2004.

Gamper, Martina, ed. *Psychiatrische Institutionen in Österreich um 1900*. Vienna: Ärzte, 2009.

García-Albea, E. "Wilder G. Penfield en la residencia de estudiantes (Madrid, 1924)." *Revista neurologia* 39 (2004): 872–878.

Garrison, Fielding. *History of Neurology*. ed. Lawrence C. McHenry Jr. Springfield, IL: Thomas, 1969.

Gauchet, Marcel. *L'inconscient cérébral*. Paris: Seuil, 1992.

Gavrus, Delia. "Men of dreams and men of action: Neurologists, neurosurgeons, and the performance of professional identity, 1920–1950." *Bulletin of the History of Medicine* 85 (2011): 57–92.

———. "'Men of strong opinions': Identity, self-representation, and the performance of neurosurgery, 1919–1950." PhD diss., University of Toronto, 2011.

———. "Mind over matter: Sherrington, Penfield, Eccles, Walshe and the dualist movement in neuroscience." MCIS Briefings, Comparative Program on Health and Society Lupina Foundation Working Papers Series 2005–6, edited by Jillian Clare Cohen and Lisa Forman, University of Toronto, 2006, 51–75.

Gelfand, Toby. "Neurologist or psychiatrist? The public and private domains of Jean-Martin Charcot." *Journal of the History of the Behavioral Sciences* 36 (2000): 215–229.

Gerabek, Werner. "Der Weg zur Bismarckschen Invaliditäts- und Altersversicherung aus medizinhistorischer Sicht." *Würzburger medizinhistorische Mitteilungen* (1992): 331–356.

Gere, Cathy. "Nature's experiment": Epilepsy, localization of brain function and the emergence of the cerebral subject." In *Neurocultures: Glimpses into an Expanding Universe*, edited by Francisco Ortega and Fernando Vidal, 235–247. Frankfurt: Peter Lang, 2011.

Gere, Cathy, and Charlie Gere. "The brain in a vat." Special issue, edited by Cathy Gere and Charlie Gere, *Studies in History and Philosophy of Science Part C: Studies in History and Philosophy of Biological and Biomedical Sciences* 35.2 (2004).

Geschwind, Norman. "Carl Wernicke, the Breslau school and the history of aphasia." In *Selected Papers on Language and the Brain*, edited by R. S. Cohen and M. W. Wartofsky, 42–61. Dordrecht: D. Reidel, 1963.

———. "Wernicke's contribution to the study of aphasia." *Cortex* 3 (1967): 449–463.

Gill, A. S. "Wilder Penfield, Pío Del Río-Hortega, and the discovery of oligodendroglia." *Neurosurgery* 60 (2007): 940–948.

Gilman, Sander, Helen King, Roy Porter, G. S. Rousseau, and Elaine Showalter, eds. *Hysteria beyond Freud*. Berkeley: University of California Press, 1993.

———. "Sigmund Freud, electrotherapy, and the voice." In *Diseases and Diagnoses: The Second Age of Biology*, 159–196. New Brunswick, NJ: Transaction, 2010.

Goetz, Christopher, Michel Bonduelle, and Toby Gelfand. *Charcot: Constructing Neurology*. New York: Oxford University Press, 1995.

Goldstein, Jan. *Console and Classify: The French Psychiatric Profession in the Nineteenth Century*. Chicago: University of Chicago Press, 1987.

———. *The Post-Revolutionary Self: Politics and Psyche in France*. Cambridge, MA: Harvard University Press, 2005.

Goldstein, Kurt. "Carl Wernicke (1848–1904)." In *Founders of Neurology: One Hundred and Forty-Six Biographical Sketches by Eighty-Eight Authors*, edited by Webb Haymaker and Francis Schiller, 531–535. Springfield, IL: Thomas, 1970.

———. "Die Restitution bei Schädigungen der Hirnrinde." *Deutsche Zeitschrift für Nervenheilkunde* 116 (1930): 2–26.

———. *The Organism: A Holistic Approach to Biology Derived from Pathological Data in Man*. New York: Zone Books, 1995.

Goldstein, Kurt, Viktor v. Weizsäcker, and Otfrid Foerster. "Aussprache zum Bericht und Schlussworte: 20. Jahresversammlung der Gesellschaft Deutscher Nervenärzte in Dresden." *Deutsche Zeitschrift für Nervenheilkunde* 116 (1930): 29–31, 42–45.

Greenberg, Steven. "The history of dermatome mapping." *Archives of Neurology* 60 (2003): 126–131.

Greenberg, Valerie. *Freud and His Aphasia Book: Language and the Sources of Psychoanalysis*. Ithaca, NY: Cornell University Press, 1997.

Greenblatt, Samuel. "Cerebral localization: From theory to practice; Paul Broca and Hughlings Jackson to David Ferrier and William Macewen." In *A History of Neurosurgery in Its Scientific and Professional Contexts*, edited by Samuel Greenblatt, 137–152. Park Ridge, IL: American Association of Neurological Surgeons, 1997.

———. "The historiography of neurosurgery: Organizing themes and methodological issues." In *A History of Neurosurgery in Its Scientific and Professional Contexts*, edited by S. H. Greenblatt, 3–10. Park Ridge, IL: American Association of Neurological Surgeons, 1997.

Griesinger, Wilhelm. *Die Pathologie und Therapie der psychischen Krankheiten für Aerzte und Studirende*. 4th ed. Braunschweig: Wreden, 1876.

———. "Ueber psychische Reflexactionen: Mit einem Blick auf das Wesen der psychischen Krankheiten." *Archiv für physiologische Heilkunde* 2 (1843): 76–113.

Grob, Gerald. *The Mad among Us: A History of the Care of America's Mentally Ill*. New York: Free Press, 1994.

Gröger, Helmut. "Zur Entwicklung der Psychiatrie in der Wiener Medizinischen Schule." In *Gründe der Seele: Die Wiener Psychiatrie im 20. Jahrhundert*, ed. Brigitta Keintzel and Eberhard Gabriel, 30–48. Vienna: Picus, 1999.

Guenther, Katja. "Exercises in therapy—neurological gymnastics between *Kurort* and hospital medicine, 1880–1945." *Bulletin of the History of Medicine* 88.1 (2014): 102–131.

———. "Recasting neuropsychiatry—Freud's 'critical introduction' and the convergence of French and German brain science." *Psychoanalysis and History* 14.2 (2012): 203–226.

Guttmann, Giselher, and Inge Scholz-Strasser, eds. *Freud and the Neurosciences: From Brain Research to the Unconscious*. Vienna: Verlag der Österreichischen Akademie der Wissenschaften, 1998.

Hagner, Michael. "*Der Geist bei der Arbeit*: Die visuelle Repräsentation cerebraler Prozesse." In Hagner, *Der Geist bei der Arbeit*, 164–194.

———, ed. *Der Geist bei der Arbeit: Historische Untersuchungen zur Hirnforschung*. Göttingen: Wallstein, 2006.

———. *Homo cerebralis: Der Wandel vom Seelenorgan zum Gehirn*. Berlin: Berlin, 1997.

———. *Geniale Gehirne: Zur Geschichte der Elitegehirnforschung*. Munich: dtv, 2007.

———. "Lokalisation, Funktion, Cytoarchitektonik: Wege zur Modellierung des Gehirns." In *Objekte, Differenzen, Konjunkturen: Experimentalsysteme im historischen Kontext*, edited by Michael Hagner, Hans-Jörg Rheinberger, and Bettina Wahrig-Schmidt, 121–150. Berlin: Akademie, 1994.

Hagner, Michael, and Cornelius Borck. "Mindful practices: On the neurosciences in the twentieth century." *Science in Context* 14.4 (2001): 507–510.

Hale, Nathan. *The Rise and Crisis of Psychoanalysis in America: Freud and the Americans, 1917–1985*. New York: Oxford University Press, 1995.

Hall, Marshall. *Memoirs on the Nervous System*. London: Sherwood, 1837.

———. "On the reflex function of the medulla oblongata and medulla spinalis." *Philosophical Transactions of the Royal Society of London* 123 (1833): 635–665.

Hannaway, Caroline, and Ann La Berge. *Constructing Paris Medicine*. Amsterdam: Rodopi, 1998.

Harrington, Anne. *The Cure Within: A History of Mind-Body Medicine*. New York: Norton, 2008.

———. *Medicine, Mind and the Double Brain: A Study in Nineteenth-Century Thought*. Princeton, NJ: Princeton University Press, 1987.

———. *Reenchanted Science: Holism in German Culture from Wilhelm II to Hitler*. Princeton, NJ: Princeton University Press, 1996.

"Hatten Sie eine Ahnung?" *Zeit am Montag* 21 (May 23, 1932).

Head, Henry, and Gordon Holmes. "Sensory disturbances from cerebral lesions." *Brain* 34.2–3 (1911): 102–254.

Heilbronner, Karl. "Nekrolog C. Wernicke." *Allgemeine Zeitschrift für Psychiatrie* 62 (1905): 881–892.

Herzberg, David. *Happy Pills in America: From Miltown to Prozac*. Baltimore: Johns Hopkins University Press, 2009.

Hess, Volker, and Eric Engstrom. "Neurologie an der Charité zwischen medizinischer und psychiatrischer Klinik." In *Geschichte der Neurologie in Berlin*, edited by Bernd Holdorff and Rolf Winau, 99–110. Berlin: De Gruyter, 2001.

Hess, Volker, and Sophie Ledebur. "Taking and keeping: A note on the emergence and function of hospital patient records." *Journal of the Society of Archivists* 32 (2011): 21–32.

Hess, Volker, and Andrew Mendelsohn, "Case and series: Medical knowledge and paper technology, 1600–1900." *History of Science* 48 (2010): 287–314.

Hesse, Günter. "Patient Lenin: Ein Übermensch?" *Deutsches Ärzteblatt* 10 (1975): 682–686, 755–760, 835–839, 3205–3207.

Hirschmüller, Albrecht. *Freuds Begegnung mit der Psychiatrie: Von der Hirnmythologie zur Neurosenlehre.* Tübingen: Edition Diskord, 1991.

———. *The Life and Work of Josef Breuer: Physiology and Psychoanalysis.* New York: New York University Press, 1989.

Hitzig, Eduard. "Neubau der psychiatrischen und Nervenklinik für die Universität Halle a. S." *Klinisches Jahrbuch* 2 (1890): 383–405.

Holdorff, Bernd. "Die Lokalisationsdiskussion vor 60 Jahren und heute (O. Foerster, K. Goldstein, V. v. Weizsäcker)." *Schriftenreihe der Deutschen Gesellschaft für Geschichte der Nervenheilkunde* 1 (1996): 139–141.

Holdorff, Bernd, and Rolf Winau, eds. *Geschichte der Neurologie in Berlin.* Berlin: De Gruyter, 2001.

Hubenstorf, Michael. "Österreichische Ärzte-Emigration." In *Vertriebene Vernunft I: Emigration und Exil österreichischer Wissenschaft 1930–1940*, edited by Friedrich Stadler, 359–415. Vienna: Jugend und Volk, 1987.

Hunkin, S. J. "Experience with Foerster's operation." *Journal of Orthopaedic Surgery* s2–11 (1913): 207–214.

Huxley, Thomas. "On the hypothesis that animals are automata and its history." *Fortnightly Review* 22 (1874): 199–245.

Jacyna, Stephen. *Lost Words: Narratives of Language and the Brain, 1825–1926.* Princeton, NJ: Princeton University Press, 2000.

———. *Medicine and Modernism: A Biography of Sir Henry Head.* London: Pickering and Chatto, 2008.

Janßen, Sandra. "Von der Dissoziation zum System: Das Konzept des Unbewussten als Abkömmling des Reflexparadigmas in der Theorie Freuds." *Berichte zur Wissenschaftsgeschichte* 32 (2009): 36–52.

Jasper, Herbert, Cosimo Ajmone-Marson, and Julius Stoll. "Corticofugal projections to the brain stem." *Archives of Neurology and Psychiatry* 67 (1952): 155–171.

Jaspers, Karl. *General Psychopathology.* Translated by J. Hoenig and Marian W. Hamilton. Chicago: University of Chicago Press, 1963.

Jeannerod, Marc. *The Brain Machine: The Development of Neurophysiological Thought.* Translated by David Urion. Cambridge, MA: Harvard University Press, 1985.

Jones, Ernest. *The Life and Work of Sigmund Freud.* Vol. 1. New York: Basic Books, 1953.

Joyce, Kelly A. *Magnetic Appeal: MRI and the Myth of Transparency.* Ithaca, NY: Cornell University Press, 2008.

Jung, Carl Gustav. "Vorbemerkung der Redaktion." In *Jahrbuch für psychoanalytische und psychopathologische Forschungen*, vol. 1, edited by Eugen Bleuler, Sigmund Freud, and C. G. Jung. Leipzig: Deuticke, 1909.

Kandel, Eric, James Schwartz, Thomas Jessel, and Steven Siegelbaum. *Principles of Neural Science*. 5th ed. New York: McGraw-Hill, 2012.

Kiejna, Andrzej, and Małgorzata Wójtowicz. *Z dziejów Kliniki Psychiatrycznej i Chorób Nerwowych we Wrocławiu: wybitni przedstawiciele i budowle—Zur Geschichte der psychiatrischen und Nervenklinik in Breslau: Bedeutende Vertreter und Bauwerke*. Wrocław: Fundacja Ochrony Zdrowia Psychicznego, 1999.

Killen, Andreas. *Berlin Electropolis: Shock, Nerves and German Modernity*. Berkeley: University of California Press, 2006.

Kleeberg, Ingrid. "Poetik der nervösen Revolution: Psychophysiologie und das politisch Imaginäre, 1750–1860." PhD diss., Konstanz University, 2011.

Kleist, Karl. "Carl Wernicke." In Kolle, *Große Nervenärzte*, 2:106–128.

Kolb, Bryan, and Ian Whishaw. *An Introduction to Brain and Behavior*. New York: Worth, 2009.

Kolle, Kurt. *Große Nervenärzte*. Stuttgart: Thieme, 1959.

Krell, David Farrell. *Of Memory, Reminiscence, and Writing: On the Verge*. Bloomington: Indiana University Press, 1990.

Kroker, Kenton. *The Sleep of Others and the Transformation of Sleep Research*. Toronto: University of Toronto Press, 2007.

Krüger, Hans-Peter, ed. *Hirn als Subjekt? Philosophische Grenzfragen der Neurobiologie*. Berlin: Akademie, 2007.

Kubie, Lawrence. "Some implications for psychoanalysis of modern concepts of the organization of the brain." *Psychoanalytic Quarterly* 22 (1953): 21–52.

Kuhlendahl, Hans. "Otfrid Foerster (9.11.1873–14.6.1941)." In *Recent Progress in Neurological Surgery: Proceedings of the Symposia of the Fifth International Congress of Neurological Surgery, Tokyo, October 7–13, 1973*, 205–209. Amsterdam: Excerpta Medica, 1974.

Kushner, Howard. "Norman Geschwind and the use of history in the (re)birth of behavioral neurology." *Journal of the History of the Neurosciences* 24 (2015): 173–192.

Lakoff, Andrew. *Pharmaceutical Reason: Knowledge and Value in Global Psychiatry*. Cambridge: Cambridge University Press, 2005.

Lamb, Susan D. *Pathologist of the Mind: Adolf Meyer and the Origins of American Psychiatry*. Baltimore: Johns Hopkins University Press, 2014.

Langer, Dieter. "Paul Ferdinand Schilder: Leben und Werk." PhD diss., Mainz University, 1979.

Laplanche, Jean. *Life and Death in Psychoanalysis*. Translated by Jeffrey Mehlman. 1970. Reprint, Baltimore: Johns Hopkins University Press, 1985. Page references are to the reprint edition.

Laplanche, Jean, and J.-B. Pontalis. *The Language of Psycho-Analysis*. Translated by Donald Nicholson-Smith. The International Psycho-Analytical Library, 94:1–497. London: Hogarth Press and the Institute of Psycho-Analysis, 1973.

Ledebur, Sophie. "Das Wissen der Anstaltspsychiatrie in der Moderne: Zur Geschichte der Heil- und Pflegeanstalten *Am Steinhof* in Wien." PhD diss., University of Vienna, 2011.

———. "Schreiben und Beschreiben: Zur epistemischen Funktion von psychiatrischen Krankenakten, ihrer Archivierung und deren Übersetzung in Fallgeschichten." *Berichte zur Wissenschaftsgeschichte* 34 (2011): 102–124.

Leppmann, Arthur. "Heinrich Neumann 1814–1884." In *Deutsche Irrenärzte: Einzelbilder ihres Lebens und Wirkens*, edited by T. Kirchhoff, 1:261–265. Berlin: Springer, 1921.

Lesky, Erna. "Carl von Rokitansky (1804–1878)." *Neue österreichische Biographie ab 1815* 12 (1969): 38–51.

———. *Die Wiener medizinische Schule im 19. Jahrhundert*. Graz: Böhlaus, 1978.

———. Introduction to *Carl von Rokitansky: Selbstbiographie und Antrittsrede*, by Carl Rokitansky, edited by Erna Lesky. Vienna: Böhlau, 1960.

———. *Meilensteine der Wiener Medizin: Grosse Ärzte Österreichs in drei Jahrhunderten*. Vienna: Wilhelm Maudrich, 1981.

———. *The Vienna Medical School of the 19th Century*. Translated by L. Williams and I. S. Levij. Baltimore: Johns Hopkins University Press, 1976. Leuschner, Wolfgang. "Einleitung." In *Zur Auffassung der Aphasien: Eine kritische Studie*, by Sigmund Freud, edited by Paul Vogel, 7–31. Frankfurt: Fischer, 2001.

Levin, Kenneth. *Freud's Early Psychology of the Neuroses: A Historical Perspective*. Pittsburgh: University of Pittsburgh Press, 1978.

Lewandowsky, Max, ed. *Handbuch der Neurologie*. Ergänzungsband, vol. 2. Berlin: Julius Springer, 1924.

Lewis, Jefferson. *Something Hidden: A Biography of Wilder Penfield*. Toronto: Doubleday, 1981.

Leys, Ruth. "Adolf Meyer's life chart and the representation of individuality." *Representations* 34 (1991): 1–28.

———. *From Sympathy to Reflex: Marshall Hall and His Opponents*. New York: Garland, 1990.

———. "Meyer, Jung, and the limits of association." *Bulletin of the History of Medicine* 59.3 (1985): 345–360.

———. "Meyer's dealings with Jones: A chapter in the history of the American response to psychoanalysis," *Journal of the History of the Behavioral Sciences* 17 (1981): 445–465.

———. *Trauma: A Genealogy*. Chicago: University of Chicago Press, 2000.

Lichtenberger, Elisabeth. *Wirtschaftsfunktion und Sozialstruktur der Wiener Ringstrasse*. Vienna: Böhlau, 1970.

Liepmann, Hugo. "Carl Wernicke." In *Deutsche Irrenärzte: Einzelbilder ihres Lebens und Wirkens*, edited by Theodor Kirchhoff, 242. Berlin: Julius Springer, 1921–24.

———. *Über Störungen des Handelns bei Gehirnkranken*. Berlin: Karger, 1905.

———. "Wernickes Einfluss auf die klinische Psychiatrie." *Monatsschrift für Psychiatrie und Neurologie* 30 (1911): 1–37.

Littlefield, Melissa, and Jenell Johnson, eds. *The Neuroscientific Turn: Transdisciplinarity in the Age of the Brain*. Ann Arbor: University of Michigan Press, 2012.

Lunbeck, Elizabeth. *The Americanization of Narcissism*. Cambridge, MA: Harvard University Press, 2014.

———. *The Psychiatric Persuasion: Knowledge, Gender, and Power in Modern America*. Princeton, NJ: Princeton University Press, 1994.

Mahony, Patrick. *Freud and the Rat Man*. New Haven, CT: Yale University Press, 1986.

Makari, George. *Revolution in Mind: The Creation of Psychoanalysis*. New York: HarperCollins, 2008.

Mann, Ludwig. "Wernickes Stellung zur Elektrodiagnostik und Elektrotherapie." *Zeitschrift für Elektrotherapie und Elektrodiagnostik* 7 (1905): 311–317.

Marinelli, Lydia, and Andreas Mayer. *Dreaming by the Book: Freud's Interpretation of Dreams and the History of the Psychoanalytic Movement*. Translated by Susan Fairfield. New York: Other Press, 2003.

———. *Forgetting Freud? For a New Historiography of Psychoanalysis*. *Science in Context*, special issue, 19.1 (2006).

Marneros, Andreas, and Frank Pillmann. *Das Wort Psychiatrie wurde in Halle geboren: Von den Anfängen der deutschen Psychiatrie*. Stuttgart: Schattauer, 2005.

Marshall, J. C., and G. R. Fink. "Cerebral localization, then and now." *Neuroimage* 20, suppl. 1 (2003): S2–S7.

Marx, Otto. "Freud and aphasia: An historical analysis." *American Journal of Psychiatry* 124 (1967): 815–825.

———. "Psychiatry on a neuropathological basis: Th. Meynert's application for the extension of his venia legendi." *Clio Medica* 6.1 (1971): 139–158.

Mayer, Andreas. *Mikroskopie des Psyche: Die Anfänge der Psychoanalyse im Hypnose-Labor*. Göttingen: Wallstein, 2002.

———. *Sites of the Unconscious: Hypnosis and the Emergence of the Psychoanalytic Setting*. Chicago: University of Chicago Press, 2013.

"McGill University, Department of Psychiatry, 50th Anniversary." *Journal of Psychiatry and Neuroscience* 18.4 (1993): 141–142.

Menninger, Anneliese. *Sigmund Freud als Autor in Villarets Handwörterbuch der Gesamten Medizin von 1888–1891*. Hamburg: Dr. Kovač, 2011.

———. "Zu den Beiträgen Sigmund Freuds in Villarets Handwörterbuch der Gesamten Medizin (1888–91)." *Luzifer Amor* 49.1 (2012): 83–105.

Métraux, Alexandre. "Metamorphosen der Hirnwissenschaft: Warum Sigmund Freuds 'Entwurf einer Psychologie' aufgegeben wurde." In *Ecce Cortex: Beiträge zur Geschichte des modernen Gehirns*, edited by Michael Hagner, 75–109. Göttingen: Wallstein, 1999.

Metzl, Jonathan. *Prozac on the Couch: Prescribing Gender in the Era of Wonder Drugs*. Durham, NC: Duke University Press, 2003.

Meyer, Adolf. *The Collected Papers of Adolf Meyer*. Edited by Eunice E. Winters. Baltimore: Johns Hopkins Press, 1950–.

———. "Dealing with mental disease." *Modern Hospital* 51 (1938): 87–89. Also in Meyer, *Collected Papers*, vol. 4.

———. "A discussion of some fundamental issues in Freud's psychoanalysis." *State Hospitals Bulletin*, n.s., 2 (1909–10): 827–848. Also in Meyer, *Collected Papers*, vol. 2.

———. "The dynamic interpretation of dementia praecox." Lecture delivered at the celebration of the twentieth anniversary of Clark University, September 1909. *American Journal of Psychology* 21 (1910): 385–403. Also in Meyer, *Collected Papers*, vol. 2.

———. "Interpretation of obsessions." *Psychological Bulletin* 3 (1906): 280–283. Also in Meyer, *Collected Papers*, vol. 2.

———. "Outlines of examinations." Privately printed, 1918, 1923. Also in Meyer, *Collected Papers*, vol. 3.

———. "Paul Ferdinand Schilder, M.D., 1886–1940." *Journal of Nervous and Mental Disease*, 93 (1941): 812–814. Also in Meyer, *Collected Papers*, vol. 3.

———. "Preparation for psychiatry." *Archives of Neurology and Psychiatry* 30 (1933): 1111–1125. Also in Meyer, *Collected Papers*, vol. 3.

———. "Psychobiology in the first year of medical school." *Journal of the Association of American Medical Colleges* 10 (1935): 365–72. Also in Meyer, *Collected Papers*, vol. 3.

Meynert, Theodor. "Anatomie der Hirnrinde als Träger des Vorstellungslebens und ihrer Verbindungsbahnen mit den empfindenden Oberflächen und den bewegenden Massen." In *Lehrbuch der psychischen Krankheiten*, edited by Maximilian Leidesdorf, 45–73. Erlangen: Enke, 1865.

———. "Beiträge zur Theorie der maniakalischen Bewegungserscheinungen nach dem Gange und Sitze ihres Zustandekommens." *Archiv für Psychiatrie und Nervenkrankheiten* 2 (1870): 622–642.

———. "Das Gesammtgewicht und Theilgewichte des Gehirnes in ihren Beziehungen zum Geschlechte, dem Lebensalter und dem Irrsinn, untersucht nach einer neuen Wägungsmethode an den Gehirnen der in der Wiener Irrenanstalt im Jahre 1866 Verstorbenen." *Vierteljahresschrift für Psychiatrie in ihren Beziehungen zur Morphologie und Pathologie des Centralnervensystems, der physiologischen Psychologie, Statistik und gerichtlichen Medizin* 2 (1867): 125–170.

———. "Das Zusammenwirken der Gehirntheile." Talk given at the Tenth International Medical Congress in Berlin, 1890. In *Sammlung von populärwissenschaftlichen Vorträgen über den Bau und die Leistungen des Gehirns*, edited by Theodor Meynert, 201–231. Vienna: Braumüller, 1892.

———. "Die acuten hallucinatorischen Formen des Wahnsinns." *Jahrbücher für Psychiatrie* 2 (1881): 181–196.

———. "Die Blosslegung des Bündelverlaufes im Grosshirnstamme." *Oesterreichische Zeitschrift für praktische Heilkunde* 11 (1865): 5–8, 25–28, 85–89, 153–156, 184–186, 437–440.

———. "Die Medianebene des Hirnstammes als ein Stück der Leitungsbahn zwischen dem Vorstellungsgebiete und den motorischen Hirnnerven." *Allgemeine Wiener medizinische Zeitung* 51 (1865): 411, 419.

———. "Ein Fall von Sprachstörung, anatomisch begründet." *Medizinische Jahrbücher* 12 (1866): 152–187.

———. "Ein Fortschritt in der psychiatrischen Diagnostik." *Mittheilungen des Wiener medicinischen Doctoren-Collegiums* 3 (1877): 249–268.

———. "Fragment aus den anatomischen Corollarien und der Physiologie des Vorderhirns." *Jahrbücher für Psychiatrie* 2 (1881): 65–91.

———. *Klinische Vorlesungen über Psychiatrie auf wissenschaftlichen Grundlagen für Studirende und Ärzte, Juristen und Psychologen.* Vienna: Braumüller, 1890.

———. "Kraniologische Beiträge zur Lehre von der psychopathischen Veranlagung." *Jahrbücher für Psychiatrie* 1 (1879): 69–93, 153–171.

———. *Psychiatrie: Klinik der Erkrankungen des Vorderhirns begründet auf dessen Bau, Leistungen und Ernährung.* Vienna: Braumüller, 1884.

———. *Psychiatry: A Clinical Treatise on Diseases of the Fore-Brain Based upon a Study of Its Structures, Function, and Nutrition.* Translated by Bernard Sachs. New York: Putnam's, 1885.

———. "Skizzen über Umfang und wissenschaftliche Anordnung des psychiatrischen Lehrstoffes," *Psychiatrisches Centralblatt* 1 (1876): 2–85.

———. "Studien über das pathologisch-anatomische Material der Wiener Irren-Anstalt." *Vierteljahresschrift für Psychiatrie* 3 (1868): 381–402.

———. "Ueber die Bedeutung der Stirnentwicklung." Talk given at the Wissenschaftlicher Club in Vienna, 1886. In *Sammlung von populär-wissenschaftlichen Vorträgen über den Bau und die Leistungen des Gehirns*, edited by Theodor Meynert, 101–110. Vienna: Braumüller, 1892.

———. "Ueber die Nothwendigkeit und Tragweite einer anatomischen Richtung in der Psychiatrie." *Wiener medizinische Wochenschrift* 18 (1868): 573–576, 589–591.

———. "Ueber Fortschritte der Lehre von den psychiatrischen Krankheitsformen." *Psychiatrisches Centralblatt* 6, 7 (1877): 53–126; 1 (1878): 1–23.

———. "Vom Gehirne der Säugethiere." In *Handbuch der Lehre von den Geweben des Menschen und der Thiere*, edited by Salomon Stricker, 1:694–808. Leipzig: Engelmann, 1872.

———. "Zum Verständniss der functionellen Nervenkrankheiten." *Wiener medizinische Blätter* 5 (1882): 481–484, 517–518.

Micale, Mark. *Approaching Hysteria: Disease and Its Interpretations.* Princeton, NJ: Princeton University Press, 1994.

———. *Hysterical Men: The Hidden History of Male Nervous Illness.* Cambridge, MA: Harvard University Press, 2008.

Miciotto, Robert J. "Carl Rokitansky: Nineteenth-century pathologist and leader of the New Vienna School." PhD diss., Johns Hopkins University, 1979.

Minkowski, M. "Die Stellung der Neurologie im medizinischen Unterricht." *Schweizer Archiv für Neurologie und Psychiatrie* 30 (1933): 159–177.

Mommsen, Wolfgang. "Das Ringen um den nationalen Staat: Die Gründung und der innere Ausbau des Deutschen Reiches unter Otto von Bismarck 1850–1980." In *Propyläen Geschichte Deutschland*, edited by Dieter Groh, vol. 7, part 1, 624–664. Berlin, Propyläen, 1993.

Müller, Johannes. *Handbuch der Physiologie des Menschen*. Vol. 1. Coblenz: Hölscher, 1833.

Munk, Hermann. *Ueber die Functionen der Grosshirnrinde*. Berlin: Hirschwald, 1881.

Neumärker, K.-J. "Carl Wernicke und Karl Kleist: Zwei Biographien—eine Richtung in ihrer Entwicklung." *Fundamenta Psychiatrica* 8 (1994): 178–184.

"The 1949 annual meeting." *Bulletin of the American Psychoanalytic Association* 5C (1949): 31–75.

Nitzschke, Bernd. *Aufbruch nach Inner-Afrika: Essays über Sigmund Freud und die Wurzeln der Psychoanalyse*. Göttingen: Vandenhoeck and Ruprecht, 1998.

Noë, Alva. *Action in Perception*. Cambridge, MA: MIT Press, 2004.

Nonne, Max. "Über Wert und Bedeutung der modernen Syphilistherapie für die Behandlung von Erkrankungen des Nervensystems." *Deutsche Zeitschrift für Nervenheilkunde* (1911–12): 166–250.

Nussbaum, Martha. *Not for Profit: Why Democracy Needs the Humanities*. Princeton, NJ: Princeton University Press, 2012.

Pantel, Johannes. "Neurologie, Psychiatrie und Innere Medizin: Verlauf und Dynamik eines historischen Streites." *Würzburger medizinhistorische Mitteilungen* 11 (1993): 77–99.

———. "Streitfall Nervenheilkunde—eine Studie zur disziplinären Genese der klinischen Neurologie in Deutschland." *Fortschritte der Neurologie und Psychiatrie* 61 (1993): 144–156.

Pantel, Johannes, and Axel Bauer. "Die Institutionalisierung der Pathologischen Anatomie im 19. Jahrhundert an den Universitäten Deutschlands, der deutschen Schweiz und Österreichs." *Gesnerus* 47 (1990): 303–328.

Penfield, Wilder. "The cerebral cortex in man. I. The cerebral cortex and consciousness." *Archives of Neurology and Psychiatry* 40 (1938): 417–442.

———. "Clinical notes from a trip to Great Britain." *Archives of Neurology and Psychiatry* 47 (1942): 1030–1036.

———. *The Difficult Art of Giving: The Epic of Alan Gregg*. Boston: Little, Brown, 1967.

———. "The electrode, the brain and the mind." *Journal of Neurology* 201 (1972): 297–309.

———. "The evidence for a cerebral vascular mechanism in epilepsy." *Annals of Internal Medicine* 7.3 (1933): 303–310.

———. *The Excitable Cortex in Conscious Man*. Springfield, IL: Thomas, 1958.

———. "Intracerebral vascular nerves." *Archives of Neurology and Psychiatry* 27 (1932): 30–44.

———. "Les effets des spasmes vasculaires dans l'épilepsie." *Union Médicale du Canada* 63 (1934): 1275–1282.

———. "Memory mechanisms." *Archives of Neurology and Psychiatry* 67 (1952): 178–198.

———. *The Mystery of the Mind: A Critical Study of Consciousness and the Human Brain*. Princeton, NJ: Princeton University Press, 1975.

———. *No Man Alone: A Neurosurgeon's Life*. Boston: Little, Brown, 1977.

———. "Orientation of scientific research to war." *American Scientist* 30.2 (1941): 116–118, 136.

———. "Posttraumatic epilepsy." *American Journal of Psychiatry* 100 (1944): 750–751.

———. "The radical treatment of traumatic epilepsy and its rationale." *Canadian Medical Association Journal* 23 (1930): 189–197.

———. *The Significance of the Montreal Neurological Institute*. In *The Neurological Biographies and Addresses: Foundation Volume Published for the Staff, to Commemorate the Opening of the Montreal Neurological Institute of the Montreal Neurological Institute*, 37–54. London: Oxford University Press, 1936.

———. "Some observations on the cerebral cortex of man." Ferrier Lecture. *Proceedings of the Royal Society of London* B 134 (1947): 329–347.

———. "Some problems of wartime neurology." *Archives of Neurology and Psychiatry* 47 (1942): 839–841.

Penfield, Wilder, and Edwin Boldrey. "Somatic motor and sensory representation in the cerebral cortex of man as studied by electrical stimulation." *Brain* 60.4 (1937): 389–443.

Penfield, Wilder, and J. Chorobski. "Cerebral vasodilator nerves and their pathway from the medulla oblongata with observations on the pial and intracerebral vascular plexus." *Archives of Neurology and Psychiatry* 28 (1932): 1257–1289.

Penfield, Wilder, and Theodore Erickson. *Epilepsy and Cerebral Localization: A Study of the Mechanism, Treatment and Prevention of Epileptic Seizures*. Springfield, IL: Thomas, 1941.

Penfield, Wilder, and J. Evans. "The frontal lobe in man: A clinical study of maximum removals." *Brain* 58.1 (1935): 115–133.

Penfield, Wilder, and Donald Hebb. "Human behavior after extensive bilateral removal from the frontal lobes." *Archives of Neurology and Psychiatry* 44 (1940): 421–438.

Penfield, Wilder, and Herbert Jasper. "Electroencephalograms in post-traumatic epilepsy: Pre-operative and post-operative studies." *American Journal of Psychiatry* 100 (1943): 365–377.

———. *Epilepsy and the Functional Anatomy of the Human Brain*. Boston: Little, Brown, 1954.

Penfield, Wilder, and P. Perot. "The brain's record of auditory and visual experience: A final summary and discussion." *Brain* 86 (1963): 595–696.

Penfield, Wilder, and Theodore Rasmussen. *The Cerebral Cortex of Man: A Clinical Study of Localization of Function*. New York: Macmillan, 1950.

Penfield, Wilder, and Lamar Roberts. *Speech and Brain Mechanisms*. Princeton, NJ, Princeton University Press, 1959.

Phelps, Scott. "Blind to their blindness: Indifference, anosognosia, and the denial of illness, 1880–1960." PhD diss., Harvard University, 2013.

Philippon, Jacques, and Jacques Poirier. *Joseph Babinski: A Biography*. Oxford: Oxford University Press, 2009.

Pick, Arnold. *Die agrammatischen Sprachstörungen: Studien zur psychologischen Grundlegung der Aphasielehre*. Vol. 1. Berlin: Springer, 1913.

Pillmann, Frank, T. Arndt, U. Ehrt, A. Haring, E. Kumbier, and A. Marneros. "An analysis of Wernicke's original case records: His contribution to the concept of cycloid psychoses." *History of Psychiatry* 11 (2000): 355–369.

Piñero, José María López. *Historical Origins of the Concept of Neurosis*. Translated by D. Berrios. Cambridge: Cambridge University Press, 1983.

Poetzl, Otto. "Julius Wagner von Jauregg." *Wiener klinische Wochenschrift* 53 (1940): 1–4.

Pogliano, Claudio. "Penfield's homunculus and other grotesque creatures from the land of if." *Annali di Storia della Scienza* 27.1 (2012): 141–162.

Popinski, Krysztof. *Studenten an der Universität Breslau 1871 bis 1921: Eine sozialgeschichtliche Untersuchung*. Translated by Thorsten Möllenbeck. Historia Academica, vol. 26 (Würzburg: Studentengeschichtliche Vereinigung des Coburger Convents, 2009).

Porath, Erik. "Vom Reflexbogen zum psychischen Apparat: Neurologie und Psychoanalyse um 1900." *Berichte zur Wissenschaftsgeschichte* 32 (2009): 53–69.

Proceedings of the First International Neurological Congress, Berne, Switzerland, August 31st to September 4th, 1931. Berne: The Congress, 1932.

Quick, Tom. "Techniques of life: Zoology, psychology and technical subjectivity (c. 1820–1890)." PhD diss., University College London, 2011.

Raz, Mical. "Between the ego and the icepick: Psychosurgery, psychoanalysis, and psychiatric discourse." *Bulletin of the History of Medicine* 82 (2008): 387–420.

———. *The Lobotomy Letters: The Making of American Psychosurgery*. Rochester, NY: University of Rochester Press, 2013.

Reicheneder, Johann. "'Lokalisation': Ein bisher unbekannt gebliebener Beitrag Freuds zu Villarets Handwörterbuch der gesamten Medizin." *Jahrbuch der Psychoanalyse* 32 (1994): 155–182.

Richter, Prof. "Eine Stätte, von wo Licht ausging." *Die Gartenlaube* 47–48 (1863): 747–750, 757–759.

Rimpau, Wilhelm. "Die Krise der Neurologie in erkenntnistheoretischer Weise: Kontroverse zwischen Viktor von Weizsäcker, Kurt Goldstein und Otfrid Foerster zum Lokalisationsprinzip 1930." *Nervenarzt* 80 (2009): 970–974.

Ringer, Fritz. *The Decline of the German Mandarins*. Middletown, CT: Wesleyan University Press, 1990.

Río Hortega, P. del, and Wilder Penfield. "Cerebral cicatrix: The reaction of neuroglia and microglia to brain wounds." *Bulletin of the Johns Hopkins Hospital* 41 (1927): 278–303.

Roazen, Paul. *How Freud Worked: First-Hand Accounts of Patients*. Northvale, NJ: J. Aronson, 1995.

Roelcke, Volker. "Unterwegs zur Psychiatrie als Wissenschaft: Das Projekt einer 'Irrenstatistik' und Emil Kraepelins Neuformulierung der psychiatrischen Klassifikation." In *Psychiatrie im 19. Jahrhundert: Forschungen zur Geschichte von psychiatrischen Institutionen, Debatten und Praktiken im deutschen Sprachraum*, edited by Eric Engstrom and Volker Roelcke, 169–188. Mainz: Schwabe, 2003.

Rokitansky, Carl. *Handbuch der allgemeinen pathologischen Anatomie*. Vienna: Braumüller and Seidel, 1846.

Rose, Nikolas. *The Politics of Life Itself: Biomedicine, Power and Subjectivity in the Twenty-first Century*. Princeton, NJ: Princeton University Press, 2007.

Rose, Nikolas, and Joelle Abi-Rached. *Neuro: The New Brain Sciences and the Management of the Mind*. Princeton, NJ: Princeton University Press, 2013.

Rose, Clifford. *History of British Neurology*. London: Imperial College Press, 2012.

Rosen, George. *The Specialization of Medicine with Particular Reference to Ophthalmology*. New York: Froben Press, 1944.

Rosenberg, Charles. "Contested boundaries, psychiatry, disease, and diagnosis." *Perspectives in Biology and Medicine* 49.3 (2006): 407–427.

———. *The Origins of Specialization in American Medicine: An Anthology of Sources*. New York: Garland, 1989.

———. "The therapeutic revolution: Medicine, meaning, and social change in 19th-century America." *Perspectives in Biology and Medicine* 20 (1977): 485–506.

Rüdin, Ernst. "Eröffnungsansprache auf der 1. Jahresversammlung der Gesellschaft deutscher Neurologen und Psychiater." *Deutsche Zeitschrift für Nervenheilkunde* 139 (1936): 5–11.

Rumpler, Helmut, and Helmut Denk, eds. *Carl Freiherr von Rokitansky (1804–1878): Pathologe, Politiker, Philosoph, Gründer der Wiener Medizinischen Schule des 19. Jahrhunderts*. Vienna: Böhlau, 2005.

Satzinger, Helga. *Die Geschichte der genetisch orientierten Hirnforschung von Cécile und Oskar Vogt in der Zeit von 1895 bis ca. 1927*. Stuttgart: Deutscher Apotheker Verlag, 1998.

Saunders, M. "'Science and futurology in the to-day and to-morrow series': Matter, consciousness, time and language." *Interdisciplinary Science Reviews* 34.1 (2009): 68–78.

Schiff, Nicholas. "Wilder Graves Penfield: Philosophy, physiology, and the mystery of mind." Senior thesis, Stanford University, Stanford, CA, April 1987.

Schilder, Paul. "Bemerkungen ueber die Psychologie des paralytischen Größenwahns." *Zeitschrift für die gesamte Neurologie und Psychiatrie* 74 (1922): 1–14.

———. *Brain and Personality: Studies in the Psychological Aspects of Cerebral Neuropathology and the Neuropsychiatric Aspects of the Motility of Schizophrenics*. New York: Nervous and Mental Disease, 1931.

———. *Das Körperschema: Ein Beitrag zur Lehre vom Bewusstsein des eigenen Körpers*. Berlin: Springer, 1923.

———. "Das Unbewusste." *Zeitschrift für die gesamte Neurologie und Psychiatrie* 80 (1923): 96–116.

———. "Der gegenwärtige Stand der Neurosenlehre." *Klinische Wochenschrift* 6.2. (1927): 49–51.

———. "Die Psychotherapie der Psychosen." *Zentralblatt für die gesamte Neurologie und Psychiatrie* 44 (1926): 182.

———. "Entwurf zu einer Psychologie der Schizophrenie und Paraphrenie." *Deutsche medizinische Wochenschrift* 49 (1923): 1433–1435.

———. "Gesichtspunkte zur allgemeinen Psychiatrie." *Archiv für Psychiatrie und Nervenkrankheiten* 59 (1918): 699–712.

———. *The Image and Appearance of the Human Body: Studies in the Constructive Energies of the Psyche*. New York, International Universities Press, 1950.

———. *Mind: Perception and Thought in Their Constructive Aspects*. New York: Columbia University Press, 1942.

———. "Probleme der klinischen Psychiatrie." *Medizinische Klinik* 21 (1925): 79–82.

———. "Problems in the technique of psychoanalysis." *Psychoanalytic Review* 17 (1930): 1–19.

———. "Psychologie der Schizophrenie vom psychoanalytischen Standpunkt." *Zeitschrift für die gesamte Neurologie und Psychiatrie* 112 (1928): 279–282.

———. "Scope of psychiatry in schizophrenia." *American Journal of Psychiatry* 11 (1932): 1181–1187.

———. "Über die amyloide Entartung der Haut." *Frankfurter Zeitschrift für Pathologie* 3 (1909): 782–794.

———. "Vita." In *Paul Schilder: Mind Explorer*, edited by Donald Shaskan and William Roller, 266–270. New York: Human Sciences Press, 1985.

———. *Wahn und Erkenntnis*. Berlin, Springer: 1918.

Schilder, Paul, and W. Haberfeld. "Die Tetanie der Kaninchen." *Mitteilungen aus dem Grenzgebiet der Medizin und Chirurgie* 20 (1909): 728–756.

Schilder, Paul, and K. Kassowitz. "Einige Versuche über die Feinheit der Empfindungen bei bewegter Tastfläche." *Pflügers Archiv für die gesamte Physiologie der Menschen und Tiere* 122 (1908): 119–129.

Schilder, Paul, F. Paul, and H. Weidner. "Zur Kenntnis symbolähnlicher Bildungen im Rahmen der Schizophrenie." *Zeitschrift für die gesamte Neurologie und Psychiatrie* 26 (1914): 201–244.

Schlesinger, Arthur. "Die Foerster'sche Operation, Sammelreferat." *Neurologisches Centralblatt* 29 (1910): 970–978.

Schmidt, Johann Baptist. "Gehörs- und Sprachstörung in Folge von Apoplexie." *Allgemeine Zeitschrift für Psychiatrie und psychisch-gerichtliche Medicin* 27 (1871): 304–306.

Schmiedebach, Heinz-Peter. "Siegeszug des somatischen Positivismus." In *Psychiatrie und Psychologie im Widerstreit: Die Auseinandersetzung in der Berliner medicinisch-psychologischen Gesellschaft, 1867–1899*. Husum: Matthiesen, 1986.

Schorske, Carl. *Fin-de-siècle Vienna: Politics and Culture*. Cambridge: Cambridge University Press, 1979.

Schott, G. D. "Penfield's homunculus: A note on cerebral cartography." *Journal of Neurology, Neurosurgery, and Psychiatry* 56 (1993): 329–333.

Schott, Heinz, and Rainer Tölle. *Geschichte der Psychiatrie*. Stuttgart: Beck, 2006.

Schwalbe, Gustav. *Lehrbuch der Neurologie, zugleich des zweiten Bandes von Hoffmann's Lehrbuch der Anatomie des Menschen*. Erlangen: Besold, 1881.

Sealey, Anne. "The strange case of the Freudian case history: The role of long case histories in the development of psychoanalysis." *History of the Human Sciences* 24 (2011): 36–50.

Seebacher, Felicitas. *"Freiheit der Naturforschung": Carl Freiherr von Rokitansky und die Wiener Medizinische Schule: Wissenschaft und Politik im Konflikt*. Vienna: Österreichische Akademie der Wissenschaften, 2006.

Seigel, Jerrold. *The Idea of the Self: Thought and Experience in Western Europe since the Seventeenth Century*. Cambridge: Cambridge University Press, 2005.

Seung, Sebastian. *Connectome: How the Brain's Wiring Makes Us Who We Are*. Boston: Houghton Mifflin Harcourt, 2012.

Shaskan, Donald, and William Roller, eds. *Paul Schilder: Mind Explorer*. New York: Human Sciences Press, 1985.

Shepherd, Gordon. *Creating Modern Neuroscience: The Revolutionary 1950s*. Oxford: Oxford University Press, 2010.

Shorter, Edward. *A History of Psychiatry: From the Era of the Asylum to the Age of Prozac*. New York: Wiley, 1997.

Sigerist, Henry. "From Bismarck to Beveridge: Developments and trends in social security legislation." *Journal of Public Health Policy* 20 (1999): 474–496. Originally published in *Bulletin of the History of Medicine* 13 (1943): 365–388.

Skoda, Josef. *Abhandlung über Perkussion und Auskultation*. Vienna: J. G. Ritter von Mösle's Witwe and Braumüller, 1839.

Smith, D. "Freud's neural unconscious." In *The Pre-Psychoanalytic Writings of Sigmund Freud*, edited by G. Van de Vijver and F. Geerardyn, 155–164. London: Karnac, 2002.

Smith, Dale. "The evolution of modern surgery: A brief overview." In *A History of Neurosurgery in Its Scientific and Professional Contexts*. Edited by Samuel Greenblatt, 11–26. Park Ridge, IL: American Association of Neurological Surgeons, 1997.

Smith, Roger. *Between Mind and Nature: A History of Psychology*. London: Reaktion Books, 2013.

———. *Inhibition: History and Meaning in the Sciences of Mind and Brain*. Berkeley: University of California Press, 1992.

Snyder, Peter J., and Harry A. Whitaker. "Neurologic heuristics and artistic whimsy: The cerebral cartography of Wilder Penfield." *Journal of the History of the Neurosciences* 22.3 (2013): 277–291.

Solms, Mark, and Michael Saling. *A Moment of Transition: Two Neuroscientific Articles by Sigmund Freud*. London: Institute of Psycho-Analysis / Karnac, 1990.

———. "On psychoanalysis and neuroscience: Freud's attitude to the localizationist tradition." *International Journal of Psycho-Analysis* 67 (1986): 397–416.

Sporns, Olaf. *Discovering the Human Connectome*. Cambridge, MA: MIT Press, 2012.

———. "The human connectome: A complex network." *Annals of the New York Academy of Sciences* 1224 (2011): 109–125.

Springer, Elisabeth. *Geschichte und Kulturleben der Wiener Ringstrasse*. Wiesbaden: Franz Steiner, 1979.

Stahnisch, Frank. "German-speaking émigré neuroscientists in North America after 1933: Critical reflections on emigration-induced scientific change." *Österreichische Zeitschrift für Geschichtswissenschaften* 21 (2010): 36–68.

Starr, Susan L. *Regions of the Mind: Brain Research and the Quest for Scientific Certainty*. Stanford, CA: Stanford University Press, 1989.

Stephan, Klaas Enno, Jorge J. Riera, Gustavo Deco, and Barry Horwitz. "The Brain Connectivity Workshops: Moving the frontiers of computational systems neuroscience." *Neuroimage* 42.1 (2008): 1–9.

Stevens, Rosemary. *American Medicine and the Public Interest*. Berkeley: University of California Press, 1998.

———. *Medical Practice in Modern England: The Impact of Specialization and State Medicine*. New Haven, CT: Yale University Press, 1966.

Strachey, James. "Editor's introduction to J. Breuer and S. Freud (1893–95) Studies on Hysteria." In Freud, *Standard Edition*, vol. 2, *Studies on Hysteria*, ix–xxvii.

———. "Freud's psycho-analytic procedure" ([1903] 1904). In Freud, *Standard Edition*, vol. 7, *A Case of Hysteria, Three Essays on Sexuality and Other Works (1901–5)*, 247–254.

Strümpell, Adolf. *Aus dem Leben eines deutschen Klinikers: Erinnerungen und Beobachtungen*. Leipzig: Vogel, 1925.

Sulloway, Frank. *Freud, Biologist of the Mind: Beyond the Psychoanalytic Legend*. Cambridge, MA: Harvard University Press, 1992.

Tallis, Raymond. "Think brain scans can reveal our innermost thoughts? Think again." *Observer*, June 1, 2013, www.theguardian.com/commentisfree/2013/jun/02/brain-scans-innermost-thoughts, accessed July 7, 2014.

Taylor, Charles. *Sources of the Self: The Making of the Modern Identity*. Cambridge, MA: Harvard University Press, 1989.

Temkin, Oswei. *The Falling Sickness: A History of Epilepsy from the Greeks to the Beginnings of Modern Neurology*. Baltimore: Johns Hopkins Press, 1971.

Tesak, Jürgen. *"Der aphasische Symptomencomplex" von Carl Wernicke*. Idstein: Schulz-Kirchner, 2005.

———. *Geschichte der Aphasie*. Idstein: Schulz-Kirchner, 2001.

Tietze, Alexander. "Die Technik der Foersterschen Operation." *Mitteilungen aus den Grenzgebieten der Chirurgie und Medizin* 20 (1909): 559–564.

Todes, Daniel. *Ivan Pavlov: A Russian Life in Science*. New York: Oxford University Press, 2014.

———. *Pavlov's Physiology Factory: Experiment, Interpretation, Laboratory Enterprise*. Baltimore: Johns Hopkins University Press, 2001.

Tomes, Nancy. *A Generous Confidence: Thomas Story Kirkbride and the Art of Asylum-Keeping, 1840–1882*. Cambridge: Cambridge University Press, 1983.

Tönnis, Wilhelm. "Zirkulationsstörungen bei krankhaftem Schädelinnendruck." *Zeitschrift für die gesamte Neurologie und Psychiatrie* 167 (1939): 462–465.

Topp, Leslie. "The modern mental hospital in late nineteenth-century Germany and Austria: Psychiatric space and images of freedom and control." In *Madness, Architecture and the Built Environment: Psychiatric Spaces in Historical Context*, edited by Leslie Topp, James E. Moran, and Jonathan Andrews, 241–261. New York: Routledge, 2007.

———. "Psychiatric institutions, their architecture, and the politics of regional autonomy in the Austrian-Hungarian monarchy." *Studies in History and Philosophy of Biological and Biomedical Sciences* 38 (2007): 733–755.

Tuchman, Arleen. *Science, Medicine and the State in Germany: The Case of Baden, 1815–1871*. New York: Oxford University Press, 1993.

Universitas Vratislaviensis. *Chronik der Königlichen Universität zu Breslau (der Schlesischen Friedrich-Wilhelms-Universität zu Breslau)*. Vols. 1–27. Breslau, 1888–1913.

Universitätsbund Breslau. *Aus dem Leben der Universität Breslau*. Breslau: Breslau Genossenschafts Buchdruckerei, 1936.

Uttal, William R. *Distributed Neural System: Beyond the New Phrenology*. Cornwall-on-Hudson, NY: Sloan, 2009.

———. *The New Phrenology: The Limits of Localizing Cognitive Processes in the Brain*. Cambridge, MA: MIT Press, 2001.

Van Gehuchten, Arthur. "La radicotomie postérieure." *Le Névraxe* 11.2–3 (1910): 245–292.

Varela, Francisco, Evan Thompson, and Eleanor Rosch. *The Embodied Mind: Cognitive Science and Human Experience*. Cambridge: MIT Press, 1991.

Verzeichniss der Vorlesungen an der Königlichen Universität Breslau. Breslau, Sommersemester 1903–Sommersemester 1935.

Vidal, Fernando. *The Sciences of the Soul*. Chicago: University of Chicago Press, 2011.

Vidal, Fernando, and Francisco Ortega, eds. *Neurocultures: Glimpses into an Expanding Universe*. Frankfurt: Peter Lang, 2011.

Villaret, Albert, ed. *Handwörterbuch der gesamten Medizin*. Stuttgart: Ferdinand Enke, 1888–91.

Vogt, Cécile, and Oskar Vogt. "Die vergleichend-architektonische und die vergleichend-reizphysiologische Felderung der Großhirnrinde unter besonderer Berücksichtigung der menschlichen." *Die Naturwissenschaften* 14.50–51 (1926): 1190–1194.

Vöhringer, Margarete, and Yvonne Wübben. "Phantome im Labor: Die Verbreitung der Reflexe in Hirnforschung, Kunst und Technik." Special issue, *Berichte zur Wissenschaftsgeschichte* 32 (2009).

Von Weizsäcker, Victor. "Otfrid Foerster 1873–1941." *Deutsche Zeitschrift für Nervenheilkunde* 153 (1942): 1–23.

Wagner-Jauregg, Julius. "Über die lanzinierenden Schmerzen der Tabetiker." *Wiener klinische Wochenschrift* 40 (1924): 187–204.

Wegener, Mai. *Neuronen und Neurosen: Der psychische Apparat bei Freud und Lacan: Ein historisch-theoretischer Versuch zu Freuds "Entwurf" von 1895*. Munich: Fink, 2004.

Weidman, Nadine. *Constructing Scientific Psychology: Karl Lashley's Mind-Brain Debates*. Cambridge: Cambridge University Press, 1999.

Weisz, George. *Divide and Conquer: A Comparative History of Medical Specialization*. New York: Oxford University Press, 2006.

Wernicke, Carl. "Aphasie und Geisteskrankheit." Talk given at the Verhandlungen des Congresses für innere Medizin, Wiesbaden 1890. In Wernicke, *Gesammelte Aufsätze*, 153–160.

———. *Atlas des Gehirns: Schnitte durch das menschliche Gehirn in photographischen Originalien herausgegeben mit Unterstützung der Königlichen Akademie der Wissenschaften in Berlin*. 3 vols. Breslau: Franck and Weigert, 1897–1903.

———. *Der aphasische Symptomencomplex: Eine psychologische Studie auf anatomischer Basis*. Breslau: Cohn and Weigert, 1874. Reprinted in Wernicke, *Gesammelte Aufsätze*, 1–70. Page references are to the reprint edition.

———. "Die Aufgaben der klinischen Psychiatrie." *Breslauer ärztliche Zeitschrift* 13 (1887). Reprinted in Wernicke, *Gesammelte Aufsätze*, 146–152.

———. "Ein Fall von Ponserkrankung." *Archiv für Psychiatrie und Nervenkrankheiten* 7 (1877): 513–538.

———. "Erkrankungen der inneren Kapsel: Ein Beitrag zur Diagnose der Heerderkrankungen." Breslau: Cohn and Weigert, 1875. Reprinted in Wernicke, *Gesammelte Aufsätze*, 167–188. Page references are to the reprint edition.

———. *Gesammelte Aufsätze und kritische Referate zur Pathologie des Nervensystems*. Berlin: H. Kornfeld, 1893.

———. *Grundriss der Psychiatrie in klinischen Vorlesungen*. Vol. 1. 1st ed. Leipzig: Thieme, 1894.

———. *Grundriss der Psychiatrie in klinischen Vorlesungen*. 2nd ed. Leipzig: Thieme, 1906.

———. "Herabsetzung der electrischen Erregbarkeit bei cerebraler Lähmung." *Breslauer ärztliche Zeitschrift* 17 (1886). Reprinted in Wernicke, *Gesammelte Aufsätze*, 258–267. Page references are to the reprint edition.

———. *Krankenvorstellungen aus der Psychiatrischen Klinik in Breslau*. 3 vols. Breslau: Schletter, 1899–1900.

———. *Lehrbuch der Gehirnkrankheiten für Aerzte und Studirende*. Kassel: Fischer, 1881–83.

———. "Stadtasyle und psychiatrische Kliniken." *Klinisches Jahrbuch* 2 (1890): 186–193.

———. "Tagesfragen." *Monatsschrift für Psychiatrie und Neurologie* 1 (1897): 1–5.

———. "Ueber das Bewusstsein." *Allgemeine Zeitschrift für Psychiatrie* 35 (1879). Reprinted in Wernicke, *Gesammelte Aufsätze*, 130–140. Page references are to the reprint edition.

———. "Ueber den wissenschaftlichen Standpunkt in der Psychiatrie." *Wiener medizinische Wochenschrift* 42 (1880): 1149–1152; 43 (1880): 1179–1182.

———. "Ueber die Irrenversorgung der Stadt Breslau." *Allgemeine Zeitschrift für Psychiatrie und psychisch-gerichtliche Medizin* 45 (1889): 432–441.

———. "Zweck und Ziel der Psychiatrischen Kliniken." *Klinisches Jahrbuch* 1 (1889): 218–223.

———. "Zwei Fälle von Rindenläsion." *Arbeiten aus der psychiatrischen Klinik in Breslau* 2 (1895): 33–52.

Westphal, Carl. "Ueber einige durch mechanische Einwirkung auf Sehnen und Muskeln hervorgebrachte Bewegungs-Erscheinungen." *Archiv für Psychiatrie und Nervenkrankheiten* 5 (1875): 803–834.

Whitrow, Magda. *Julius Wagner-Jauregg (1857–1940)*. London: Smith-Gordon, 1993.

Wieseltier, Leon. "Crimes against humanities: Now science wants to invade the liberal arts: Don't let it happen." *New Republic*, September 3, 2013.

Winter, Alison. *Memory: Fragments of a Modern History*. Chicago: University of Chicago Press, 2012.

Wittels, Fritz. "Paul Schilder—1886–1940." *Psychoanalytic Quarterly* 10 (1941): 131–134.

Wittelshöfer, Leopold. "Notiz." *Wiener medizinische Wochenschrift* 27 (1870): 578–579.

Wübben, Yvonne. *Verrückte Sprache: Psychiatrie und Dichter in der Anstalt des 19. Jahrhunderts*. Konstanz: Konstanz University Press, 2012.

Young, Allan. *The Harmony of Illusions: Inventing Post-Traumatic Stress Disorder*. Princeton, NJ: Princeton University Press, 1995.

Young, Robert. *Mind, Brain, and Adaption in the Nineteenth Century: Cerebral Localization and Its Biological Context from Gall to Ferrier*. New York: Oxford University Press, 1990.

Young-Bruehl, Elisabeth. *Anna Freud: A Biography*. New York: Summit Books, 1988.

Zülch, Klaus-Joachim. *Die geschichtliche Entwicklung der deutschen Neurologie*. Berlin: Springer, 1987.

———. *Otfrid Foerster, Arzt und Naturforscher, 9.11.1873–15.6.1941*. Berlin: Springer, 1966.

———. *Otfrid Foerster, Physician and Naturalist, November 9, 1873–June 15, 1941*. Translated by Adolf Rosenauer and Joseph P. Evans. Berlin: Springer, 1969.

———. "Otfrid Foerster und die Lokalisationslehre." *Leopoldina* 9 (1973): 164–177.

———. "The place of neurology in medicine and its future." *Handbook of Clinical Neurology* 1 (1969): 1–44.

Index

Page numbers in italics refer to figures.